The Python-Based Laboratory

The Python-Based Laboratory

A Hands-On Guide for Scientists and Engineers

JOHN ESSICK

OXFORD
UNIVERSITY PRESS

Great Clarendon Street, Oxford, OX2 6DP,
United Kingdom

Oxford University Press is a department of the University of Oxford.
It furthers the University's objective of excellence in research, scholarship,
and education by publishing worldwide. Oxford is a registered trade mark of
Oxford University Press in the UK and in certain other countries

Published in the United States of America by Oxford University Press
198 Madison Avenue, New York, NY 10016, United States of America

British Library Cataloguing in Publication Data
Data available

Library of Congress Control Number: 2024939879

ISBN 9780198896982
ISBN 9780198896999 (pbk.)

DOI: 10.1093/9780191998478.001.0001

Printed and bound by
CPI Group (UK) Ltd, Croydon, CR0 4YY

The manufacturer's authorised representative in the EU for product
safety is Oxford University Press España S.A. of el Parque Empresarial San
Fernando de Henares, Avenida de Castilla, 2 – 28830 Madrid
(www.oup.es/en).

To my favorite Pythonista, Devon

Preface

The Python-Based Laboratory: A Hands-On Guide for Scientists and Engineers provides a learn-by-doing approach to acquiring the Python programming skills needed to implement computer-controlled experimental work. This book is not a manual-like presentation of Python. Rather, *The Python-Based Laboratory* leads its readers to mastery of the popular, open-source Python computer language in its role as a powerful laboratory tool by carrying out interesting and relevant projects that explore the acquisition, production, analysis, and presentation of digitized waveforms. Readers, who are assumed to have no prior computer programming or Python background, begin writing meaningful programs in the first few pages.

The Python-Based Laboratory can be used as a textbook for science and engineering instructional laboratory students who are being taught up-to-date Python-based experimental skills. The book also works well as a self-study guide for professional laboratory researchers, industrial engineers, hobbyists, and electronics enthusiasts seeking to automate tasks using Python. Topics covered include the control of data-acquisition devices (including multifunction data-acquisition hardware and IEEE 488.2-interfaced stand-alone instruments), data file storage and presentation, digitized data concepts (such as resolution, sampling frequency, and aliasing), and data analysis techniques (curve fitting and fast Fourier transform). As readers work their way through the book, they build several computer-based instruments, including a DC voltmeter, digital oscilloscope, DC voltage source, waveform generator, blinking LED array, digital thermometer, spectrum analyzer, and frequency meter. Each chapter concludes with a Do It Yourself project and a Use It! example as well as a healthy selection of homework-style problems, allowing readers to test their understanding and further develop their Python-based experimentation skills.

Python has become a very popular programming choice for laboratory and industrial automation in recent years, often replacing the use of the industry-standard languages LabVIEW (product of NI, formerly, National Instruments) and MATLAB (product of MathWorks). Reasons for Python's growth in this area include the following:

- Python is free and open source, in contrast to the commercial products LabVIEW and MATLAB, making it an attractive option in academic, research, and industrial laboratories.
- Python is widely adopted in education. Because it is relatively easy to learn and transfers well to other domains such as scientific modeling, web development, and data science, it is not unusual for today's students to be taught and then use Python in one or more of their courses.
- Python has a large and active user community, which has developed an abundance of software libraries to carry out standard computing tasks needed in laboratory applications.
- Python software drivers are increasingly being provided by equipment manufacturers for their instruments.

The Python-Based Laboratory takes advantage of Python add-ons (called packages or libraries). Some of these Python add-ons facilitate graphical user interfaces (GUIs) and data presentation, one of the strengths of LabVIEW programming, while other Python add-ons enable array-based mathematical calculations, a powerful feature of MATLAB. By utilizing these add-ons in *The Python-Based Laboratory*, readers will be able to write Python programs that incorporate the favorable features of both LabVIEW and MATLAB programming.

To serve the largest possible audience, *The Python-Based Laboratory* employs only free and open-source software, namely, Python itself, the IDLE Integrated Development Environment (for program development and execution), the Matplotlib library (for plotting), the Tkinter package (for creating GUIs), the NumPy and SciPy packages (for mathematical functions and array-based mathematical operations), and the PyVISA and NI-VISA packages (for IEEE 488.2 control of stand-alone instruments). Likewise, only data-acquisition (DAQ) devices with free Python driver software are used, specifically, NI DAQ devices with the NI-DAQmx driver and nidaqmx Python package and Measurement Computing Corporation (MCC) DAQ devices with the mcculw package.

The progression of topics in *The Python-Based Laboratory* is as follows:

Chapter 1: Python Program Development. This chapter focuses on writing a simple program to test the parity (i.e., evenness or oddness) of an integer. Central features of Python are presented, including its programming environment (terminal session and program editor), data types, conditional branching, exception (error) handling, and program documentation. In addition, Python's support of object-oriented programming is introduced.

Chapter 2: Graphical User Interface Using Tkinter. Readers use Tkinter, the graphical user interface toolkit included as part of the Python Standard Library, to equip their integer-parity-testing program with a GUI. In the process of creating this program, Python's support of both object-oriented programming (via classes) and functional programming (via functions) is discussed and implemented. Readers carry out the procedure for creating their own custom-made function.

Chapter 3: Counted Looping and Waveform Plots. The properties of a digitized waveform are studied as readers write a sine-wave plotting program. Plotting of the digitized sine-wave data is facilitated by Python's Matplotlib library, while the digitized data are generated by three counted-looping methods—for loop, list comprehension, and vectorization. The latter method is realized through the array-based mathematics of the NumPy library. Python's iterable data structures called list and string are discussed, along with the various ways prewritten software can be imported into a Python program. Finally, the procedure for placing a Matplotlib plot on a GUI is implemented.

Chapter 4: Conditional Looping and Real-Time Plots. Readers learn to use Python's conditional looping structure— the while Loop—to create a real-time plot of a sine wave. The advantage of using Matplotlib's animation module, along with its blitting feature, for such plots is explored and the method for placing an animation on a GUI is executed. Indexing and slicing of iterable data structures is explained and the tuple data structure is defined.

Chapter 5: GUI-Equipped Waveform Simulator. Readers write a digitized waveform simulation program with advanced GUI features, including numeric input controls, a pull-down menu, and an embedded waveform plot. The completed program creates waveforms with sine, cosine, DC level, square, sawtooth, triangle, and user-defined shapes. In writing this program, three Python "utility" modules are created to carry out the following tasks: creation of waveform data with a desired shape, creation of a graphical user interface with interactive controls, and creation of a GUI-embedded plot to display waveform data. These three modules are used in programs created throughout the remainder of the book.

Chapter 6: Introduction to Data-Acquisition Device Features. This chapter assumes that readers have access to a multifunction DAQ device from either NI or MCC. Both NI and MCC are manufacturers of popular, high-quality DAQ devices, where MCC devices are the lower priced option with (satisfactorily) scaled-down specifications. Using the interactive program appropriate for their device (MAX for NI, DAQami for MCC), readers familiarize themselves with the capabilities of a multifunction DAQ device, including analog-to-digital conversion, digital-to-analog conversion, digital input/output operations, and digital pulse counting. In the course of this work, explanations are given of the following important digitized data concepts: analog input modes, range, resolution, sampling frequency, aliasing, digital input/output levels, and digital triggering. The NI USB-6002 and MCC USB-202 DAQ devices are highlighted, but the programs developed apply to most other NI and MCC multifunction DAQ devices.

Chapter 7: Data Acquisition Using NI DAQ Device. After their Chapter 6 orientation to the NI DAQ device's capabilities via MAX, readers write Python programs that execute analog-to-digital, digital-to-analog, and digital input/output tasks on the NI DAQ device as they construct several computer-based instruments, including a DC voltmeter, digital oscilloscope, DC voltage source, waveform generator, and blinking LED array. The necessity for placing GUI and DAQ operations on separate execution threads is explored as well as the need for triggering in an oscilloscope application.

Chapter 8: Data Acquisition Using MCC DAQ Device. After their Chapter 6 orientation to the MCC DAQ device's capabilities via DAQami, readers write Python programs that execute analog-to-digital, digital-to-analog, and digital input/output tasks on the MCC DAQ device as they construct several computer-based instruments, including a DC voltmeter, digital oscilloscope, DC voltage source, waveform generator (for upgraded MCC models), and blinking LED array. The necessity for placing GUI and DAQ operations on separate execution threads is explored as well as the need for triggering in an oscilloscope application.

Chapter 9: Data Files and Character Strings. Readers learn how to use NumPy functions to store data in, and read data from, one- and two-dimensional spreadsheet-formatted computer files. Numeric formatting, escape sequences, and the Unicode character coding scheme are covered. The chapter concludes by giving the NumPy-based method for writing and reading binary-formatted files.

Chapter 10: Data Analysis: Curve Fitting. A thermistor is a semiconducting device commonly used as a temperature sensor. A physical model for a thermistor's temperature-dependent resistance is presented and then readers write GUI-equipped Python programs that fit a thermistor's temperature versus resistance calibration data to various functional forms consistent with the physical model. First, the NumPy polynomial curve-fitting function is used to fit the data to a third-order polynomial. Next, SciPy's more general nonlinear fitting function is used to fit the data to several alternate functions, including the widely used Steinhart–Hart equation. Goodness of the fitted curve is judged via residual plots. After obtaining the fitted calibration curve for a thermistor, a computer-based digital thermometer is constructed.

Chapter 11: Data Analysis: Fast Fourier Transform. Using SciPy functions, readers write Python programs to investigate the proper use of the fast Fourier transform (FFT). Readers apply SciPy's complex Fourier transform function to various sinusoidal inputs, demonstrating that this function produces the expected complex-amplitude spectra. In addition, a GUI-equipped program that simply plots the magnitude of the complex-amplitude as a function of frequency is used to explore the effects of spectral leakage, windowing, and aliasing. With that understanding in place, a computer-based spectrum analyzer is constructed.

Chapter 12: Control of Stand-Alone Instruments Using VISA. Using PyVISA and NI-VISA, Python-based control of an IEEE 488.2-interfaced stand-alone instrument over the Universal Serial Bus (USB) as well as the General Purpose Interface Bus (GPIB) is studied. The Agilent 34410A digital multimeter (which has the same command set as digital multimeters from other manufacturers) is used to demonstrate the central concepts of IEEE 488.2 interface bus communication between a computer and stand-alone instrument, including synchronization methods. Readers write GUI-equipped Python programs to control the instrument's data-taking functions. In the latter portion of the chapter, reader create an instrument driver for the digital multimeter using the Python dictionary data structure as a lookup table for instrument specifications.

Appendix A: Installing Python and IDLE. Instructions are given to assist readers in loading Python and IDLE onto their computer. IDLE is used throughout the book to create and execute Python programs.

Key features of *The Python-Based Laboratory* include its emphasis on real-world problem solving, its early introduction and routine use of data-acquisition hardware, its Do It Yourself projects and Use It! examples at the end of each chapter, and its healthy offering of back-of-the-chapter homework problems.

Real-world problem solving: Chapter topics and exercises provide examples of how commonly encountered problems in the laboratory are solved by scientists and engineers. Python features, along with relevant mathematical background, are introduced in the course of solving these problems. The "best practice" strategies of modularity and code reuse are emphasized so that readers optimize their use of Python. In addition, many of the book's programs are equipped with a graphical user interface, a desired feature in computer programs that control experiments.

Data-acquisition usage throughout: Exercises involving DAQ hardware appear early and then routinely in *The Python-Based Laboratory.* Of particular note, following the book's first five software-only chapters that teach the fundamentals of the Python programming language, data acquisition using a DAQ device is covered in Chapters 6 and, depending on the manufacturer of your DAQ device, 7 or 8. For a professor or self-learner who wishes to devote only three or four weeks to instruction in Python-based data acquisition, Chapters 1 through 7 (or 8) will provide the needed instructional materials. For those planning a more comprehensive study of Python-based lab skills, a digital thermometer and spectrum analyzer are constructed in Chapters 10 and 11, respectively. In Chapter 12, data are acquired remotely from a stand-alone instrument using the USB and/or GPIB interface bus. Additionally, commonly used interfacing circuits consisting of low-cost integrated circuits are presented. Circuits include anti-aliasing filters, an audio amplifier, and a thermocouple signal conditioner.

Do It Yourself projects: Each of the book's chapters concludes with a Do It Yourself project that poses an interesting problem and directs readers in finding a solution by applying the chapter's material. In some chapters, this project involves writing a program that carries out a statistical analysis of flipping coins (Chapter 3), creates a square wave from a combination of sine waves (Chapter 5), or writes and reads a binary-formatted data file (Chapter 9); in other chapters, the reader constructs a computer-based instrument including a millisecond-resolution stopwatch (Chapter 6), digital thermometer (Chapter 10), and a spectrum analyzer (Chapter 11).

Use It! examples: Ready-to-use example programs, which carry out common tasks encountered in laboratory work, are presented at the end of each chapter. Some of these examples involve programming solutions, for example, showing how to document programs, test-run Python coding as "main" programs in order to find programming bugs, and save data during runtime. Other examples are low-cost hardware solutions, including anti-aliasing through the use of an eighth-order Butterworth lowpass filter, amplification and cold-junction compensation for a thermocouple temperature measurement, and configuration of a LAN/Ethernet interface.

Back-of-the-chapter homework problems: A selection of homework-style problems is included at the end of each chapter so that interested readers can further develop their Python-based skills. In some of these problems, readers test their understanding by applying the chapter topics to new applications (e.g., lock-in detection); in others, readers use programs written within the chapter to explore important experimental issues (e.g., frequency resolution of a fast Fourier transform). Finally, a number of problems introduce readers to features of Python relevant to, but not included in, the chapter's text (e.g., code-execution timer).

To aid readers in creating their Python programs, the following conventions are used throughout the book:

- *Italic* text signals the first-time use of important terms and concepts and highlights character strings to be entered on a program's graphical user interface using the keyboard.
- **Bold** text designates pull-down menu selections and graphical user interface label names.
- `Constant Width` text is used for the Python programs to be written by readers as well as within paragraphs to refer to Python program elements. The programs themselves are within a gray background.
- **Straight Bold** text designates the names of programs written by readers as well as the path within the computer file system to those programs.

Any suggestions or corrections are gladly welcomed and can be sent to John Essick at jessick@reed.edu.

Updates, answers to frequently asked questions, and ancillary materials for *The Python-Based Laboratory* (including Python code for all programs within the book as well as solutions to the problems) are available at http://academic.reed.edu/physics/faculty/essick.

Additionally, the book's Python programs and problem solutions can be downloaded at http://www.oup.com/us/essick.

I'd like to express my gratitude to the Physics Department at the University of Newcastle, Australia who inspired me to undertake this project through a Fulbright Specialist visit, especially to Vicki Keast for arranging my visit and Lachlan Rogers for sharing his deep experience with Python in teaching and in research. Also, much thanks to David Horn for his support with Digilent products and many thanks to Devon Essick and Jennifer Heath for reading portions of the book's manuscript and providing invaluable feedback.

For their support and assistance in preparing *The Python-Based Laboratory*, I thank Dan Taber and David Lipp of Oxford University Press and Thilagavathi Ezhumalai of Integra Software Services. For their helpful comments and suggestions, I express my appreciation to the four anonymous reviewers who provided early comments on the book as part of the OUP peer review process.

Finally, to my wife Katie: Thank you for your love and support while I worked on this project.

John Essick
Portland, Oregon

Contents

CHAPTER 1

Python Program Development

In this chapter, we will run our first Python code, both in a terminal session and in a Python program. So, first, check that a recent version of Python (3.6 or later) is installed on your computer. If you need to install Python, please refer to Appendix A for instructions on how to install it.

To create our Python program, we will use a code editor. As described in Appendix A, out of the many possible choices for a Python code editor, IDLE will be used throughout the book. IDLE is present on your computer as it is included when you install Python. If you favor (and have experience with) another Python code editor, please feel free to use that as you work through the book. The book's instructions should be easily adaptable to your preferred editor.

1.1 Python Programming Environment

Welcome to the world of Python-based laboratory automation. In the following pages, we will explore how the popular, general-purpose programming language Python is employed by scientists and engineers to create a computer-controlled laboratory. The Python-based skills that you develop through this hands-on study will enable you to write software programs that perform all of the laboratory tasks required for your state-of-the-art research and industrial applications, including instrument control, data acquisition, data analysis, data presentation, and data storage.

When you install Python on your computer, the installation includes the essential programming elements of the Python language (e.g., basic arithmetic operations, if-else statements, while Loops), along with the *Python Standard Library*, an extensive library of handy programming elements such as constants, functions, operators, and much more. A Python program (also called a *script*, although I use the term *program* throughout the book) is a sequence of text statements constructed using these programming elements and is created using a text-based code editor. Once the program is complete, the editor can be used to instruct the *Python Interpreter* (which is installed as part of the Python installation process) to convert your sequence of Python statements into the lower-level code that enables your computer to execute your program.

To begin developing your skill in Python programming, in this first chapter, I will guide you through the steps of creating a program using the Python programming elements installed on your computer. Together, we will explore key features of the Python programming environment as we write a program that detects the *parity* of a given integer, that is, whether it

The Python-Based Laboratory. John Essick, Oxford University Press. © John Essick (2024). DOI: 10.1093/9780191998478.003.0001

is even or odd. The parity of an integer is defined by the following test: Using long division, divide the integer by 2. If the remainder is 0 (that is, it "divides evenly"), the integer is even; if the remainder is 1, the integer is odd (by convention, the remainder is assumed to be positive when dividing a negative integer by a positive integer). According to this definition, for example, the integers 0, 4, and −4 are even, while 3 and −3 are odd.

1.2 Python Terminal Session

In the program we will create, the following steps will be carried out: An input integer is divided by 2 and the value of the remainder is then used to determine whether the integer is odd or even. In computer programming, the remainder R that results when x is divided by y is given by the *modulo operator*, which is included as one of Python's core arithmetic operations and is denoted by the percent sign character (`%`). Thus, in Python, the remainder R that results when x is divided by y is calculated using the statement `x % y`.

When writing a Python program, it's often helpful to first test programming statements in an *interactive command-line session*, also called a *terminal session*. To start a terminal session in Windows, (depending on your version of Windows) enter *cmd* in the search box next to the bottom-left Windows icon or in the **Start>>Run** input box. Then, in the **Command Prompt** window, type *python* in lowercase and press <*Enter*>. On the macOS, click on **Terminal.app** in the **Applications>>Utilities** folder. Then, at the prompt, type *python3* in lowercase.

If you have Python properly installed, the Python Interpreter will be started, allowing you to run Python commands one-at-a-time in the terminal session. The session will initiate with a message detailing information about the installed version of Python on your computer. Below the version information you will see a *Python prompt* (>>>), where a Python statement can be entered and then executed. If, at any time, you need to exit the Python Interpreter, type `exit()`, then press <*Enter*>.

Verify the proper command syntax of the Python modulo operator by typing `4 % 2` at the Python prompt, then press <*Enter*>. The result of this command will be displayed on the next line, followed by a new line with another Python prompt, as shown:

```
>>> 4 % 2
0
>>>
```

Next, try the statements `3 % 2`, `0 % 2`, `-3 % 2`, and `-4 % 2` to confirm the results are as expected.

1.3 Object-Oriented Programming

The Python language supports *object-oriented programming*. In this programming approach, items in your program are *objects*, which can have properties (called *attributes*) and associated functions (called *methods*). An object is created by using the *assignment* operator, denoted by the equal sign character (=).

We can demonstrate some features of object-oriented programming with the following example involving a *string*. In Python, a string is a sequence of (Unicode) characters surrounded by either single or double quotation marks (we'll choose to use single quotation marks throughout the book). Thus, a simple assignment operation is `x = 'python'`. Here, an object is created and given the name `x` and the string `python` is assigned to the object. The attribute of `x` is the string `python`, and since `x` is a string-type object, there are many (over 40) available string methods that can be made to act on it. One such method is `capitalize()`, which will return a copy of the string object with the first character capitalized and the rest of the string in lowercase. Note that parentheses are included as part of this method name. Since methods are functions that act on an object, their names always end with a set of parentheses. To implement a method upon an object, the dot notation `object_name.method_name()` is used. In your terminal session, demonstrate the effect of this method by the following sequence of commands:

```
>>> x = 'python'
>>> x.capitalize()
'Python'
>>>
```

1.4 Developing Parity-Test Code in Terminal Session

For our parity-testing program, we will need to determine whether the resulting value of the modulo operation is 0 or 1. In order to carry out this test, we will assign the modulo result to be the attribute of an object named `remainder` via the following statement: `remainder = 4 % 2`. Use your terminal session to verify that the `remainder` object has been properly assigned, where the value of `remainder` is displayed using Python's `print()` function:

```
>>> remainder = 4 % 2
>>> print(remainder)
0
>>>
```

Within a program, we can determine whether `remainder` equals 0 by using Python's *Equal to* comparison operator, denoted by two equal sign characters (==). A comparison operator makes its relevant comparison of two quantities and produces a *Boolean* object, i.e., an object whose value is one of two possible values, *True* or *False*. Use your terminal session to determine whether `remainder` is equal to zero as follows:

```
>>> remainder = 4 % 2
>>> remainder == 0
True
>>>
```

To report the determined parity of the input integer, we'll have our program choose to display the appropriate message—**Integer is Even** or **Integer is Odd**. In a Python program, such decisions are made using the *if. . .elif. . .else* conditional branching statement (*elif* is Python's abbreviation for *elseif*). The *if. . .elif. . .else* structure allows commands to be executed conditionally, depending on the result of one or more logical tests (that is, tests that produce Boolean values of either `True` or `False`). The form of the *if. . .elif. . .else* statement is as follows:

```
if <logical test 1>:
    <statement 1>
elif <logical test 2>:
    <statement 2>
else:
    <statement>
```

Here, if *<logical test 1>* produces a `True` value, *<statement 1>* is executed. Otherwise, if *<logical test 2>* produces a `True` value, *<statement 2>* is executed. Additional *elif* conditions can be added, if needed. If none of the logical tests preceding the (optional) *else* produce a `True` value, *<statement>* is executed. Note that the lines beginning with *if*, *elif*, and *else* all end in a colon (`:`). If you forget to place a colon, Python will produce an error message at runtime to remind you. Also, as we will discuss later, the indentations of the statement lines are significant, with the standard indents being four spaces. Use your terminal session to carry out a trial run of the following conditional-branching code:

```
>>> remainder = 4 % 2
>>> if remainder == 0:
...     print('Integer is Even')
... elif remainder == 1:
...     print('Integer is Odd')
...
Integer is Even
>>>
```

Remember the colons at the end of the *if* and *elif* lines and indent the `print()` lines by four spaces. Press *<Enter>* after the final line with three dots (`. . .`) to alert the Python Interpreter that your *if. . .elif. . .else* statement is complete.

Finally, at the top of our program, we will ask the user to input a desired integer for the parity test. Such data entry is accomplished by Python's `input(prompt)` command, which reads a user input and returns the entered value as a string. The command's optional `prompt` parameter is a message string that is displayed before the user enters an input. Demonstrate the functioning of the `input(prompt)` command in your terminal session by first typing the following:

```
>>> x = input('Enter Integer:')
Enter Integer:
```

Next, type an integer value, such as *6*, after the colon and press *<Enter>*:

```
>>> x = input('Enter Integer:')
Enter Integer:6
>>>
```

This code has assigned the returned value of the `input()` command to the object named `x`. The value of `x` can be displayed by a `print()` command:

```
>>> x = input('Enter Integer:')
Enter Integer:6
>>> print(x)
6
>>>
```

As mentioned above, the `input()` command returns the entered value as a string. This fact can be verified by the `type()` command, which reports the data type of an object (e.g., `'str'` for string, `'int'` for integer, `'bool'` for Boolean):

```
>>> x = input('Enter Integer:')
Enter Integer:6
>>> print(x)
6
>>> type(x)
<class 'str'>
>>>
```

Being a string, the object `x` cannot be directly used in a mathematical calculation such as the modulo operation. Instead, `x` must first be recast into a numeric data type, for example, an integer by using the `int()` command:

```
>>> x = input('Enter Integer:')
Enter Integer:6
>>> print(x)
6
>>> type(x)
<class 'str'>
>>> y = int(x)
>>> type(y)
<class 'int'>
>>>
```

One final note on terminology: In the definition of a function like `input(prompt)`, the quantity `prompt` in parentheses of this definition is an example of a *parameter*, i.e., a piece of information the function needs to do its work. When a particular value is assigned to the parameter as in `input('Enter Integer:')`, that particular value (in this case, `'Enter Integer:'`) is called an *argument*. You may find people use the terms *parameter* and *argument* interchangeably when referring to a function's variables, but technically these terms do have (slightly) different meanings.

We now have explored the needed components for our integer-parity program and are ready to write our first Python program. Close the terminal session by typing `exit()` and pressing *<Enter>*:

1.5 Creating and Running Parity-Test Program

Search for the IDLE application on your computer and then start it. When IDLE starts up, you will be presented with an *IDLE Shell window*, where you will supply inputs to your program and view its outputs. In the **File** drop-down menu at the top of the IDLE Shell window, create a new program by selecting **New File**. An *IDLE Editor window* will appear, where you will enter your Python program.

From now on, a sequence of choices like selecting **New File** from the **File** drop-down menu will be indicated as follows: **File>>New File**. Also, henceforth, the IDLE Shell window and the IDLE Editor window will be simply called the *Shell* and *Editor*, respectively. While we will create, save, and run our Python programs in the Editor, the Shell will be used to display program output (until we learn how to create a graphical user interface for our programs). As we will find below, a terminal session can also be run in the Shell.

In the Editor, enter the following sequence of statements, where the function of each statement will be familiar from your work in the terminal session; remember that the lines beginning with *if* and *elif* both end in a colon and that the lines with `print()` statements must be indented by four spaces:

```
entry_string = input('Enter Integer:')

entry_integer = int(entry_string)

remainder = entry_integer % 2

if remainder == 0:
    print('Integer is Even')
elif remainder == 1:
    print('Integer is Odd')
```

This Python program detects the parity of an integer, so let's save it in a computer file with a descriptive name. By convention, Python file names are all lowercase with words separated by underscores. The extension **.py** is used to identify the file as a Python program. In the Editor, save your program by selecting **File>>Save**, which will activate the **Save As** dialog window. Because you will be writing and saving many programs in your study of Python, let's first create a folder in which to store this work. In the **Save As** window, navigate to a desired storage place on your computing system (e.g., **Desktop**, **Documents** folder, flash drive), and then click on the **New folder** button. In the highlighted box that appears next to the folder icon, use your name to entitle the new folder **YourNamePythonWork** and then click **Open** to open this folder. Finally, in the **File name:** box, type **integer_parity_detector.py** and then press **Save**.

Now that your program is saved, execute it by selecting **Run>>Run Module** in the Editor. In the Shell, type your choice of integer after the prompt **Enter Integer:** and verify that its parity is correctly determined. Run your program with a few other choices of positive and negative integers to confirm it works for positive, negative, and zero integer inputs.

Now that you've successfully gone through the process of writing a Python program, I'll point out two important features of Python programming. First, in broad terms, Python is a high-level interpreted language as opposed to a high-level compiled language. In high-level compiled languages like C, C++, and Java, user-written ("human understandable") programs can only be run after they are first translated into low-level ("machine understandable") machine code through a process called *compilation*. Upon successful compilation, an executable file is generated, which can then be run by a computer. In contrast, programs written in the high-level interpreted Python language do not need to be compiled prior to running. Rather, in a multistep process during program execution, program statements are converted to low-level instructions, which allow the Python Interpreter to execute the program line-by-line. For this reason, a Python program is commonly called a *script*, as it provides the sequence of statements the Python Interpreter follows during runtime. As a Python programmer, you will encounter a consequence of this line-by-line runtime execution in the following way: Say you have made a programming error somewhere near the end of your Python program. When you run this program, you'll find that it runs correctly up to the point of the error, producing valid outputs along the way, but when the line with the error is reached, the Python Interpreter will stop running the program and generate an error statement. That statement will identify the line in the program with the error and provide some hints that hopefully help you to identify and correct the problem. Another consequence of the interpreted mode of execution is this: When coding the same algorithm, Python programs will typically take longer to run than programs written in a compiled language like C. This extra execution time can be understood by an analogy: A conversation with someone in your shared native language will be quicker than the same conversation in which an interpreter is translating your language "on the fly" into the different language of your listener.

Second, in Python, indentation has meaning. The structure of a computer program typically includes blocks of related code, for example, a block consisting of multiple lines of commands below an *if* statement, which are to be executed when the *if* statement's logical test is `True`. Some programming languages such as C and Java use a set of curly braces to mark off such blocks of code, while other languages use a set of keywords such as **begin** and **end**. In such languages, a common practice is to use consistent indentation to enhance the readability of a program, but the indentation has no actual programming function. In contrast, the developers of Python decided that indentation itself was enough to define the structure of a program and, with that decision, eliminated the need for curly braces and keywords for that purpose. Thus, in our **integer_parity_detector.py** program, the *if. . .elif* statement was constructed as follows:

```
if remainder == 0:
    print('Integer is Even')
elif remainder == 1:
    print('Integer is Odd')
```

Here, the line with the first `print()` statement is tied to the line with the *if* statement because of its indentation, while indentation ties the line with the second `print()` statement to the *elif* statement. To indent a line, spaces should be used (as opposed to tabs), and it is preferred Python practice to indent by four spaces. You may find that your Python editor provides this four-space indentation for you automatically or that the Tab key actually produces four separate spaces (rather than a single tab). For those with experience in other programming languages, Python's functional use of indentation may seem odd at first. But, after spending some time writing Python code, it becomes natural and saves a lot of curly-brace and keyword typing.

1.6 Exceptions and Exceptions Handling

Now for some bad news. Try running **integer_parity_detector.py** and entering a non-integer value such as *3.6* when prompted by **Enter Integer:**. You'll find that this entry results in the following error message (called a *traceback*):

```
Traceback (most recent call last):
  File "integer_parity_detector.py", line 3, in <module>
    entry_int = int(entry_string)
ValueError: invalid literal for int() with base 10:'3.6'
```

The last line of the traceback identifies the problem as a *ValueError* when `'3.6'` is entered for the argument of the `int()` function.

As often happens in programing, we've discovered a bug that appears during runtime, which produces an error message that stops execution of our program. In Python, such a runtime error is called an *exception*. Since our program works perfectly as long as only integers are used as input, one solution to the exception we have uncovered is to simply modify the `input()` function's prompt to the cautionary (and confidence-sapping) message such as **Enter Integer: (WARNING: Program Crashes for Non-Integer Input)**. However, we can do better than that. Let's fix the bug by anticipating its possible presence and building *exception handling* into our code.

To investigate this problem, note that the last line in the Shell contains a Python prompt (>>>), meaning the Shell has activated the Python Interpreter and an interactive terminal session can be carried out in this window. At the Python prompt within the Shell, try entering `3.6` as the argument for the `int()` function:

```
>>> x = int(3.6)
>>> print(x)
3
```

You'll find that no traceback results. To understand what is going on, we must first realize that the value `3.6` used above is a *floating-point number* (called a *float*, for short), a numeric data type we have not encountered up to this point. In Python, real numbers, i.e., numbers that include a decimal point, are formatted as double-precision floating-point values. Under this formatting, eight bytes of computer memory are used to represent real numbers in the range of $\pm 1.79 \times 10^{308}$.

Above, we have provided the `int()` function with a real number argument and it is performing as designed, converting `3.6` to the integer `3` by truncating the real number at its decimal point.

Now recall that, when running **integer_parity_detector.py**, the entry from the `input()` function is formatted as a string data type. Thus, in our program, the `int()` function is evaluating the string `'3.6'`and not the floating-point number `3.6`. At the Python prompt within the Shell, try entering `'3.6'`as the argument for the `int()` function (remember that the surrounding single quotation marks format the quantity inside as a string):

```
>>> x = int('3.6')
```

You'll find that reproduces the traceback we observed when running **integer_parity_detector.py**.

When an error-producing bug like this is identified, Python provides the *try. . .except* statement for handling it. The form of this statement is as follows:

```
try:
    <block of code>
except <error_name>:
    <block of code to execute if error found in try block>
```

Note the important four-space indentation for the blocks of code associated with the *try* and *except* lines.

With **integer_parity_detector.py** open in the Editor, use **File>>Save As . . .** to create a new program named **integer_parity_detector_exception.py**. Then, modify the program as follows and save your result:

```
entry_string = input('Enter Integer:')

try:
    entry_integer = int(entry_string)
    remainder = entry_integer %2

    if remainder == 0:
        print('Integer is Even')
    elif remainder == 1:
        print('Integer is Odd')

except ValueError:
    print('Entry is Not Integer')
```

Run **integer_parity_detector_exception.py** by entering the value of *3.6* and demonstrate that the `ValueError` exception is handled by producing the output message **Entry is Not Integer**, rather than terminating the program with a traceback. Try entering alphabetic letter, like *A*, to see what happens.

Do It Yourself

Each of the book's chapters concludes with a *Do It Yourself* (DIY) project that poses an interesting problem and directs you in finding a solution by applying the chapter's material. A suggested name for the Python-program solution that you will write is given first; programming elements used in constructing the program will be explained. For this first DIY project, a completed solution program is given and you are asked to simply understand and replicate the program as shown, then demonstrate that your code can be run successfully. In subsequent chapters, the DIY projects will often be more loosely directed, requiring you to fill in some of the programming details.

parallel_resistors.py For two resistors R_1 and R_2 in parallel, the equivalent resistance R is given by $R = \frac{1}{\frac{1}{R_1}+\frac{1}{R_2}}$. Write a program that asks a user to input the resistor values R_1 and R_2, calculates the equivalent parallel resistance R, and then displays this value in the Shell.

The `input(prompt)` command returns the single string entered at the Shell's `prompt`. If this single string is a concatenation of two substrings separated by a specified `separator` (such as a comma), the string method `split(separator)` can be used to break the string produced by `input(prompt)` into the two substring values as shown in the code below. Note that the two substrings can be accessed by putting two variable names to the left of the equal sign (i.e., the assignment operator). The complete **parallel_resistors.py** program is as follows. Here `float()` converts the string representation of a number to its floating-point representation:

```
prompt = 'Enter Two Comma-Separated Resistor Values (Ohm):'
r1_string, r2_string = input(prompt).split(',')

r1 = float(r1_string)
r2 = float(r2_string)

r = 1/(1/r1 + 1/r2)
r_integer = int(r)

print('Equivalent Parallel Resistance (Ohm)=', r_integer)
```

Save your work, then run the program, inputting two resistance values. For $R_1 = 100\,\Omega$ and $R_2 = 400\,\Omega$, the input and output in the Shell will appears as shown here:

```
Enter Two Comma-Separated Resistor Values (Ohm):100,400
Equivalent Parallel Resistance (Ohm)= 80
```

A final note about the `print()` command: As shown above, if you'd like to print the value of an object named `item`, along with a text message, the syntax is as follows:

```
    print('The Value of Item =', item)
```

For future reference, if you'd like to print the values of multiple objects, use the following the syntax:

```
    print(item_1, item_2,...)
```

Use It!

Each of the book's chapters also concludes with a *Use It!* example, which is a useful tool for your everyday work in the laboratory. Use It! examples are based on (and, in some cases, extend) the Python programming techniques presented in their associated chapters. In later chapters, some of the Use It! examples will highlight hardware circuitry commonly employed in computer-based data acquisition applications. Completed Python programs are given in Use It! examples, along with explanations of how they function. It is left to you to employ your skills to replicate the examples.

Comments for Program Documentation It is good programming practice to provide comments within your code to explain (to other programmers and to your future self) the objective of the different code blocks within your program. Comments are ignored by the Python Interpreter and have the sole purpose of enhancing readers' understanding of the code. Python offers two forms of program commenting: *docstring* and *single-line comment.*

A docstring appears at the top of a program and is a (typically multiline) text statement, surrounded on each side by three single or (most commonly) double quotation marks. This statement provides an explanation of what the program does and may provide useful information about relevant quantities (for example, in the case of yet-to-be-discussed *functions*, definitions of input parameters and output return values).

A single-line comment begins with a hash symbol (#) and a space, followed by text that explains the code block on the line below it. Ideally, comments are concise, complete English sentences. If the comment is multiline, each line must begin with a hash symbol and space. The purpose of commenting is to explain *why* a code block does what it does; comments are not needed to simply explain *what* it does (e.g., a `print()` statement is self-explanatory and does not require a comment).

As an example, here is the **parallel_resistors.py** program from this chapter's Do It Yourself project with a docstring and single-line comments included:

```
"""
Given two resistor values as input, the equivalent parallel
resistance is calculated and displayed. Resistance values
are in the unit of Ohm.
"""

# Single string is input with two resistance values
# separated by a comma. Single string is split at comma
# into two strings, each representing one of the resistor
# values.
prompt='Enter Two Comma-Separated Resistor Values (Ohm):'
r1_string, r2_string = input(prompt).split(',')

# Convert string to float data type in preparation for
# calculation.
r1 = float(r1_string)
r2 = float(r2_string)

# Calculate equivalent parallel resistance as float and
# then truncate to integer.
r = 1/(1/r1 + 1/r2)
r_int = int(r)

print('Equivalent Parallel Resistance (Ohm)=',r_int)
```

Problems

To test your programming skill, every chapter ends with a selection of homework-style problems. Each problem begins with a suggested descriptive name (including the **.py** extension) for the solution program that you will write. Programming tips (e.g., suggested functions for use in the solution program) are given at the end of many of the problem statements. In this chapter, programs are written in the Editor and output is printed in the Shell.

1. **celsius_to_fahrenheit.py** Write a program that, given a temperature T_C in Celsius, converts the temperature to its Fahrenheit scale equivalent T_F, and then prints this value. The formula to convert between the two temperature scales is $T_F = \frac{9}{5}T_C + 32$. Use your program to show that 20 °C is equivalent 68 °F.
2. **account_balance.py** If an investor deposits an initial amount P ("principal") into a bank account with a fixed annual percentage interest rate I, the accrued amount B ("balance") in the account after N years is given by $B = P \times (1 + I/100)^N$. Write a program that prompts the user to input the three values of P, I, and N. Given these three input values, the program calculates the account balance, and then prints that resulting balance. Show that a $1000 initial deposit becomes $1480 after 10 years, if the interest rate is 4%. Programming tip: In Python, x^y is written as `x**y`.
3. **equivalent_integer_detector.py** Assume that a floating-point number with all zeros to the right of its decimal point can be termed "equivalent to an integer" (e.g., *3.000*... is equivalent to the integer *3*). Create a program that, when given a float in response to the prompt **Enter Floating-Point Number:**, determines if that float is equivalent to an integer and then prints the appropriate message, either **Entry is Equivalent to Integer** or **Entry is Not Equivalent to Integer**. Programming tip: `float()` converts its string or integer argument to its floating-point representation.
4. **area_of_circle.py** The area of a circle is $A = \pi r^2$. Here, r is the radius of the circle and it is a positive number. Create a program that prompts the user to input a value for a circle's radius. The program then uses this value (as a float) to calculate the circle's area and prints the result as **Area of Circle = XXX**, where XXX is the calculated area. However, if the user enters a negative value at the prompt, the message **Radius Must Be Positive Number** is printed. Programming tip: In Python, x^y is written as `x**y` and the *less than* operator is the "<" symbol. Also, use the approximation $\pi = 3.1416$.
5. **and_gate.py** Write a program that accepts two Boolean values (i.e., `True` or `False`) as *operands* and performs the logical AND operation on these operands. Print the *resultant* Boolean value of the AND operation. Verify that the only time the resultant is `True` is when both operands are `True`. Programming tip: Given two operands `x` and `y`, the AND operation is performed by the Python command `x and y`.
6. **python_program_detector.py** The file name of a Python program has the form **file_name.py**; i.e., the extension is **.py**. Write a program that prompts a user for a file name (including the extension) and then determines whether the file is a Python program. Report the result by printing **File is Python Program** or **File is Not Python Program**. Use your program to show that **and_gate.py** is a Python program and **voltage_data.txt** is not.

7. **factor_of_3_and_4.py** Write a program that, given an integer as input, tests whether that integer can be divided evenly (i.e., remainder equals zero) by both 3 and 4. Alert the user of the test result by printing either **Does Divide Evenly By 3 and 4** or **Does Not Divide Evenly By 3 and 4**. Programming tip: The Python AND operation `x and y` is `True` only if both `x` and `y` are `True`.

8. **length_converter.py** A length has a *magnitude* (number denoting its size) and a *unit*. Write a program that when first run, presents the user with the message **Enter Length in Unit of cm or in (with Space Separating Magnitude and Unit):**. Then, if the user enters a length in centimeters (e.g., *10 cm*), the program converts the length to inches; if a length in inches is entered (e.g., *10 in*), the program converts the length to centimeters. Print the converted value XXX via the appropriate message **Length in Inches = XXX** or **Length in Centimeters = XXX**. If the user does not include a single space between the entered magnitude and unit, a `ValueError` exception will be raised. If this error occurs, make the program print the message **Entry Must Have Single Space Between Magnitude and Unit**. Show that when *10 cm* (*10 in*) is entered in your program, the reported conversion is 3.94 in (*25.4 cm*) and also if *10cm* or *10in* is entered, the error message is printed. 1 in = 2.54 cm.

CHAPTER 2

Graphical User Interface Using Tkinter

Throughout this book, we will use Python's default code editor IDLE to create our programs. As shorthand, the IDLE Shell window and IDLE Editor window will be called the *Shell* and *Editor*, respectively. If you favor (and have experience with) another Python code editor, please feel free to use that as you work through the book. The book's instructions should be easily adaptable to your preferred editor.

Additionally, as outlined in Chapter 1's Use It!, it is good programming practice to document your programs with comments. However, since the programs presented in this book are explained in the surrounding text and programming objects are given descriptive names, comments will not be included in the book's programs. To develop this discipline, it's highly recommended that you include your own comments as you replicate the book's programs on your computer.

2.1 Tkinter GUI Toolkit

In Chapter 1, we found that inputs to, and outputs from, an executing Python program can be managed through a terminal session or the Shell. For the laboratory-based programs we will be writing, though, a *graphical user interface* (abbreviated as *GUI* and pronounced "gooey") will be a much more effective means for handling experimental inputs and outputs. In this chapter, you will learn how to create a custom-made GUI, composed of an operating-system window with interactive inputs and outputs that perform such tasks as controlling experimental parameters and monitoring acquired data. With these skills in place, you will build a GUI for your **integer_parity_detector.py** program.

There are several toolkits (aka *frameworks*) available for building GUIs for Python programs, each with its own strengths and learning-curve steepness as well as possible requirements for licensing. For our work, we will use the *Tkinter* toolkit (pronounced "tee-kay-inter"), as it is the GUI toolkit included (for free) in the Python Standard Library and it has a gentler learning curve than most other options. Learning to use the Tkinter toolkit will provide a good introduction to the concepts you must master to create GUIs for your Python programs. While not up to the sheen of the GUIs of commercial products like Excel or Photoshop, our Tkinter-based GUIs will be equipped with all of the functionality we need for our laboratory programs.

The Python-Based Laboratory. John Essick, Oxford University Press. © John Essick (2024). DOI: 10.1093/9780191998478.003.0002

If, in the future, you wish to create more sophisticated GUIs for your Python applications, popular options are the PyQt-related toolkits.

Tkinter is a *Python binding*, a form of software that we will encounter several other times in our work. A bit of history will help explain the meaning of the technical term *binding*. In the late 1980s when GUIs were first becoming widely available, as an aid for programming tools used in engineering research, a University of California, Berkeley professor developed the open-source programming language *Tool Command Language (Tcl)*, which included a GUI toolkit named *Tk*. Because of its speed and simplicity, the combination of Tcl with its Tk toolkit (denoted as *Tcl/Tk*) rapidly became popular among open-source programmers in many fields. When the developers of Python decided to include a GUI toolkit as part of the Python Standard Library, instead of writing the required code from scratch, they simply wrote software that served as a Python interface to the Tk GUI toolkit. That is, this software takes inputs from a set of Python commands and translates them into the Tcl language in order to access the functionality of the Tk toolkit. Since this software serves as a connection between Python and Tk, it is called the *Tk interface*, or Tkinter for short. In software development, code that maps commands in one language to those of another language is termed a *binding*. Thus, Tkinter is a Python binding. Quite commonly, the term *wrapper* is used interchangeably with the term binding (but there is disagreement about whether this is strictly valid, with some saying a wrapper only interfaces separate code written in the same language).

Tkinter provides the basic building-block components needed for assembling an interactive GUI. The foundational element of the GUI is an operating-system window (commonly called the *root window*), which is created by using the command `Tk()`. This root window then acts as a *container* that is populated with your choice of other useful elements called *widgets*. Some of the useful available widgets we will become familiar with are *label*, *entry*, *optionMenu*, and *button*. In addition, the *frame* widget can help organize the GUI layout.

2.2 Tkinter Basics

To use Tkinter, we must first deepen our understanding of the Python Standard Library, which was loaded onto your computer when you installed Python. A file containing Python code is termed a *module* and the Python Standard Library is an extensive (well over 200) collection of modules, each designed to carry out a common task required by programmers; for complete documentation of this resource, use the search term "Python Standard Library." The library's modules divide into two types: *built-in* and *importable*. The built-in modules are automatically available to you as your write a Python program and include functions (e.g., the `int()` function we used in Chapter 1), constants (such as the Boolean `True` and `False`), data-type operations (including the *Equal to* comparison we have used), and exceptions (for error handling). In contrast, the other modules in the library are only available for use in your programs if you first use the `import` command. Of present interest to us, the `tkinter` module is an importable module (note, by convention, Python module names are all lowercase). So, to use this module in your program, at the top of your program the statement `import tkinter` must appear. By loading into memory at runtime only those importable modules that have been explicitly imported, Python minimizes memory usage and increases performance.

Let's first explore a few features of Tkinter in an interactive terminal session within the Shell. As we explore these features, it's important to keep in mind that Tkinter implements object-oriented programming, where a command is used to create an object. The properties of that object (called *attributes*) can then be set by associated functions (called *methods*).

Start IDLE and, at the Python prompt in the Shell, import the `tkinter` module with the following (lowercase) command:

```
>>> import tkinter
```

Next, use the assignment operation, along with the `Tk()` command, to create an operating-system window object with the name `root`. To alert Python that the `Tk()` command is found in the `tkinter` module, use the dot notation `module_name.command_name()` as below:

```
>>> import tkinter
>>> root = tkinter.Tk()
```

A root window with the default name **tk** will appear on your computer screen.

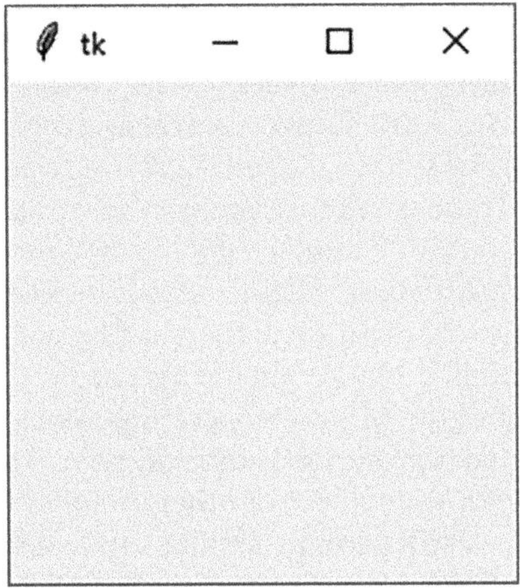

There are a few methods associated with this object, including `geometry()` and `title()`, which control two of its attributes, pixel size and title-bar label, respectively. Try test-driving these methods, using the dot notation `object_name.method_name()`:

```
>>> import tkinter
>>> root = tkinter.Tk()
>>> root.geometry('750x250')
>>> root.title('My Window')
```

The window will now have a size of 750-by-250 pixels with the title **My Window** in its title bar.

Create a text-label object called *label* using the Tkinter command `Label()`, which has the form

```
tkinter.Label(container, text='<text to be displayed>')
```

Note that, as with all Tkinter commands, the name of the command is written in *camel case*, where the first letter of a word is capitalized, the rest of the word's characters are lowercase. As another example, Tkinter's multiword command for the *option menu* widget is written as `OptionMenu()`. Add the label-creation statement shown below:

```
>>> import tkinter
>>> root = tkinter.Tk()
>>> root.geometry('750x250')
>>> root.title('My Window')
>>> label = tkinter.Label(root, text='This is a Label')
```

Note that after you pressed <*Enter*> and the label object was created, it did not appear as yet in its designated container (i.e., the `root` window). To place a Tkinter widget into its designated container, there are three possible ("geometry manager") methods for controlling its placement—`pack()`, `grid()`, and `place()`. The `pack()` method is the easiest to implement as it simply organizes widgets in only vertical and horizontal positions within the widget's container. Hence, placement options are limited to just `TOP` (default), `BOTTOM`, `LEFT`, and `RIGHT`, where the placement option is chosen using a parameter called *side*. The uppercase directional constants are contained in the `tkinter` module so, for example, placement at the left side of the container is chosen via `side=tkinter.LEFT`. Use this method to place the label object at the bottom of the `root` window as follows:

```
>>> import tkinter
>>> root = tkinter.Tk()
>>> root.geometry('750x250')
>>> root.title('My Window')
>>> label = tkinter.Label(root, text='This is a Label')
>>> label.pack(side=tkinter.BOTTOM)
```

The window will now appear as shown here:

Finally, the `root` window has a method called `mainloop()`, which causes the Tkinter-based application to run in an infinite loop, constantly checking to see if an event such as a button press or data entry has occurred. Constantly checking for such events is necessary as they usually trigger a program to branch off to some important operation. This process is termed *event handling*. However, the event handling of a GUI can monopolize a lot of your computer processing time and end up blocking other important processes like communication with a data-acquisition device. In later chapters, we will come back to this problem and solve it using a technique called *threading*. For now, complete your session by entering `root.mainloop()` as shown:

```
>>> import tkinter
>>> root = tkinter.Tk()
>>> root.geometry('750x250')
>>> root.title('My Window')
>>> label = tkinter.Label(root, text='This is a Label')
>>> label.pack(side=tkinter.BOTTOM)
>>> root.mainloop()
```

Your program is now in an infinite loop, waiting to do something once an event (such as a button press) is detected. There's no event-producing widget in our window, so the program will be stuck in this infinite loop until we close the window. One indication of this preoccupation with event handling is the lack of a Python prompt in the Shell after `mainloop()` was activated. Stop this program by closing the `root` window with **File≫Close Window** or clicking on the **close button** in the window's title bar.

Before we move on, here's some good news. Above, we imported the `tkinter` module using the command `import tkinter` and so, for example, when we used its `Tk()` function, the required dot notation was `tkinter.Tk()`. However, Python allows the following labor-saving form of the import command: `import tkinter as tk`. Then, employing the declared abbreviation `tk`, the command to use `Tk()` becomes the much-shortened `tk.Tk()`.

2.3 GUI-Equipped Parity-Test Program

Now let's write a Python program that tests the parity of an integer and is equipped with an interactive GUI for supplying input and viewing outputs. With IDLE open, in the Editor, create a file using **File≫New File** and then save this file in the **YourNamePythonWork** folder under the name **integer_parity_detector_gui.py** via **File≫Save**.

Then, in the Editor, write the following GUI-producing code, which takes advantage of the abbreviated form of `tkinter`. We'll explore the meaning of the new commands after you finish writing this code and saving it.

```
import tkinter as tk

root = tk.Tk()
root.title('Integer Parity Detector')
root.geometry('300x120')
```

```
lbl_1 = tk.Label(root, text='Enter Integer in Box Below:')
var = tk.StringVar()
ent = tk.Entry(root, textvariable=var)
lbl_2 = tk.Label(root, text='', width=20)
btn = tk.Button(root, text='Parity Test', command=None)

lbl_1.pack()
ent.pack()
lbl_2.pack()
btn.pack()

root.mainloop()
```

Save your work. The first five lines are familiar from our terminal session. The `tkinter` module is imported and assigned the abbreviation `tk`, while the next three lines create a root window with a certain size and name in its title bar. The fifth line implements the `Label()` command that we explored in our terminal session. This command creates a *prompt label* in the `root` window that reads **Enter Integer in Box Below:**. The label object is assigned the (commonly used) abbreviated name `lbl_1` (shortly, there will be a `lbl_2`).

The next four lines are composed of mostly new commands. These commands create four GUI-related objects—string variable, entry widget, label widget, and button widget—that are assigned the (commonly used) abbreviated names `var`, `ent`, `lbl_2`, and `btn`, respectively.

The first two objects are related and create what we'll call the *entry box*. The entry widget named `ent` created by `Entry()` is a rectangular box in which a user can input a single line of text (i.e., a string) from the keyboard, while `StringVar()` creates the *string variable* object `var` that stores the input text. To create the entry widget `ent`, the form of the `Entry()` command is as follows:

```
ent = tkinter.Entry(container, textvariable)
```

Here, `container` designates the GUI container (in our program, the `root` window) in which to place the widget and `textvariable` specifies the variable in which the entered text is to be stored (for us, that's `var`).

Next, a `Label()` command creates an object `lbl_2` we'll call the *output label*. As before, the first parameter of `Label()` designates `root` as the label's container. However, there are two more parameters, one of which we haven't encountered before. The option `text=''` (there is no space between the two quotation marks) initializes the label as empty of text, while the option `width=20` specifies the horizontal width of the rectangular box housing the label to be 20 *text units* (i.e., not 20 pixels). A text unit is the width of a character in your system's default font; if that font is not monospaced, the text unit is taken to be the width of the number-zero character (0).

For future reference, the background color of a label widget (i.e., the color behind the displayed text) can be controlled by the *background* parameter (`bg`, for short) of `Label()`. So, for example, to obtain a white background, include `bg='white'` as an argument of `Label()`. Some of the other possible choices are `'black'`, `'red'`, `'green'`, and `'blue'`. In the above program, we have not included the parameter `bg` when calling `Label()`; thus, the label's background color will be your computing system's default color, which is the same (default) background color as the root window.

Finally, a push button widget named `btn` is created using the `Button()` command, which has the following form:

```
btn = tkinter.Button(container, text, command)
```

In our program, the specified container is `root`. The option `text='Parity Test'` places this text within the GUI button, while the `command` option specifies an action to take when the button is clicked by the mouse; i.e., this option provides event handling. At present, in our program, we have `command=None`, which obviously means no action is triggered by a button press.

The sequence of four `pack()` commands place the prompt label, entry box, output label, and button in the `root` window, in that order. Since the `side` option is not specified, the default value `tk.TOP` is used. Thus, the prompt label is placed first at the top (center) of the root window. The entry box is then placed at the topmost location that remains available, which is just below the prompt label. The output label again goes to the remaining topmost location, i.e., just below the entry box. Finally, the button goes just below the output label, which is the last remaining topmost location. So, our GUI is composed of a stack of four widgets, ordered from top to bottom as prompt label, entry box, output label, and button.

The last command in our program is `root.mainloop()`, which executes the program in an infinite loop as it waits for an event. A press of the button on our GUI is the event for which `root.mainloop()` awaits, and each button press will trigger the action specified by the `command` option within `Button()`, after which `root.mainloop()` will return the program to infinite-loop execution as it awaits the next event. Execution of the infinite loop is terminated when the `root` window is closed.

Let's run your (saved) program now. In the Editor, select **Run≫Run Module**. If all goes well, a GUI window named **Integer Parity Detector** will appear as shown below, with the four-widget stack of prompt text above a rectangular entry box, followed by a rectangular output label, and finally a push bottom labeled **Parity Test** at the bottom.

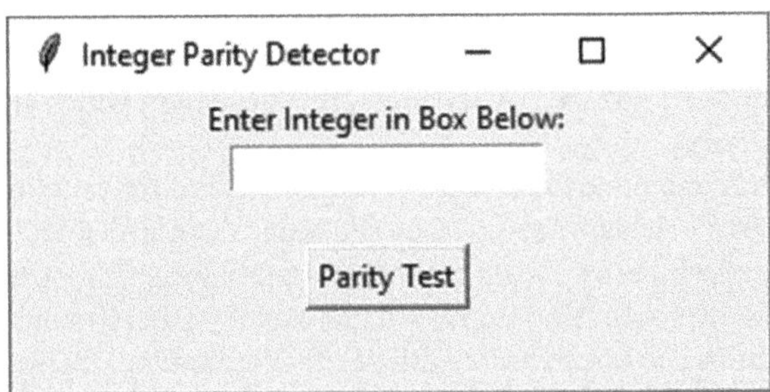

The rectangular output label will not be evident because it has the same background color as the root window. As our work unfolds, we will find that the rectangular output label is, in fact, where we have placed it. If you want to check now that it really is there, temporarily change its background color to blue with this modification to your program:

```
lbl_2 = tk.Label(root, text='', width=20, bg='blue')
```

Then, run the program and the output label will come out of hiding.

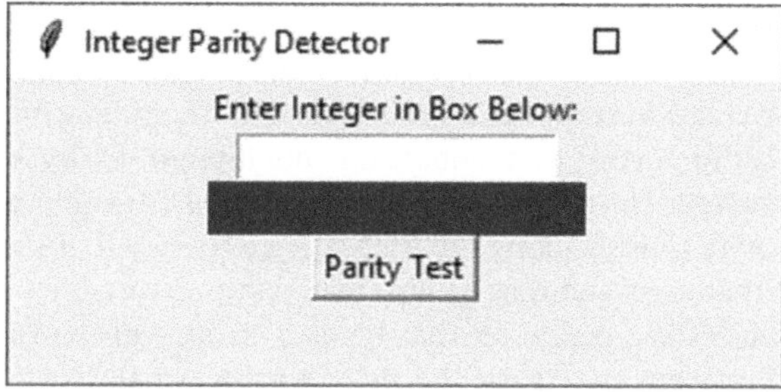

Delete the argument `bg='blue'` in `Label()`, then save and rerun the program. Click on the entry box with your mouse and type in an integer, say, *5*. Then, click on the **Parity Test** button. You will find that nothing happens, a result of requesting no action when the button is pressed by setting the `Button()` argument as `command=None`.

To trigger a useful action when the **Parity Test** button is pressed, we need to explore a new programming element—the Python *function*. While we are at it, we'll also probe further into a related element from the realm of object-oriented programming that we've already been implementing—the Python *class*.

2.4 Functions and Classes

A well-written Python program is hierarchical in nature. It consists of a top-level ("main") program, which calls lower-level subprograms. These subprograms may call even lower-level subprograms, which in turn may call still lower-level subprograms. There is no limit to the level of subprogram layering in a Python program. In this hierarchical structure, the main program acts as a supervisor, coordinating the execution of its collection of subprograms. Within this modular approach to programming, a subprogram is written to carry out each component task required for the successful completion of the greater program's overall mission. Modularity means that, when needed, each subprogram can be executed as a *stand-alone program* (i.e., run independent of the top-level program), simplifying the programmer's inevitable job of program debugging. Additionally, since subprograms are designed to perform well-defined tasks, they can be reused as components in other top-level programs.

Python has two types of subprograms, termed *functions* and *classes*. Many of these subprograms are built-in or importable elements of the programming language, available as prewritten code for your use. The definition of a Python function is analogous to the definition of a function in mathematics. A Python function is a block of code that, when provided input data, performs a specific task and returns a resulting output. The `input()` command we used in Chapter 1 is an example of one of Python's built-in functions.

A Python class is a more complex programming element than a function. It is the central tool of object-oriented programming, one of the programming paradigms supported by the Python language. In object-oriented programming, programming elements are defined to be *objects* with associated *attributes* and *methods*, which are properties of the object and functions that act on the object, respectively. A class is a generic blueprint for creating a specific type of object (e.g., a Tkinter label widget) with associated attributes (e.g., `width`, `height`, `font`, `text`) and methods (e.g., `config(text)` can be used to change text within the label object). Many of Python's built-in programming elements are classes and we have already been using several in our work. For example, in the terminal session of Chapter 1, hidden from our view, the assignment command `x = 'python'` used the string class named `str()` to create an *instance* of a string object called x with the attribute python. The str() class provided us access to its method capitalize() through use of the dot notation object.method(). In addition, all of Tkinter's widget-creation commands are classes and their definitions are contained in the tkinter module. By importing that module into our **integer_parity_detector_gui.py** program, we were able to access those widget-creation classes. For example, the command var = tkinter. StringVar() is invoking the StringVar() class to *instantiate* (i.e., create) a particular instance of that class and that newly created string-variable object is assigned the name var. The var object is a container that holds a string and the value of that string is an attribute of var. If we want to retrieve the string's value, we utilize the get() method of the StringVar() class via use of the dot notation, i.e., value = var.get(). On the other hand, the set() method of the StringVar() class allows one to change the value of the string held by var as follows: var.set(new_string).

Python's libraries are filled with an extensive collection of built-in functions and classes. In addition, Python gives us the ability to write our own custom-written versions of these types of programming elements as well. Writing a custom-made class is an involved topic and so, in this book, we will only use the prewritten classes in Python libraries. However, creating a custom-made function is straightforward and we will write quite a few. It is this latter skill on which we now wish to focus our attention.

To create your own function, use the following form to define it:

```
def function_name(<parameter_1>, <parameter_2>, ...):
    <statements>
    return <output_1>, <output_2>, ...
```

Here, def is the keyword for creating a function, and that keyword is followed by the function name, which (by convention) is all lowercase. If the function is given a multiword name, the words are separated by underscores. Remember the all-important colon at the end of this first line and the four-space indentation for the body of the function. Data are input from the calling program via the <parameter> values, which are then processed by the <statements>. Processing results are returned to the calling program in the <output> values.

The syntax for calling a function and receiving its return value(s) is as follows:

```
<output> = function_name(<parameters>)
```

where <output>=<output_1>, <output_2>, ... and <parameters>= (<parameter_1>, <parameter_2>, ...

Going forward, many of the classes and functions we will encounter have longish names and/or quite a few parameters and arguments. To better fit such a class or function on the book's page as well as to render its multiple parameters and arguments more readable, we will follow the Python convention of wrapping parentheses-enclosed arguments and parameters onto multiple lines, with the wrapped elements (typically) aligned with the element next to the opening parenthesis of the class or function. For example, by this convention, the statement

```
lbl = tk.Label(root, text='Enter Integer in Box Below:')
```

may be rewritten as

```
lbl = tk.Label(root,
               text='Enter Integer in Box Below:')
```

By convention, within its parentheses, neighboring parameters and arguments of a class or function are separated by a comma followed by a space. Additionally, arguments such as `text='Enter Integer in Box Below:'` are called *keyword arguments*. For a keyword argument, no spaces should be placed around the equal sign.

2.5 Adding Trial Function to Parity-Test Program

Going back to our program **integer_parity_detector_gui.py**, we can now code in a trial function to perform the following action when the button is pressed: After a user has entered a string into the entry box, that string is displayed in the output label when the user clicks the button.

To accomplish our goal, we will implement methods associated with Tkinter's built-in classes `StringVar()` and `Label()`. First, included in our program is the statement `var = tk.StringVar()`. Here, the `StringVar()` class is being used to create a string variable named `var`, which is then used to store the text a user has entered in the entry box. Our program can access the value of `var` via the `get()` method of `StringVar()`, i.e., through the dot-notation command `var.get()`. Second, within our program, the `tk.Label()` class is used to create the output label named `lbl_2`, which is initialized to display an *empty string* (i.e., a string devoid of text). As the program is executed, the value of the text displayed in `lbl_2` can be changed via the `config(text)` method of `Label()`, i.e., through the dot-notation command `lbl_2.config(text)`. Thus, the text entered in the entry box can be used to update the text displayed in the label by the following function called `update_label()`:

```
def update_label():

    entry_string = var.get()
    message = entry_string
    lbl_2.config(text=message)

    return
```

Note that this function has no input parameters from, or output values to, the calling program. However, a set of parentheses is still required as part of the name in order to identify it as a function.

Add this function to your program as shown below. Note that now within `Button()`, we have changed `command=None` to `command=update_label` (without the parentheses) so that this function is executed when the button is pressed. Also note that, within the program, the function definition must appear before the function is called.

```
import tkinter as tk

def update_label():

    entry_string = var.get()
    message = entry_string
    lbl_2.config(text=message)

    return
```

```
root = tk.Tk()
root.title('Integer Parity Detector')
root.geometry('300x120')

lbl_1 = tk.Label(root, text='Enter Integer in Box Below:')
var = tk.StringVar()
ent = tk.Entry(root, textvariable=var)
lbl_2 = tk.Label(root, text='', width=20)
btn = tk.Button(root, text='Parity Test',
                command=update_label)

lbl_1.pack()
ent.pack()
lbl_2.pack()
btn.pack()

root.mainloop()
```

Save your work, then run the program and see if it performs as expected.

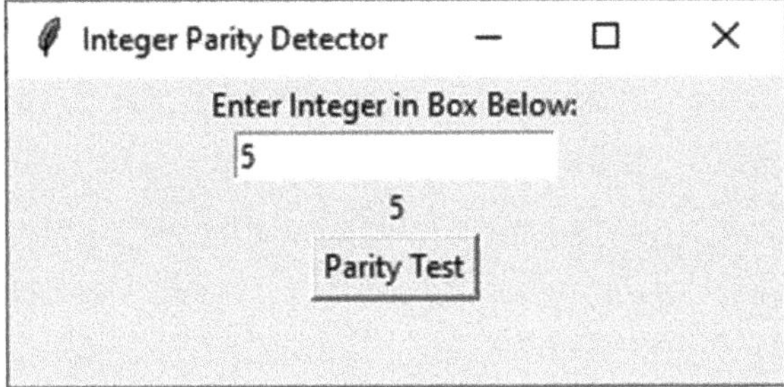

Note that, by default, the text within an entry widget and a label widget is left justified and center justified, respectively. If you'd like the text within the entry box to be center justified in order to match the justification of text within the output label, include the `justify` keyword argument in the `Entry()` class as shown here:

```
ent = tk.Entry(root, textvariable=var, justify=tk.CENTER)
```

The options for `justify` are `tk.LEFT` (default), `tk.CENTER`, and `tk.RIGHT`.

2.6 Adding Parity-Test Function

Now let's write a function that determines the parity of an input integer. The function will be used as a subprogram to our main program. In the Editor, open **integer_parity_detector_exception.py**. Then, using **File≫Save As...**, create a new program named **parity_test_function.py** in the **YourNamePythonWork** folder. Define a function named `even_or_odd()` by modifying the program as shown next. Save your work.

```
def even_or_odd(entry_string):

    try:
        entry_int = int(entry_string)
        remainder = entry_int % 2

        if remainder == 0:
            message = 'Integer is Even'
        elif remainder == 1:
            message = 'Integer is Odd'

    except ValueError:
        message = 'Entry is Not Integer'

    return message
```

A Python file with the extension **.py** such as **parity_test_function.py** is termed a module and code from any module can be placed within another Python program using the `import` command. As shown below, **integer_parity_detector_gui.py**, which is our top-level ("main") program, can be completed by importing the module **parity_test_function.py** via the command `import parity_test_function as ptf` (this import statement appears at the top of the main program). Then, **parity_test_function**'s parity-detecting function `ptf.even_or_odd()` can be used within the main program's `update_label()` function, where the object named `message` is assigned the return value of `ptf.even_or_odd()`. With **integer_parity_detector_gui.py** open in the Editor, complete the program as follows (note the text within the entry box is now center justified):

```
import tkinter as tk
import parity_test_function as ptf

def update_label():

    entry_string = var.get()
    message = ptf.even_or_odd(entry_string)
    lbl_2.config(text=message)

    return

root = tk.Tk()
root.title('Integer Parity Detector')
root.geometry('300x120')

lbl_1 = tk.Label(root, text='Enter Integer in Box Below:')
var = tk.StringVar()
ent = tk.Entry(root, textvariable=var, justify=tk.CENTER)
lbl_2 = tk.Label(root, text='', width=20)
btn = tk.Button(root, text='Parity Test',
                command=update_label)
```

```
lbl_1.pack()
ent.pack()
lbl_2.pack()
btn.pack()

root.mainloop()
```

Save your work, then run **integer_parity_detector_gui.py** and enjoy the results!

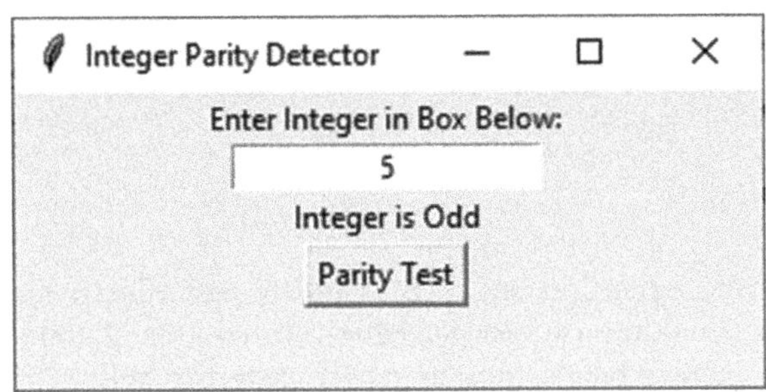

2.7 Adding a GUI Frame

In designing a graphical user interface, it's worth devoting some extra effort to improving its visual appeal and overall user-friendliness. Tkinter offers a useful widget for this purpose—the *frame*. The frame is a rectangular-shaped container, with an optional border, in which to place one or more widgets, allowing one to organize the GUI layout. In addition, *padding* around the perimeter of a frame can be used to provide spacing between neighboring GUI elements.

To create a Tkinter frame named `frame` inside of a specified `container` (e.g., the `root` window) with some selected `options` (such as `height`, `width`, `background color`, `border width`), the command is as follows:

```
frame = tkinter.Frame(container, options)
```

Once created, the `frame` object is placed within the specified `container` using the command `frame.pack(side, padx, pady)`. The (familiar) parameter `side` determines the placement of a widget within its container via `tk.TOP` (default), `tk.BOTTOM`, `tk.LEFT`, or `tk.RIGHT`. The (new-to-us) parameters `padx` and `pady` specify the amount of horizontal and vertical padding, respectively, to leave outside of each side of the frame (or, for a bordered frame, outside of its border), where the padding is measured in pixels. Finally, the following `options` produce a blue-bordered frame: `bg='blue'` and `bd=2`. These options set the background (and border) color of the frame to blue and place a border around the frame that is two pixels thick, respectively. Thus, to place such a blue-bordered frame in the `root` window, with zero padding beyond the border, the code is as shown here:

```
frame = tk.Frame(root, bg='blue', bd=2)
frame.pack(padx=0, pady=0)
```

With **integer_parity_detector_gui.py** open in the Editor, modify the program as follows, so that the prompt label, entry box, and output label are placed within a blue-bordered frame (i.e., their `container` is `frame`) and the button is positioned below the frame (i.e., its `container` is `root`):

```
import tkinter as tk
import parity_test_function as ptf

def update_label():

    entry_string = var.get()
    message = ptf.even_or_odd(entry_string)
    lbl_2.config(text=message)

    return

root = tk.Tk()
root.title('Integer Parity Detector')
root.geometry('300x120')

frame = tk.Frame(root, bg='blue', bd=2)
frame.pack(padx=0, pady=0)


lbl_1 = tk.Label(frame, text='Enter Integer in Box Below:')
var = tk.StringVar()
ent = tk.Entry(frame, textvariable=var, justify=tk.CENTER)
lbl_2 = tk.Label(frame, text='', width=20)
btn = tk.Button(root, text='Parity Test',
                command=update_label)

lbl_1.pack()
ent.pack()
lbl_2.pack()
btn.pack()

root.mainloop()
```

Save your work, then run the program. As show below, a few problems are apparent. By setting the frame's background color to blue, its entire interior is filled with that color and, because the widgets placed within the frame are not wide enough, some of the frame's background blue color is uncovered. In addition, the button is a bit close to the bottom of the frame.

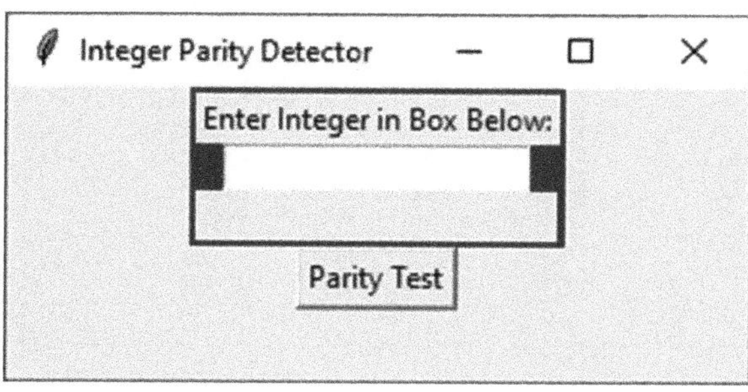

These two problems can be solved as follows. First, for the three widgets within the frame, include the argument `fill=tk.X` in their `pack()` functions. This option will cause these objects to completely fill the frame in the horizontal (`X`) direction. Second, to provide some vertical separation between the bordered-frame and the button, simply set `pady` in the frame's `pack()` function to a nonzero (pixel) value, say, `4`. Make these changes to **integer_parity_detector_gui.py**:

```
import tkinter as tk
import parity_test_function as ptf


def update_label():

    entry_string = var.get()
    message = ptf.even_or_odd(entry_string)
    lbl_2.config(text=message)

    return

root = tk.Tk()
root.title('Integer Parity Detector')
root.geometry('300x120')

frame = tk.Frame(root, bg='blue', bd=2)
frame.pack(padx=0, pady=4)

lbl_1 = tk.Label(frame,text='Enter Integer in Box Below:')
var = tk.StringVar()
ent = tk.Entry(frame, textvariable=var, justify=tk.CENTER)
lbl_2 = tk.Label(frame, text='', width=20)
btn = tk.Button(root, text='Parity Test',
                command=update_label)

lbl_1.pack(fill=tk.X)
ent.pack(fill=tk.X)
lbl_2.pack(fill=tk.X)
btn.pack()

root.mainloop()
```

Save your work, then run your completed GUI-equipped program.

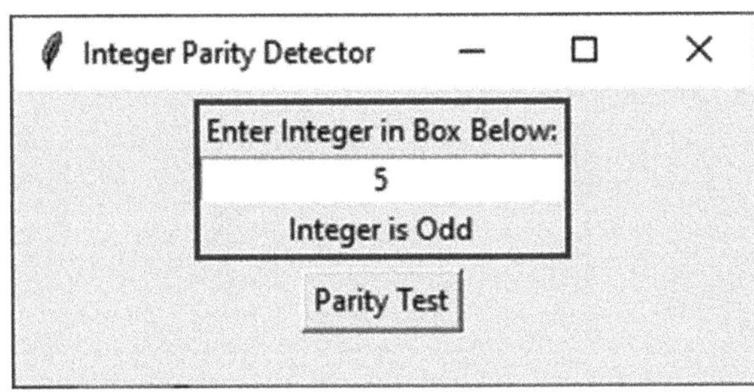

Our GUI is intended to mimic the front panel of a laboratory instrument, whose features are designed to input experimental parameters via *controls* (e.g., knobs, buttons, switches) and output acquired data via *indicators* (e.g., LEDs, numeric and graphical displays). Thus, going forward, an *entry box* and an *output label* will be called a *control* and *indicator*, respectively.

Do It Yourself

fahrenheit_to_celsius.py Write a GUI-equipped program to do the following: Given the temperature T_F in Fahrenheit, use $T_C = (T_F - 32) \times \frac{5}{9}$ to convert the temperature to its Celsius scale equivalent T_C. Display the Celsius temperature with one digit of precision, i.e., one digit to the right of the decimal point. The function `round(number, ndigits)` rounds a float named `number` to `ndigits` of precision. Once completed, your program should appear as shown here, where we see that it correctly converts 68 °F to 20 °C.

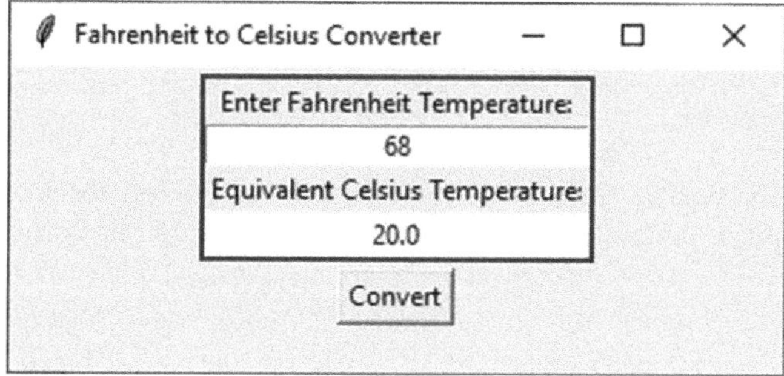

Use It!

Executing Module as Main Program In our **integer_parity_detector_gui.py** program, we carried out the parity check in a function called `even_or_odd(entry_string)` contained in a separate module named **parity_test_function.py**. In this programming approach, **integer_parity_detector_gui.py** is designated the "main" program and the module **parity_test_function.py** is subordinate to it; i.e., **parity_test_function.py** is behaving as a subroutine and is only being run when it is "called" by the main program. However, when developing a module like **parity_test_function.py**, it's often helpful to run it as a stand-alone program in order to debug it. By putting the following block of code at the bottom of a module (which we'll call a *test block*), the module can be run as a stand-alone program with the *<set of statements>* executed:

```
if __name__ == '__main__':

    <set of statements>
```

Here, both `__name__` and `__main__` include two underscores on either side of the term within (i.e., `name` and `main`) and are examples of special Python quantities called *dunder attributes*, where *dunder* stands for double underscore.

Open **parity_test_function.py** in the Editor and add the test-block coding at the bottom of the program as shown:

```
def even_or_odd(entry_string):

    try:
        entry_int = int(entry_string)
        remainder = entry_int % 2

        if remainder == 0:
            message = 'Integer is Even'
        elif remainder == 1:
            message = 'Integer is Odd'

    except ValueError:
        message = 'Entry is Not Integer'
    return message

if __name__ == '__main__':

    entry_string = input('Enter integer:')
    message = even_or_odd(entry_string)
    print(message)
```

Save **parity_test_function.py**, then run it. You will find that the test-block coding causes the program to execute as a stand-alone program, allowing you to verify that it performs as intended.

Problems

For the reason mentioned at the end of the chapter text, in these problems, an *entry box* and an *output label* will be called a *control* and *indicator*, respectively.

1. **gui_message.py** Write a program that produces this GUI:

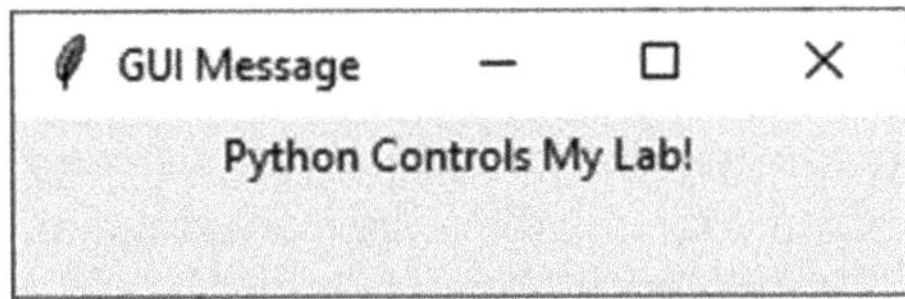

2. **exit_button.py** Write a program with the following GUI: Configure the program so that, when the **Exit** button is pressed, the window closes. Programming tip: Once created, a `root` window is closed by the command `root.destroy()`.

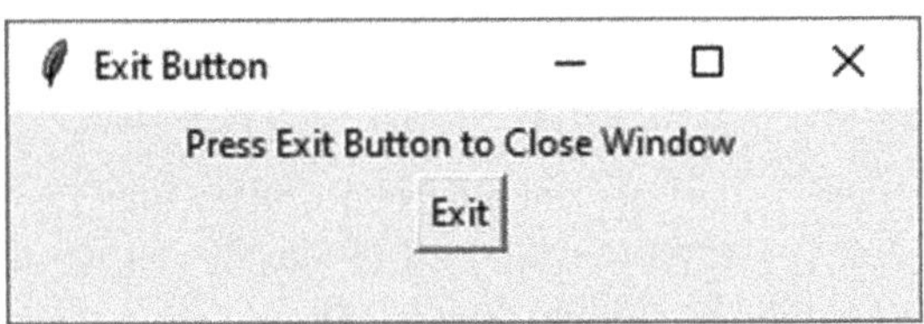

Next, try changing `root.destroy()` to `root.quit()`. When you run the program now, the infinite loop produced by the `root.mainloop()` command quits execution. If there was code below the `root.mainloop()` command, program execution would move to that code. However, if there is no code below `root.mainloop()`, the program ends its execution, while the root window remains.

3. **widget_order.py** Create the following window with two labels `lbl_1` and `lbl_2` whose text is **I am Label A** and **I am Label B**, respectively:

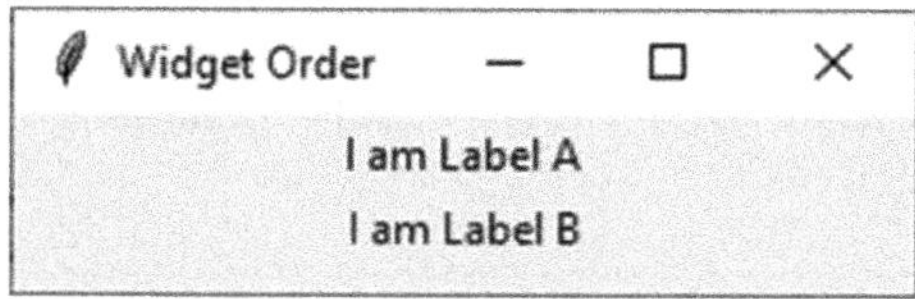

While keeping the text for labels `lbl_1` and `lbl_2` still as **I am Label A** and **I am Label B**, respectively, make the needed change in your program so that the window now appears as follows:

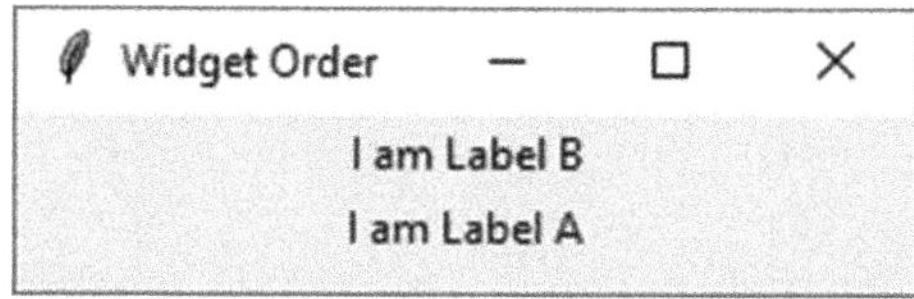

4. **grid_placement.py** The `grid()` geometry manager divides a container (e.g., root window or frame) into rows and columns. You then control the placement of a widget by specifying row and column indices as arguments of the `grid()` method. To understand how this geometry manager works, complete the following code:

```
import tkinter as tk

root = tk.Tk()
root.geometry('400x80')
root.title('Placement Using Grid')

lbl_1 = tk.Label(root, text='Number of Samples')
lbl_2 = tk.Label(root, text='Sampling Frequency (Hz)')
ent_1 = tk.Entry(root)
ent_2 = tk.Entry(root)

lbl_1.grid(row=0, column=0)
lbl_2.grid(row=1, column=0)
ent_1.grid(row=0, column=1)
ent_2.grid(row=1, column=1)

root.mainloop()
```

Save **grid_placement.py**, then run it. We haven't assigned string variables to the controls, so this program only demonstrates the GUI layout that can be produced using the `grid()` geometry manager.

5. **compound_interest.py** Assume an investment account starts with \$1.00 and pays 100% interest per year. If the interest is credited annually, i.e., once at the end of the year, the accrued amount B ("balance") in the account at the end of the year end will be \$2.00. If, instead, the interest is compounded over N equally spaced time intervals during the year (for example, $N = 12$ for monthly compounding), then the balance at the end of the year will be $B = (1 + 1/N)^N$. Write a GUI-equipped program as shown below, with a control labeled **Number of Intervals:** to accept the value of N and an indicator to display the calculated balance at the end of the year. Use your program to find the balance when interest is compounded monthly as well as when N is large. Interestingly, the latter case converges to Euler's number (= 2.7182818284 . . .) as $N \to \infty$. Programming tip: In Python, x^y is written as `x**y`.

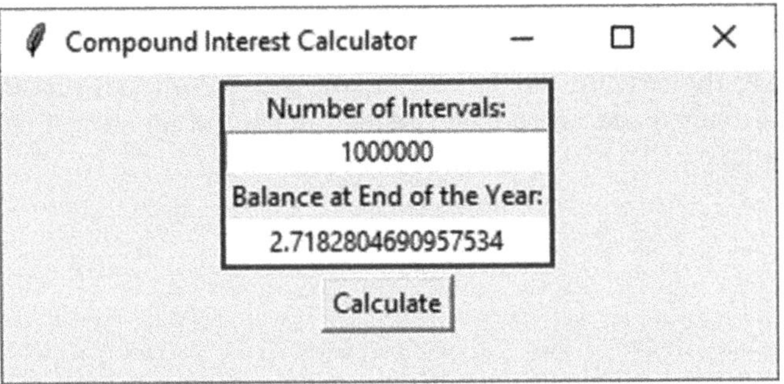

6. **repeat_n_times.py** Write a GUI-equipped program that displays an input character (or characters) repeated N times. As shown below, the GUI has two controls labeled **Enter Character(s):** and **Enter Number of Times to Repeat:** and, within the program, these controls should have associated string variables `var_1` and `var_2`, respectively. Below the two controls is an indicator with its own label **Display:**, where the repeated character(s) are viewed. In this GUI, there are two controls, three labels, one indicator, and one button, each of which needs to be placed in its designated container using a `pack()` command. Programming tip: In Python, the string `'R'` can be repeated 7 times with the command `'R'*7`; i.e., `'R'*7` produces the string `'RRRRRRR'`, where 7 is an integer (not a string).

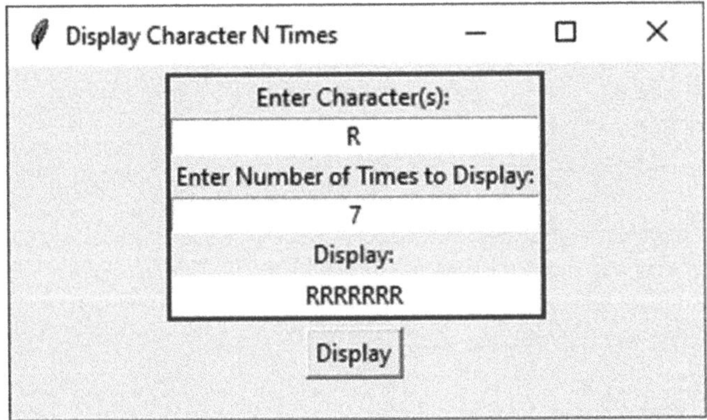

CHAPTER 3

Counted Looping and Waveform Plots

The programs presented in this book are rendered on a gray background and, for those programs, we will adhere to the following Python convention: Within its parentheses, neighboring parameters and arguments of a class or function are separated by a comma followed by a space. The class or function used in a program will often be explained in the paragraphs adjacent to the program and, to save book space, only a comma will be used to separate neighboring parameters and arguments there. For example, `tkinter.Frame(container, options)` will appear within the gray-background program, while this same class will be written as `tkinter.Frame(container,options)` in the surrounding explanatory paragraphs.

In addition, it is assumed that all of your written programs are stored in the folder named **YourNamePythonWork** (or, starting in Chapter 5, its subfolder named **Utilities**). From now on, instead of writing "the file in the **YourNamePythonWork** folder", I'll use the shorthand statement "the file in **YourNamePythonWork**".

3.1 Programming Structures and Python Packages

Python is a general-purpose, full-featured programming language. As such, it provides the means for creating programs to carry out a variety of computationally intensive tasks in areas such as scientific simulation, data analytics, and machine learning. In addition, through the ongoing development and ready availability of application-specific modules, Python is well suited for programming the laboratory and industrial tasks of concern to scientists and engineers, including computer-based instrument control, data acquisition, data analysis, and reporting.

As in all programming languages, Python programs implement tasks by using a combination of *programming structures*, each of which carries out one of the following three functions: *sequencing*, *branching*, or *looping*. Sequencing refers to controlling the ordered execution of instructions, that is, ensuring that the correct instructions are performed in the correct order. By design, Python simply executes a program line by line, starting at the first statement and progressing to the last statement. However, in writing a program, a function must be defined above the statement that calls it during execution (or else the Python Interpreter will raise a *NameError* exception). Thus, in a program, function definitions precede the block of

The Python-Based Laboratory. John Essick, Oxford University Press. © John Essick (2024). DOI: 10.1093/9780191998478.003.0003

statements being sequentially executed when the program is run; i.e., it is not unusual for the location of the first statement to be executed to be somewhat below the very top of the program. We will discover later that sequential execution can cause problems in data-acquisition applications, where a computer is commonly asked to communicate with a device streaming data from an experimental setup, while at the same time displaying that data on a monitor. We will find that Python offers a *threading* module, which solves that problem through parallel execution of multiple program threads.

Previously, we found that the *if... elif...else* statement is Python's branching structure, which allows a program to follow alternative paths of execution, depending on the resulting Boolean value of a logical test.

We now turn our attention to looping. For this category of operation, a looping structure executes a given task repeatedly until a certain condition is met. Python provides two such looping structures—the *for loop* and the *while loop*. The `for` loop, which is explored in this chapter, repeatedly executes a task for an initially specified number of times. Thus, the `for` loop is an example of a *counted looping* structure. As we will find, Python provides several equivalent methods for carrying out this type of repetitive task. In contrast, a `while` loop, the subject of Chapter 4, repeatedly executes a task until an associated Boolean quantity changes its value from `True` to `False`. In contrast to the `for` loop's counted looping, the `while` loop's mode of operation is termed *conditional looping*.

An essential requirement for many of our laboratory-based programs is access to a comprehensive numerical computing toolkit and a data display library. The core Python programming language and the Python Standard Library do not provide these functionalities. However, an online repository called the *Python Package Index* will provide the resources to fulfill our needs in the form of two Python-based software packages—*NumPy* and *Matplotlib*. NumPy (pronounced "num-pie") provides a large collection of mathematical (and other) functions that carry out operations on numeric arrays, while Matplotlib offers the programmer several easy-to-implement, yet powerful graphing tools.

In this chapter, you will write sine-wave plotting programs based on `for` loop-style (i.e., counted looping) calculations. In the process of writing these programs, you will explore Python's method of creating and storing numeric data in the form of a NumPy array and learn to produce waveform plots using Matplotlib. Before studying the `for` loop, let's learn how to place NumPy and Matplotlib on your computer.

3.2 Installing NumPy and Matplotlib from PyPI

Python programming has been adopted in wide-ranging fields of academia and industry. In most cases, though, "basic" Python—i.e., the "core" Python programming language, along with its Standard Python Library—is not sufficient to meet the needs of a particular field. Thus, to fulfill application-specific requirements, open-source programmers have created *third-party* software libraries that extend Python's functionality in ways that are useful in particular fields. For scientists and engineers, two essential extensions are in the areas of array-based numerical computation and data visualization and are realized by the third-party libraries *NumPy* and *Matplotlib*, respectively. These application-specific libraries are not included with the original download of basic Python onto your computer, but they are available for free at an online repository called the *Python Package Index* (abbreviated as *PyPI* and pronounced "pie-pea-eye"). PyPI is the official, and best vetted, repository of third-party Python libraries, but there are several other Python repositories including the popular *GitHub*.

To use NumPy and Matplotlib in your programs involves the following two-step process: First, install (i.e., download from PyPI) these libraries onto your computer, then place the `import numpy` and `import matplotlib` statements at the top of your programs. Python provides an easy-to-use third-party package installer named *pip* (shorthand for *Preferred Installer Program*) that, by default, installs packages from PyPI.

To install (lowercase) `numpy` using `pip`, start a terminal session (see Section 1.2), then type the following command on Windows:

```
pip install numpy
```

On macOS, your command should be modified to the following:

```
pip3 install numpy
```

To find the directory into which NumPy has been installed, enter the following command in your terminal session (change `pip` to `pip3`, if needed):

```
pip show numpy
```

Similarly, install `matplotlib` using `pip`.

Troubleshooting tip: The response to the `pip show numpy` and `pip show matplotlib` commands will show you the file path to the folder into which NumPy and Matplotlib have been installed. One (or more) of the folders in that path will be labeled with the Python version number (e.g., 3.12) associated with the install process. For your work in this book, make sure that the IDLE application you use (and the version of Python being run) has this same version number. When running a program in which you have imported NumPy and/or Mathplotlib, if you receive an error message to the effect that those modules cannot be found, you may have more than one version of Python installed on your computing system and you are not using the version possessing the folders into which NumPy and Mathplotlib were installed.

3.3 Python for Loop

Let's now explore the first of Python's two looping programming structure—the `for` loop. The `for` loop structure is used to *iterate* (i.e., repeat) a given set of statements for an initially specified number of times. The `for` loop is a common feature in computer programs, and it takes on varying forms in each. Quite often, an integer index i is attached to each iteration of the loop, starting at $i = 0$ for the first iteration and ending at $i = N-1$ at the last iteration, for a total of N iterations. Then, the form of the `for` loop is equivalent to the pseudocode

```
for i = 0 to N-1
    execute set of statements
    increment i by 1
```

In Python, the structure of the `for` loop does not conform to the above pseudocode. Instead, the iteration of the loop is over an *iterable object*, which can be a set of integer indices from $i = 0$ to $N-1$, but can also be, say, the set of characters in a string. Thus, the form of the Python `for` loop is as follows, where the first line ends in a colon (`:`) and all of the `statements` to be executed are indented by four spaces relative to the first line:

```
for item in iterable object:
    execute set of statements
```

We see that a key to understanding the Python `for` loop is to understand the meaning of the term *iterable object*.

3.4 Iterable Objects: String and Lists

So, what is an iterable object in Python? Well, here's the definition: An iterable object is an ordered collection of elements, where the elements can be sequenced through, in order and one element at a time, by a `for` loop. In computer programming, a *data structure* is a specified format for organizing data so that it can be accessed efficiently in a given situation. The Python programming language has several data structures that are iterable objects, including strings, lists, tuples, dictionaries, and sets. Save for strings, we have not encountered any of these data structures as yet. For now, let's just focus on strings and lists.

From our previous work, we know that a string (e.g., `'python'`) is an ordered collection of (Unicode) characters, surrounded by a set of single (or double) quotation marks. Since the collection of characters is ordered, a string is an iterable object. In addition, due to the ordering, each character can be assigned an implicit index. For a string with N characters, the first character has index-0, the second character has index-1, and so on, up to the last character, which has index-($N-1$). To demonstrate that a string is an iterable object, in the Shell, run the following `for` loop. Note that `for` is lowercase and that the `for` statement ends with a colon (`:`). Also, the line with the `print(letter)` statement is indented by four spaces and you must press `<Enter>` twice after this statement to alert the Shell that the `for`-loop coding is complete:

```
>>> for letter in 'python':
...     print(letter)
...
...
    p
    y
    t
    h
    o
    n
```

Here, the iterable object is `'python'`. On the first iteration of the `for` loop, `letter` is assigned to the index-0 character (`p`) and printed, on the second iteration, `letter` is assigned to the index-1 character (`y`) and printed on the next line, and so on, until the sixth iteration when `letter` is assigned to the index-5 character (`n`) and printed on the final line. The `for` loop has then completed its execution.

A Python *list* is another example of an iterable object. This data structure is defined as follows: A list is a collection of elements in a particular order, where the elements are separated by commas and the list is surrounded by a set of square brackets ([ ]). The syntax for creating a list is as follows:

```
list_name = [element_0, element_1, element_2, ...]
```

The range of possible Python lists is quite large in that the elements in a list can be objects of any data type, including integers, floats, strings, other lists, and even mixtures of data types. So, for example, all of the following are legal Python lists: `[0, 1, 2, 3]`, `[0.1, 0.2, 0.3, 0.4]`, `['apple','banana','orange']`, `['apple', 2, 0.3]`. Since the elements in a list are ordered, again there is implicit indexing of each element, where the first element has index-0, the second element has index-1, and so on. Demonstrate that an integer list is an iterable object by running the following `for` loop in the Shell:

```
>>> for element in [0, 1, 2, 3]:
...     print(element)
...
...
    0
    1
    2
    3
```

To demonstrate that the data type within the list does not matter, run this `for` loop:

```
>>> for element in ['apple', 'banana', 'orange']:
...     print(element)
...
...
    apple
    banana
    orange
```

Another handy feature of an iterable object is that, when needed, a particular element can be accessed by *indexing*. Here's how to accomplish that operation: Write the name of the iterable object, followed by the integer index of the element that you wish to access enclosed within square brackets. Try these indexing examples in the Shell, remembering that the first element of an iterable has index-0, the second element has index-1, and so on:

```
>>> mylist = [100, 200, 300, 400]
>>> mylist[0]
    100
>>> mylist[3]
    400
>>> mystring = 'python'
>>> mystring[0]
    'p'
```

```
>>> mystring[5]
    'n'
>>>
```

A computer-controlled experiment often consists of measuring the response of an experimental setup under a sequence of N different values of an applied experimental parameter (e.g., level of applied voltage or intensity of illuminating light). Schematically, the following for loop-based program can be used to carry out this experiment:

```
for i in [0, 1, 2,..., N-1]:
    apply unique value of experimental parameter
    acquire resulting response of setup
```

Here, the iterable object is the N-element list of integers. This mode of `for` loop operation is so common that Python provides a shortcut for obtaining the needed iterable object—the `range()` function. Using this shortcut, the above program is written as follows:

```
for i in range(N):
    apply unique value of experimental parameter
    acquire resulting response of setup
```

Interestingly, although this code will execute as if `range(N)` was producing the full list of integers `[0,1,2,..., N-1]`, it doesn't actually produce any list. Instead, `range()` is what's known as an *iterator*; that is, it iterates the `for` loop by feeding it with the proper sequence of integers one at a time, but after a particular integer has done its job in iterating the `for` loop during its particular iteration, `range()` erases that integer and creates the next one in the sequence. In this way, an iterable list of N integers never has to be created, which can be a big savings in computer memory usage when N is large. In the Shell, test drive `range()` with the code

```
>>> for i in range(5):
...     print(i)
...
...
    0
    1
    2
    3
    4
```

The full selection of options for the `range()` function is as follows:

```
range(start, stop, step)
```

Here, `start` is the first integer in the sequence (defaults to *0*, if not given explicitly), `stop` is one integer-step value above the last integer in the sequence, and `step` (defaults to *1*, if not given explicitly) is the integer-step value between an integer in the sequence and the next largest integer. In our usage above, `range(N)` did not include `start` and `step` explicitly and so we have used the equivalency

```
range(N) = (start=default, stop=N, step=default)
```

Finally, if you need to make a list of integers, the results of `range(N)` can be converted to an actual list using Python's `list()` function. Give the following a try in the Shell:

```
>>> integer_list = list(range(5))
>>> print(integer_list)
    [0, 1, 2, 3, 4]
```

3.5 Sine-Wave Plot Using for Loop and List Comprehension

As an example of the use of the `for` loop and Matplotlib, let's write a program that creates a Python list of data points that represents a few cycles of the sine function and then presents these data points on a waveform plot.

This exercise will also serve as an introduction to digitized data, which is the type of data generated in a computer-controlled experiment. As an example of digitized data, assume that a physical system under investigation is producing a pure sine-wave signal with a time-varying *displacement* given by $y(t) = A \sin(2\pi f t)$, where A and f are the sine wave's amplitude and frequency, respectively. Furthermore, assume that this analog signal is input to a digitizing instrument that *samples* (i.e., measures) the signal at N equally spaced times, where the time interval between neighboring samples is Δt. Defining y_i to be the sample taken at time t_i, the resulting data set will be given by the two relations

$$t_i = i\,\Delta t \tag{3.1}$$

$$y_i = A \sin(2\pi f t_i), \tag{3.2}$$

where $i = 0, 1, 2, \ldots, N-1$. Note that the first and last data samples are taken at times $t_0 = 0$ and $t_{N-1} = (N-1)\,\Delta t$, respectively. Our goal then is to write a Python program that creates the data set described by Eqs. (3.1) and (3.2) and thus simulates the data output of the above-mentioned digitizing equipment.

To evaluate the two given equations at a succession of N different times, Python's `for` loop structure is perfectly suited for the job. In addition, the Python Standard Library includes a module named `math`, which provides access to advanced mathematical operations (i.e., operations beyond basic arithmetic and comparisons) such as trigonometric, logarithmic, and power functions. The `math` module also includes useful mathematical constants such as pi (π = 3.141592...) and Euler's number (e = 2.718281...). For the `for` loop-based program we are about to write, we will evaluate the sine function in Eq. (3.2) by using the sine function available in the `math` module. Later we will write a similar program that instead uses the sine function available in the NumPy module, in order to highlight the advantages of NumPy operations.

Here are the details of our planned Python program: Acquire N = 100 samples of a sine function with amplitude A = 5 cm and frequency f = 3 Hz, where the time interval between samples is Δt = 0.01 seconds. One hertz (Hz) equals 1 cycle per second. Then according to Eq. (3.1), the sampling occurs between the times t_0 = 0 and t_{99} = 0.99 seconds and so, over this (almost) one-second time span, the 3-Hz sine wave should undergo three cycles.

Use **File≫New File** to create a new program in the Editor, then use **File≫Save** to store it under the name **sinewave_plot_for_loop.py** in **YourNamePythonWork**. Begin coding the program by importing the needed modules. The `math` module is part of the Python Standard Library, so it is automatically present on your computer (i.e., it does not need to be installed from an online repository), but it does need to be imported at the top of the program. Using `pip`, we have already installed the Matplotlib library. A module named `pyplot` within `matplotlib` is a collection of functions that provide a user-friendly interface to the Matplotlib plotting routines. Since this module is within the Matplotlib library, i.e., the path to the module is **matplotlib≫pyplot**, it is imported using the dot notation `matplotlib.pyplot` and, as is conventional, given the abbreviated name `plt`:

```
import math
import matplotlib.pyplot as plt
```

Next, the parameters of our program are defined, including the sine-wave amplitude `a` and frequency `f` as well as the number of samples `num` and sampling time interval `delta_t`. Conforming to Python convention, all of these variable names are lowercase and, if consisting of more than one word, words are connected by underscores.

```
import math
import matplotlib.pyplot as plt

a = 5
f = 3

num = 100
delta_t = 0.01
```

Our program will produce two 100-elements lists named t and y that store the succession of times t_i and associated sine-wave displacements y_i, respectively. To create such a list in Python, the list is first initialized as an *empty list* by the command `list_name= []`. This command calls upon the `list()` class, which creates a list object and then gives access to a set of list methods, one of which is called `append()`. An element `new_element` can then be added to an existing list named `list_name` through the following dot-notation command `list_name.append(new_element)`. First, add code to initialize the two lists:

```
import math
import matplotlib.pyplot as plt

a = 5
f = 3

num = 100
delta_t = 0.01

t = []
y = []
```

Now, add a `for` loop structure that iterates 100 times, where each iteration i evaluates Eqs. (3.1) and (3.2) at time t_i and then appends the values t_i and y_i to their respective lists. Since the sine function (with argument in units of radians) and value of pi come from the and value of pi come from the `math` module, they are accessed via the dot-notation commands `math.sin()` and `math.pi`:

```
import math
import matplotlib.pyplot as plt

a = 5
f = 3

num = 100
delta_t = 0.01

t = []
y = []

for i in range(num):
    t_i = i*delta_t
    y_i = a*math.sin(2.0*math.pi*f*t_i)

    t.append(t_i)
    y.append(y_i)
```

Finally, a command calling the `plot(x,y)` function in the `pyplot` module produces a plot of y versus x. This command must be followed by a command calling the `show()` function in the `pyplot` module in order for the plot to be displayed. Complete your program with these two plot-related commands as shown:

```
import math
import matplotlib.pyplot as plt

a = 5
f = 3

num = 100
delta_t = 0.01

t = []
y = []

for i in range(num):
    t_i = i*delta_t
    y_i = a*math.sin(2.0*math.pi*f*t_i)

    t.append(t_i)
    y.append(y_i)

plt.plot(t, y)
plt.show()
```

Save your work, then run your program. You will find that Matplotlib opens a new window in which three cycles of the sine wave are displayed.

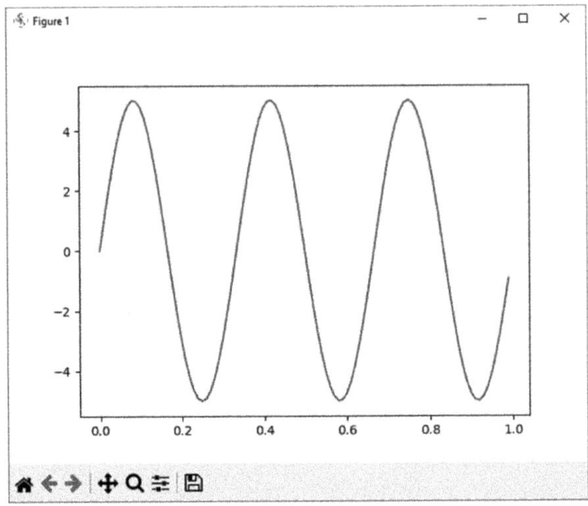

In using a `for` loop to generate a list in the program above, we found that several lines of code were required for the following steps—first, initialize the list; then, in each iteration of the `for` loop, calculate an element and append it to the list. Such list-building is a common task in programming, so Python offers a streamlined alternate coding structure called *list comprehension.*

A list comprehension combines a `for` loop and creation of new elements into a single line, and automatically initializes the list and appends each new element. The form of the list comprehension statement is as follows:

```
list_name = [expression for item in iterable object]
```

where `expression` is the relation that generates an element in the list.

Using list comprehension (after assigning `num = 100`), the code generating the 100-element *time* list in the above program can be replaced by the single line

```
t = [i*delta_t for i in range(num)]
```

Once created by this single-line command, each element in the list named `t` has a unique index *i* in the range from 0 to 99 and the value of the *i*th element can be accessed using the indexing operation `t[i]`. Thus, once the `t` list is known, the *sine-wave displacement* list can be calculated using list comprehension as follows:

```
y = [a*math.sin(2.0*math.pi*f*t[i]) for i in range(num)]
```

With **sinewave_plot_for_loop.py** open in the Editor, use **File≫Save As. . .** to create a new program named **sinewave_plot_list_comprehension.py** and save it in **YourNamePythonWork.** Modify this sine-wave plotting program using list comprehensions to generate the time `t` and sine-wave displacement `y` lists as shown:

```
import math
import matplotlib.pyplot as plt

a = 5
f = 3

num = 100
delta_t = 0.01

t = [i*delta_t for i in range(num)]
y = [a*math.sin(2.0*math.pi*f*t[i]) for i in range(num)]

plt.plot(t, y)
plt.show()
```

Save your work and run the program. It should produce the same three-cycle sine-wave plot as that produced by **sinewave_plot_for_loop.py**.

In your career as a Python programmer, you are sure to hear a critic deem certain code as *Unpythonic* and praise some coding solution as *more Pythonic* than another. In such conversations, the term Pythonic is used to describe a coding style that leverages Python's unique features in order to write code that is readable and, in the eyes of the critic, beautiful. The two programs we wrote in this section provide a good example. To a longtime Python programmer, the first program straightforwardly implements a `for` loop and looks like the work of a programmer who has a background in another computer language that has a `for` loop structure. The program carries out the required task, but it feels like a literal translation from the programmer's native programming language. The list comprehension feature is not unique to Python, but Python's version of this feature is distinctive and very readable. Thus, the second program takes advantage of this distinguishing Python feature and, to an experienced Python programmer, will appear easy to read and more stylish than the `for`-loop-based program. Most Python programmers will agree that the second program is more Pythonic.

The Python community has produced a set of Python style guides called *PEPs*, short for *Python Enhancement Proposals*. PEP 20 is called *The Zen of Python* and is a poetic collection of nineteen statements that describe Python coding philosophy. The Zen of Python can be found by entering `import this` at the Python prompt in the Shell. *PEP 8* is the official style guide for Python and is the source of the conventions used in this book such as four-space indentation and all lowercase variable names. To read the full style guide, use the search term "PEP 8." As you develop your skills in Python and learn to write Pythonic code, you will become a true *Pythonista*.

3.6 Importing Modules

Let's take a moment to discuss `import`, the Python command we have used several times to make code in a module available for use in our main program (remember, a module is any file that contains Python code and has an extension **.py**). This approach to programming allows a program's various tasks to be separated into different blocks of code, making the overall logic of the program clearer. In addition, once a module is developed for one program, it can be reused in a future program that requires its functionality.

Several variations of the `import` command are available in Python, two of which we used in Section 3.5. The most basic form is `import module_name`. When employing this form of the import statement, every item (e.g., class, function, constant) within a module named **module_name.py** is available for use in the main program through the following syntax: `module_name.item_name`. We used this form of the `import` command when writing `import math` in our sine-wave plotting program and using its `sin()` function and value for `pi` in our main program via `math.sin()` and `math.pi`. While a somewhat cumbersome way to identify an item, the `module_name.item_name` syntax self-documents the program, clearly identifying which module contains the item being called. This approach preserves the *namespace* of a module, that is, keeping the name of a module item attached solely to its module.

The second form of the import command is just a slight tweak on the first, allowing the programmer to assign an *alias* to the module name and save some keyboard work. This form of the command is `import module_name as mn`. Again, the entire contents of **module_name.py** is imported into the importing program, but now module items can be called using the easier-to-write syntax `mn.item_name`. In Section 3.5, we implemented this form when writing `import matplotlib.pyplot as plt` (`pyplot` is a module in the `matplotlib` library and so identified by the dot-notation name `matplotlib.pyplot`) and we used this module's `plot()` function in our main program via the command `plt.plot()`.

The third version of the import command is `from module_name import item_name_1, item_name_2,...`, a form we have not encountered to this point. As an example of when one might use this form of the `import` command, consider this: In our sine-wave plotting program, we only needed to use the function `sin()` and constant `pi` from the `math` module in our main program. Thus, using `from math import sin, pi` would make only those two items available in the main program, but with the added bonus that the dot notation declaring the module name will not be needed when calling these items. So, with this form of the import command in our main program, we can, for example, streamline our evaluation of the sine function to `a*sin(2.0*pi*f*t[i])`, rather than having to write the verbose expression `a*math.sin(2.0*math.pi*f*t[i]` required when using the `import math` form of the import command.

The fourth form of the import command is `from module_name import *`. The asterisk is a *wildcard character* indicating that all items in `module_name` are to be imported into the main program and are available for use without need of the dot notation to declare the module with which the items are associated (i.e., items are called in the main program in the same fashion as the third version of the import command). While seemingly convenient, use of this version of the import command can inject chaos into your program and so is not recommended. To illustrate the problem with this version of the `import` command, execute the following code in the Shell:

```
>>> from math import *
>>> d = 3
>>> e = 4
>>> d + e
    7
```

Next, execute the following code in the Shell:

```
>>> d = 3
>>> e = 4
```

```
>>> from math import *
>>> d + e
    5.718281828459045
```

What...? What happened here is called *namespace pollution*. Because the asterisk version of `import` does not require us to declare which module is defining a particular variable name, the most recent definition for the variable `e` is the one that is used. In the first case, the most recent version of `e` was in the main program, where the statement `e = 4` assigned the value of `4` to this variable. In the second case, the `math` module contains the variable `e` and assigns the value of Euler's number (2.718...) to it. Since the assignment `e = 4` occur first, followed by assignment `e = 2.718...` by the `import math` command, the `math` module assignment is the most recent version of `e`. Hopefully, this example demonstrates the perils of namespace pollution, providing a cautionary tale of the usage of `from module_name import *`.

The `import module_name as mn` form of the import command requires prepending the abbreviated module name `mn` with each usage of functions and classes that are imported from this module. Because this approach has the benefit of self-documenting programs with the source of imported programming elements, while at the same time limiting the required keyboard work, I will use the `import module_name as mn` form of the import command throughout the book.

3.7 Sine-Wave Plot Using NumPy and Vectorization

We now move on to NumPy, the third-party software library that is one of Python's big selling points for scientists and engineers. The power of NumPy derives from its fundamental data structure, a multidimensional array called *ndarray* (for N-dimensional array). In the special case of $N = 1$, a ndarray is a one-dimensional row of values (i.e., mathematically, a vector) and is simply called an *array*. Each element of an array is uniquely indexed by an integer. Higher dimensional ndarrays are mathematical matrices; for example, a two-dimensional ndarray is a table of values arranged in rows and columns, where each element is indexed by a pair of integers (a Python data structure called a *tuple*). Unlike other Python data structures such as a list, all elements of a ndarray must be numerics of the same numeric type (e.g., integer or float).

The NumPy library includes a large collection of high-level functions that can perform mathematical (as well as other useful) operations on a ndarray. The NumPy implementation of these mathematical operations and algorithms has two main advantages over the basic Python programming elements we have used so far such as the `for` loop and mathematical functions from the `math` module. First, although appearing as Python commands to the programmer, "under the hood," NumPy functions are precompiled C code and so NumPy-based operations approach the fast execution speeds of a program written in C itself. Second, NumPy supports *element-wise operations* (often called *vectorization*), where a single command can instruct a given operation to be carried out all at once on each element of an entire array. Vectorization is in contrast to the one-at-a-time operation we programed in previous sections, where a looping structure carries out the given operation on a single element in the array during each loop iteration. Thus, NumPy's element-wise operations streamline programming by eliminating the need for explicit looping and indexing.

To gain understanding about the meaning of element-wise operation, in the Shell, create a one-dimensional array named `x` using NumPy's `array()` command with a Python list as its argument as shown:

```
>>> import numpy as np
>>> x = np.array([1, 4, 9, 16, 25])
```

Next, take the square root of each element of `x` all at once by using the array `x` as the argument of NumPy's square root function:

```
>>> import numpy as np
>>> x = np.array([1, 4, 9, 16, 25])
>>> y = np.sqrt(x)
```

Verify that the square root has been taken of each element of `x` by printing `y`:

```
>>> import numpy as np
>>> x = np.array([1, 4, 9, 16, 25])
>>> y = np.sqrt(x)
>>> print(y)
    [1. 2. 3. 4. 5.]
```

Note that the NumPy array `y` is surrounded by square brackets (like a Python list), but there are no commas between the elements (unlike a Python list). Show that `y` is a (one-dimensional) NumPy ndarray by using the `type()` command:

```
>>> import numpy as np
>>> x = np.array([1, 4, 9, 16, 25])
>>> y = np.sqrt(x)
>>> print(y)
    [1. 2. 3. 4. 5.]
>>> type(y)
    <class 'numpy.ndarray'>
>>>
```

Let's now write our sine-wave plotting program, this time with the help of NumPy. In the Editor, use **File≫New File** to create a new program named **sinewave_plot_numpy.py** and save it in **YourNamePythonWork**. As before, start the program by importing the needed modules—this time, `numpy` and `matplotlib.pyplot` with the defined abbreviation `np` and `plt`, respectively. Here, at the top of our program, let's also define the amplitude and frequency of our desired sine wave:

```
import numpy as np
import matplotlib.pyplot as plt

a = 5
f = 3
```

Next, we need to specify the sequence of *time* values over which we want to evaluate the sine function. However, instead of using a Python list as the container for this sequence, we will use a NumPy array named `t`. With this decision, we can then take advantage of the NumPy function `linspace()`, which creates an array and fills it with an evenly spaced sequence of numeric values, starting at `start`, and ending at `stop`, where the number of numeric values in the sequence is `num`. The numeric values can be either integers or floats, and their data type is determined by whether the numbers are specified with or without a decimal point. To create the array `t`, the form of this NumPy function is

```
t = numpy.linspace(start, stop, num)
```

Our previous programs plotted 100 samples of a sine wave at evenly spaced times from times $t_0 = 0$ to $t_{99} = 0.99$ seconds. Create this same sequence of time values in an array named `t` as follows:

```
import numpy as np
import matplotlib.pyplot as plt

a = 5
f = 3

start = 0
stop = 0.99
num = 100

t = np.linspace(start, stop, num)
```

Now we are ready to evaluate the sine function at all of these time values `t` and produce a sequence of sine-wave displacement values we'll call `y`. Here is where the power of NumPy's element-wise mode of operation will be evident. We simply input the `t` array as part of the sine function's argument and NumPy will produce a `y` array whose *i*th element is the result of evaluating the sine function using the *i*th element of the `t` array. Using NumPy's sine function `sin()` and value for `pi`, include this coding as follows:

```
import numpy as np
import matplotlib.pyplot as plt

a = 5
f = 3

start = 0
stop = 0.99
num = 100

t = np.linspace(start, stop, num)
y = a*np.sin(2*np.pi*f*t)
```

Let's pause again to appreciate that NumPy's vectorization has eliminated the explicit use of a `for` loop as well as array indexing in our program. These operations are taking place, of course. They are just "behind the scenes" in optimized, precompiled C code. Vectorized code has many advantages. Such code is more concise and easier to read, has fewer lines of code and so fewer opportunities for bugs, and closely resembles standard mathematical notation, making it easier to code mathematical relations correctly.

In addition, note that Matplotlib can adapt to plotting data arrays, whether in the form of Python lists or NumPy ndarrays.

To complete the program, add the `plt.xlabel()` and `plt.ylabel()` commands as shown in order to include descriptive labels on the two plot axes (assuming time and sine-wave displacement are in units of seconds and centimeters, respectively):

```
import numpy as np
import matplotlib.pyplot as plt

a = 5
f = 3

start = 0
stop = 0.99
num = 100

t = np.linspace(start, stop, num)
y = a*np.sin(2*np.pi*f*t)

plt.plot(t, y)
plt.xlabel('Time (s)')
plt.ylabel('Sine-Wave Displacement (cm)')
plt.show()
```

Save your work. Run your program and demonstrate that NumPy coding results in the same three-cycle plot of the sine wave, complete with axis labeling.

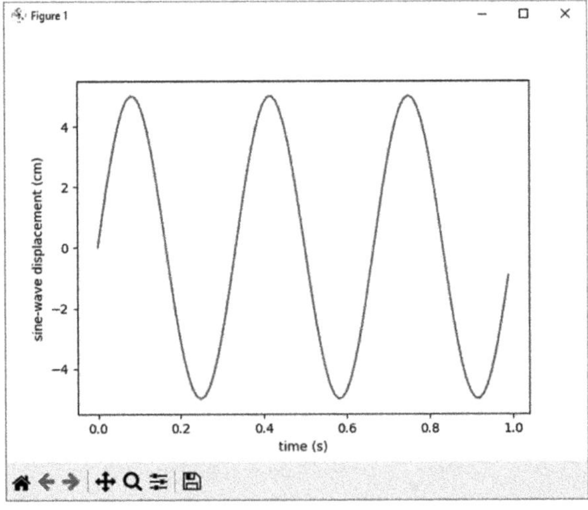

3.8 Properties of Digitized Data

Our **sinewave_plot_numpy.py** program provides a good playground for exploring issues related to digitized data. Modify the plotting portion of your program as shown below. Here, the addition of the `marker='o'` argument (the character between the quotation marks in a lowercase letter o) in `plt.plot(t,y,marker='o')` places a filled circle at the location of each data sample, while the `plt.ylim(bottom,top)` command fixes the *y*-axis scale to span the range from `bottom=-a` to `top=+a`, where `a` is the amplitude of the plotted waveform:

```
plt.plot(t, y, marker='o')
plt.xlabel('Time (s)')
plt.ylabel('Sine-Wave Displacement (cm)')
plt.ylim(-a, +a)
plt.show()
```

Run your program with `f = 3`. At 3 Hz (= 3 cycles per second) and 100 samples over the course of (close to) one second, the sine wave will complete three cycles with about 33 samples over the course of a cycle. As you can see, this level of sampling provides a good representation of the actual analog signal.

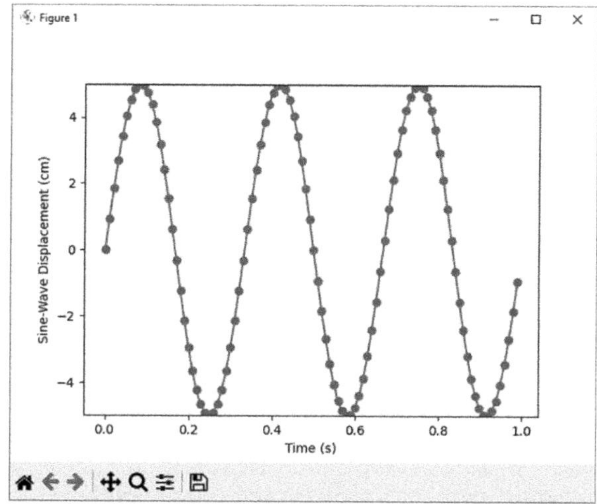

Change the frequency of your sine wave to 20 Hz by making `f=20`. Save your program and then run it. At 20 Hz and 100 samples over the course of one second, the sine wave will complete twenty cycles with 5 samples over the course of a cycle. With this level of sampling, a waveform of frequency 20 Hz is still being represented faithfully, but the degree of exactness in its representation of the actual analog signal is degraded.

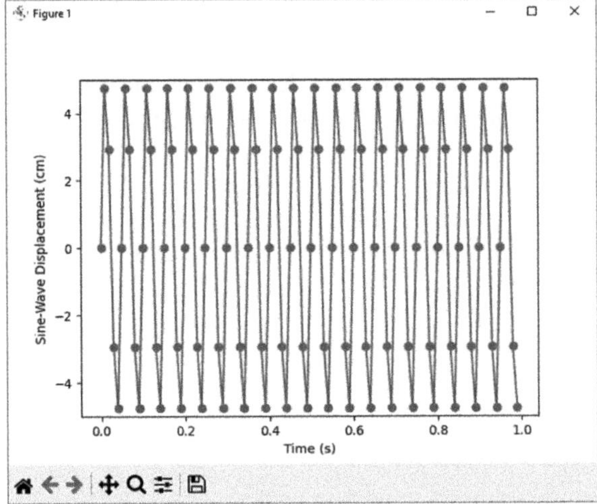

Next, change the frequency of your sine wave to 50 Hz by making `f = 50`. Save your program and then run it. Surprisingly, there is no sine wave displayed, only a string of zeroes at every sampling time.

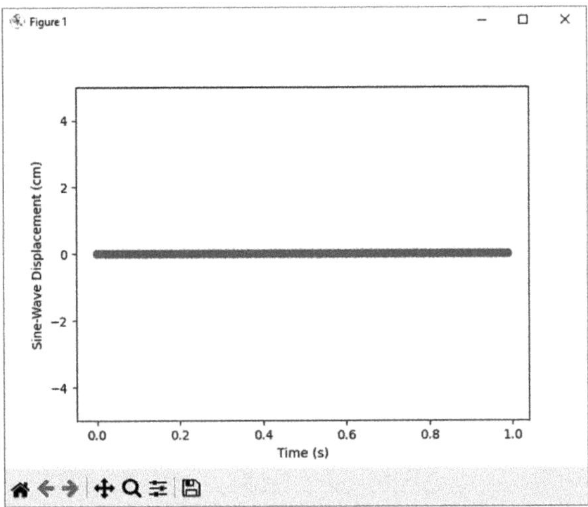

To explain what is happening, note that, at 50 Hz and 100 samples over the course of one second, the sine wave will complete fifty cycles with 2 samples over the course of each cycle. For the sine wave we are sampling, this waveform happens to be at its two zero-crossing at the moment of sampling, yielding a sample value of zero every time. If we had sampled a 50-Hz cosine wave instead, it would be at its peak and then at its trough at the moment of the two samples. Use NumPy's `cos()` function to demonstrate this outcome, if interested.

Finally, to investigate what happens when a sine wave is sampled less often than two times per cycle, change the frequency of your sine wave to 97 Hz by making `f = 97`. Save your program and then run it. Now, a nice-looking sine wave is displayed, but note that it only completes 3 cycles over the course of one second, rather than the expected 97 cycles.

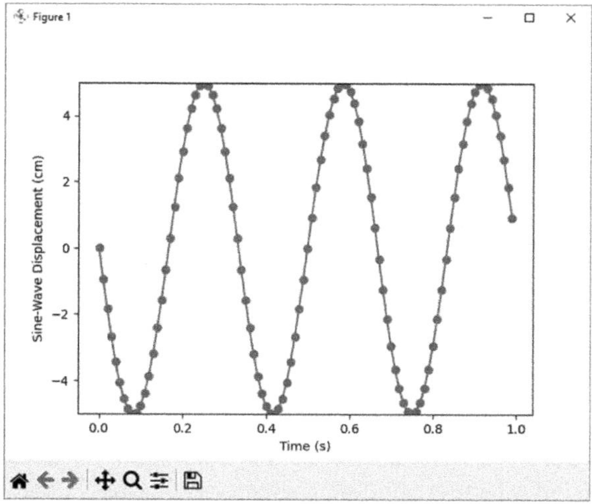

Thus, the digitized 97-Hz sine wave is being mistaken for a 3-Hz sine wave (also notice that there is a 180° phase shift in this digitized waveform compared to the actual 3-Hz sine wave plotted a few moments ago). Here, we have stumbled on a hidden danger of digitizing an analog signal called *aliasing*. Aliasing occurs when a digitizer's speed is too slow, resulting in less than two samples per cycle being taken of an analog sine wave. Stated more precisely, if one data sample is taken per time interval Δt, the *sampling frequency* is defined to be

$$f_s = \frac{1}{\Delta t}. \tag{3.3}$$

Then, to avoid aliasing, the sine wave being digitized must have a frequency less than or equal to one-half of the sampling frequency. This threshold frequency is defined to be the *Nyquist frequency* and is given by

$$f_{\text{nyquist}} = \frac{f_s}{2}. \tag{3.4}$$

For our program, 100 samples of the waveform are taken at evenly spaced times from times $t_0 = 0$ to $t_{99} = 0.99$ seconds; thus, the time between neighboring samples is $\Delta t = 0.01$ seconds and the sampling frequency is $f_s = 1/(0.01\text{ s}) = 100$ Hz. So, as we discovered, the Nyquist frequency is 50 Hz (i.e., one-half of the sampling frequency), and aliasing occurs if the signal being digitized has a frequency greater than this threshold frequency. We will revisit the phenomenon of aliasing again in later chapters.

3.9 Embedding Graph in Tkinter Window

Matplotlib is a very versatile, widely used Python library for data visualization. The library is extensive, providing the means for producing any type of plot, which can then be displayed on a computer monitor or output as an image for publication. The Matplotlib library itself is a collection of modules, each of which contains object-oriented code that define classes. When called, each class can produce a plot object (such as an axis, data-point marker, or plot legend), which together become the building blocks for creating a complete graph.

While this low-level control over plot features can produce sophisticated graphs to precise specifications, often plotting needs are simple and do not require such full-fledged control. Thankfully, Matplotlib includes the `pyplot` module, which contains a collection of convenient functions that carry out common plotting tasks. That is, `pyplot` provides user-friendly, high-level commands that, when executed, are translated into the low-level Matplotlib coding needed to carry out the requested task. In this chapter, we have previously used the `pyplot` functions `plot()` and `show()` to create a simple plot of *sine-wave displacement* versus *time* data.

In using the `pyplot` functions `plot()` and `show()`, we found that the resulting graph was displayed in a pop-up window; i.e., these `pyplot` functions created a graphical user interface for the plot. In our work ahead, we will instead want Matplotlib to place its graph in our own custom-made GUI created by Tkinter. In this section, we will learn how to accomplish that feat.

Matplotlib is equipped with several classes that allow graphs to be embedded inside Tkinter-produced GUIs. In a Tkinter window, the `Canvas()` class is used to produce a `canvas` object, which is a rectangular area in which graphical objects can be displayed. On the other hand, in Matplotlib, a graph is contained in a `figure` object that is created by the `Figure()` class. To embed a Matplotlib graph in a Tkinter window, we will use the Matplotlib class named `FigureCanvasTkAgg()`. This class acts like an interface between Matplotlib and Tkinter, taking the `figure` object produced in Matplotlib and placing that object within a `canvas` it creates in a Tkinter window. The syntax for this process is as follows:

```
canvas = FigureCanvasTkAgg(figure_name, window_name)
```

To import the `FigureCanvasTkAgg()` class, you need to know the following path to it: The `matplotlib` library contains a module named `backends`, which contains the needed behind-the-scenes coding for integration with popular *frontend* (i.e., user-interfacing) software libraries like Tkinter. The `backends` module for integrating Matplotlib with Tkinter is named `backend_tkagg`, and this is where `FigureCanvasTkAgg()` is found. Thus, the path to the needed module is **matplotlib>>backends>>backend_tkagg**, and so we will use the import command

```
import matplotlib.backends.backend_tkagg as bt
```

Begin writing our GUI-equipped sine-wave plotting program as follow: With **sinewave_plot_numpy.py** open in the Editor, use **File≫Save As...** to create a new program named **sinewave_plot_gui.py** and save it in **YourNamePythonWork**. Modify the top portion of the program by adding the `backend_tkagg` import statement discussed above as well as the import statement `import tkinter as tk` as show next. Also, assign `a=5` and `f=3`:

```
import numpy as np
import matplotlib.pyplot as plt
import matplotlib.backends.backend_tkagg as bt
import tkinter as tk
```

```
a = 5
f = 3

start = 0
stop = 0.99
num = 100

t = np.linspace(start, stop, num)
y = a*np.sin(2*np.pi*f*t)
```

As mentioned earlier, to embed a Matplotlib graph in a Tkinter window, we need Matplotlib to generate a `figure` object (which will then be inserted in a `canvas` in the Tkinter window). In addition, in order to place a plotted curve on that graph, we need Matplotlib to produce an `axes` object. These two objects, `figure` and `axes`, are created (i.e., instantiated) by Matplotlib's `Figure()` and `Axes()` classes, respectively. The `figure` object is the rectangular-shaped container for a graph and can be used to control the container size and, e.g., place a border around the container. The `axes` object, meanwhile, is used to produce the actual graph and, through application of the methods of the `Axes()` class, can control the graph's coordinate axes, tick marks, labels, plot-line style, data-point markers, and so on. While it may sound like we are in for some difficult programming, `pyplot` comes to our rescue. The following single `pyplot` command calls the `Figure()` and `Axes()` classes and creates the needed `figure` and `axes` objects, which we will give the shorthand names `fig` and `ax`, respectively, as shown here (with `pyplot` represented with its conventional abbreviation `plt`):

```
fig, ax = plt.subplots(layout='constrained')
```

The argument `layout='constrained'` prevents portions of the plot (such as labeling) from being clipped.

In your program, below the two lines that produce the `t` and `y` arrays, place the following pair of three-line code blocks. The first block should be familiar; it creates a Tkinter `root` window with a specified size and label. In the second block, the `plt` command described above is first used to instantiate the `fig` and `ax` objects. Then, the `FigureCanvasTkAgg()` class is used to create the `canvas` object and embed `fig` in it. Finally, to actually make canvas display a graph, the method `get_tk_widget().pack()` is executed.

```
root = tk.Tk()
root.title('Sine Wave Plot')
root.geometry('800x600')

fig, ax = plt.subplots(layout='constrained')
canvas = bt.FigureCanvasTkAgg(fig, root)
canvas.get_tk_widget().pack()
```

Complete the remaining portion of the program with the following code. Since the `ax` object controls the data displayed on the graph, the command `ax.plot(t,y)` will place the data plot on the graph. Also, applying the axis-labeling methods of the `Axes()` class to the `ax` object, the `ax.set_xlabel()` and `ax.set_ylabel()` statements provide labels on each axis. In place of the command `plt.show()` we have used previously, this time the command `canvas.draw()` instructs the graph to be displayed.

```
ax.plot(t, y)
ax.set_xlabel('Time (s)')
ax.set_ylabel('Sine-Wave Displacement (cm)')

canvas.draw()

root.mainloop()
```

The complete **sinewave_plot_gui.py** program is as follows:

```
import numpy as np
import matplotlib.pyplot as plt
import matplotlib.backends.backend_tkagg as bt
import tkinter as tk

a = 5
f = 3

start = 0
stop = 0.99
num = 100

t = np.linspace(start, stop, num)
y = a*np.sin(2*np.pi*f*t)

root = tk.Tk()
root.title('Sine Wave Plot')
root.geometry('800x600')

fig, ax = plt.subplots(layout='constrained')
canvas = bt.FigureCanvasTkAgg(fig, root)
canvas.get_tk_widget().pack()

ax.plot(t, y)
plt.xlabel('Time (s)')
plt.ylabel('Sine-Wave Displacement (cm)')

canvas.draw()

root.mainloop()
```

Save this finalized program and then run it. If all goes well, you will see three cycles of the sine wave displayed in your custom-made GUI.

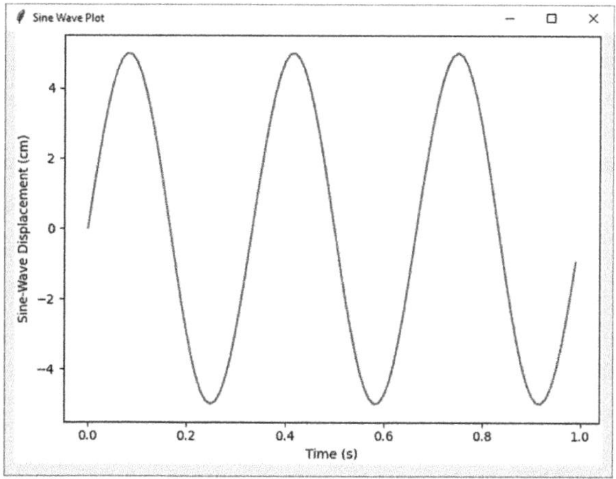

Do It Yourself

flip_four_coins.py A flipped coin can land in one of two states: *heads* or *tails*. Write a program that simulates simultaneously flipping four coins and determines how many of these coins land in the heads state.

The `random` module of the Python Standard Library has a function named `random()` that, when called, returns a random floating-point number x in the range $0 \leq x < 1.0$. Model the flip of each of the four coins using the `random.random()` function. When this function is executed, if its return value is greater than or equal to 0.5, the corresponding coin is assigned an output value of 1 (corresponding to the heads state); otherwise, that coin is assigned an output value of 0 (corresponding to the tails state). Begin the program by coding this definition of the `flip_four_coins()` function. The `for` loop performs a trial in which the four coins are flipped one at a time and the number of coins that land on a head is tallied by the object named `heads`. As shown, >= is Python's *greater than or equal to* operator.

```
import random

def flip_four_coins():

    heads = 0
    for coin in range(4):
        if random.random() >= 0.5:
            heads = heads + 1

    return heads
```

Note that there are five possible outcomes for `heads`: 0, 1, 2, 3, and 4. To verify that your code is working correctly, add the following two "test" lines below the definition:

```
heads = flip_four_coins()
print(heads)
```

Save your work, then run the program a few times for the verification. Then delete the two "test" lines.

The `flip_four_coins()` function simulates one *trial* of simultaneously flipping the four coins. Now simulate repeating such a trial over and over again. One option is to implement a `for` loop to execute N trials of the `flip_four_coins()` function and store the value of `heads` from each trial (through the `append()` method) in a list called `trials`. Perhaps a more Pythonic approach is to use the following list comprehension to create the list:

```
trials = [flip_four_coins() for trial in range(num_trials)]
```

Add code to your program to simulate `num_trials` of flipping the four coins and recording each resulting value of `heads` in a list named `trials`.

Finally, plot a histogram of the data in the `trials` list, which displays the number of trials whose outcome was 0, 1, 2, 3, and 4 heads. To create a histogram, use the `hist()` function in the `pyplot` module by including `import matplotlib.pyplot as plt` at the top of your program, then using this code:

```
plt.hist(trials, bins=5)
plt.xlabel('Number of Heads')
plt.ylabel('Number of Trials')
plt.show()
```

As shown, the number of histogram bins is set to *5* because there are five possible outcomes for `heads`.

Chose a value for `num_trials`, then save your work. Run the program. An example of the resulting histogram is shown next:

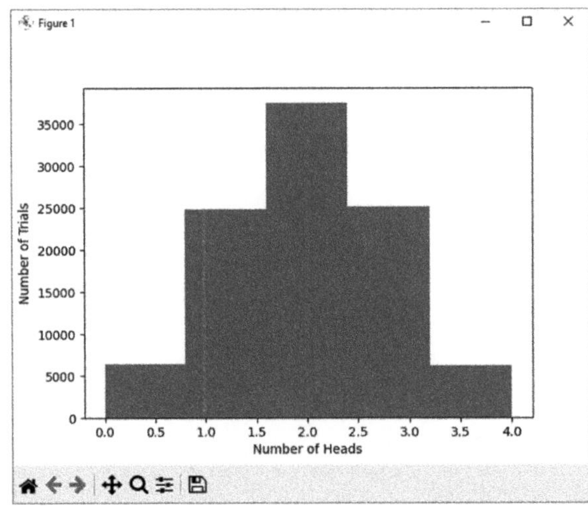

Run **flip_four_coins.py** with several choices of `num_trials`. From statistical theory, it is expected that when the number of trials N is large, the number of trials with an outcome of 0 or 4 heads should equal $N(\frac{1}{2})^4 = \frac{1}{16}N$ and the number of trials with an outcome of 2 heads should be $N\frac{4!}{2!2!}(\frac{1}{2})^4 = \frac{3}{8}N$, where $2! = 2$ and $4! = 24$. Above what (approximate) value of N are your results consistent with this expectation?

Use It!

Summation Using for Loop A famous story goes like this: In the late 1700s, a German elementary schoolteacher wanted to keep the kids in his unruly classroom occupied for 30 minutes so he gave them the following "busy work" assignment: Find the sum of all the integers from 1 to 100; that is, determine $S = 1 + 2 + 3 + \cdots + 98 + 99 + 100$.

Let's show that this assignment would have taken the kids much less than 30 minutes if Python had been available to them. Write the following program and save it under the name **sum_to_100.py** in **YourNamePythonWork**.

```
total = 0

for integer in range(1, 101):
    total = total + integer

print(total)
```

Here, the `for` loop is accumulating the sum of all the integers (in the quantity called `total`) as it iterates from `integer = 1` to `integer = 100`, i.e., the span of integers produced by the `range()` function. Since statements of the form `total = total + integer` occur often in programs, Python allows the following shorthand form of this statement: `total += integer`.

Another approach is to use the `range()` function to create the list of integers, then use the `sum()` function to sum all of the elements of the list. Add the following code to **sum_to_100.py** and show that it produces the same answer for `total`:

```
integer_list = list(range(1, 101))
total = sum(integer_list)

print(total)
```

The end of our famous story is this: Unfortunately for the elementary schoolteacher, the young mathematical genius Karl Gauss was one of his students. In less than a minute after receiving the integer-summing assignment, Gauss raised his hand and announced the correct answer. The flabbergasted teacher asked how he had arrived at the answer so quickly and Gauss explained his reasoning as follows. First, he noted that the series from 1 to 100 can be divided into two 50-element sets: 1 to 50 and 51 to 100. Next, starting at the beginning of the first set and the ending of the second, 50 pairs of integers can be formed, where the sum of each pair equals 101. That is, $1 + 100 = 101$, $2 + 99 = 101$, $3 + 98 = 101$, and so on. Thus, given 50 pairs of integers, where each pair sums to 101, the sum of all of the pairs is

$$S = 50 \times 101 = 5050.$$

Hopefully, this is the same answer your **sum_to_100.py** program gave.

By the way, Gauss's trick works for summing any series of integers $I = 1, 2, 3, \ldots, N$ as long as N is even (so that all the integers in the series can be paired). Then, one has $N/2$ pairs of integers, with each pair summing to $N + 1$, so the sum S of all of the integers is

$$S = \sum_{I=1}^{N} I = \left(\frac{N}{2}\right)(N+1) = \frac{N(N+1)}{2}.$$

Even better than that, because of a little bit of mathematical luck, $S = \sum_{I=1}^{N} I = N(N+1)/2$ is valid even if N is odd. You might have fun proving that fact.

Problems

1. **total_monthly_pay.py** I have agreed to hire you for a 30-day month. I pay you 1 penny on the first day, 2 pennies on the second day, and continue to double your daily pay on each subsequent day up to (and including) day 30. How many total dollars do you earn by the end of the month?
   (a) Write a `for`-loop-based program that determines and prints the answer to this question. (One dollar equals 100 pennies.) Programming tip: Represent each day using `range()` and the amount earn that day by `x**y`.
   (b) As an alternate solution, use list comprehension to create a 30-element list, whose elements record the number of pennies paid each day. Then, use the `sum()` function to find the total of the list's elements and convert to dollars. Print the result and verify that it agrees with the result found in part (a).
2. **finite_geometric_series.py** A finite geometric series of N terms is given by $y_N = 1 + x + x^2 + x^3 + \ldots + x^{N-1}$, where $|x| < 1$. It is well known that when $N = \infty$, $y_\infty = \frac{1}{1-x}$.
   Write a `for`-loop-based program that, given N and x as inputs, determines y_N and calculates its percentage deviation from y_∞. As output, print y_∞, y_N, and the percentage deviation. Then, use your program to answer the following question: If $x = 0.5$ and one wishes y_N to deviate from $y_\infty = 2.000\ldots$ by less than 0.1%, what is the minimum required value of N? Programming tip: Use `range()` and `x**y`.
3. **factorial.py** (a) The factorial of integer N is defined to be $N! \equiv 1 \times 2 \times 3 \times \ldots \times (N-1) \times N$ with $0! \equiv 1$. Write a program that, given N, uses a `for` loop to calculate $N!$ To check that your program works, show that $10! = 3{,}628{,}800$ and $15! = 1{,}307{,}674{,}368{,}000$. In Python 3, there is no maximum limit on the size of an integer (except those imposed by available computer memory). Programming tip: The statement `x = x*y` can be written in shorthand as `x *= y`.
   (b) **stirling_test.py** Stirling's approximation states that, for large N,

$$\ln(N!) \approx N \ln(N) - N$$

   Within your program, define a function `calculate_factorial(integer)` that calculates and returns the `factorial` of the input `integer`. In the body of your program, use a `for` loop that create a list called `differences` that contains the percentage deviation of Stirling's approximation for $\ln(N!)$ from the actual value of $\ln(N!)$ over a range of values for `integer`. Let the range of integer values be contained in a list named `integers`. Then, use `plt.plot(integers,differences)` to produce a plot and, from that plot, determine the minimum value of N so that the approximation $N\ln(N) - N$ deviates from the actual value $\ln(N!)$ by less than 1%. Programming tip: The natural logarithm of `x` can be calculated using the NumPy function `np.log(x)`, where `x` needs to be a float (an error results if `x` is an integer).

4. **sum _of_three_sines.py** A unity-amplitude square wave $y(t)$ can be approximated by the following sum of three sine waves:

$$y(t) \approx \frac{4}{\pi}\left[\sin(x) + \frac{1}{3}\sin(3x) + \frac{1}{5}\sin(5x)\right],$$

where x is in radians. Write a program that evaluates this sum at 300 equally spaced x-values in the range from $x = 0$ to $x = 6\pi$ (i.e., for three cycles of the first sine function in the sum) and then plots the resulting y versus x using `plt.plot(x,y)`, whose x-axis is calibrated in radians. When the program is run, it should produce the plot shown next. Programming tip: Use the NumPy functions `np.linspace()`, `np.pi`, and `np.sin()`.

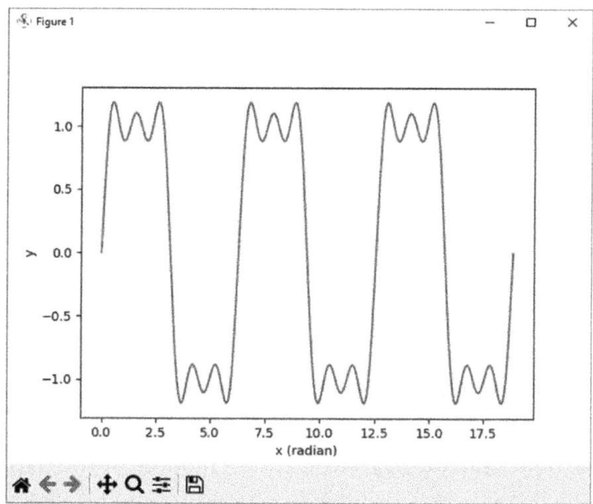

5. **loop_time_comparison.py** Write a program that compares the time it takes three methods—`for` loop, list comprehension, and `linspace()`—to create a sequence of integers from 0 to 999. To carry out this timing operation, use the `timeit(stmt)` function in Python's `timeit` module, which executes a specified code `stmt` (in our case, a function) one million times and return the total execution time. Thus, the time value reported is numerically equal to the number of microseconds per single execution of the function. Code the following program, where `for_loop_list()` and `comp_list()` create the integer sequence as a list and `numpy_array()` creates the sequence as a NumPy array:

```
import timeit as ti
import numpy as np

num = 1000

def for_loop_list():
    for_list = []
    for i in range(num):
        for_list.append(i)
    return

def comp_list():
    comp_list = [i for i in range(num)]
    return

def numpy_array():
    numpy_array = np.linspace(0, num-1, num)
    return

for_time = ti.timeit(stmt=for_loop_list)
for_time = round(for_time,1)
print('for loop time (microsecond) =', for_time)

comp_time = ti.timeit(stmt=comp_list)
comp_time = round(comp_time,1)
print('list comprehensive time (microsecond) =', comp_time)

numpy_time = ti.timeit(stmt=numpy_array)
numpy_time = round(numpy_time, 1)
print('numpy time (microsecond) =', numpy_time)
```

Save your work, then run the program. Since `timeit(stmt)` executes each function one million times, the program will take a noticeable time to run. Is the advertised fast execution speed of NumPy apparent?

6. **distribution_of_random_function.py** The `random` module of the Python Standard Library has a function named `random()`, which is a random number generator. Each time it is called, `random()` generates a double-precision floating-point number in the range from 0 to (but not including) 1, where the number is equally likely to be anywhere within this range (i.e., the probability distribution is uniform).
   (a) Demonstrate that the distribution of values produced by `random()` is uniform as follows: Write a program that employs `random.random()` to generate a list of N random numbers and then uses `plt.hist()` to plot the histogram distribution of these values. The histogram plot is shown below, where it has been run with N = 1,000,000 and the random-number range from 0 to 1 has been divided into 100 bins for the histogram plot.

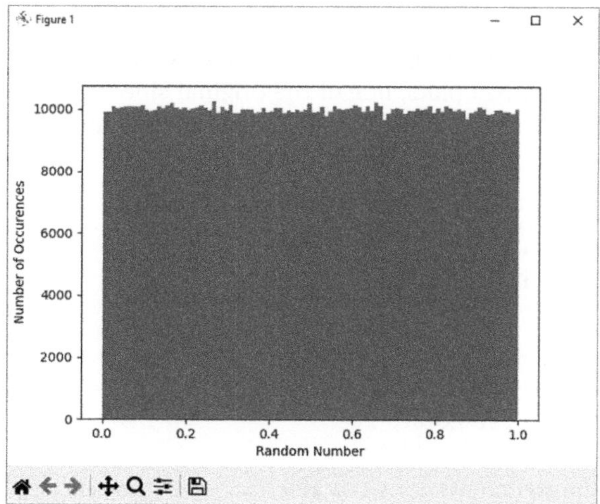

(b) During a run in which `random.random()` produces N random numbers in the range $0 \leq x < 1$ with a uniform probability distribution, the number of occurrences of a random number in the range from x to $x + dx$ is given by $dN = \rho\, dx$, where ρ is a constant. Given that $\int_{x=0}^{x=1} dN = N$, show that $\rho = N$. Then, from the definition of the distribution's mean $\bar{x} = \frac{\int_{x=0}^{x=1} x\, dN}{N}$ and standard deviation $\sigma = \sqrt{\frac{\int_{x=0}^{x=1} (x-\bar{x})^2\, dN}{N-1}}$, show that $\bar{x} = 0.5$ and $\sigma \approx 1/\sqrt{12}$ (assuming $N >> 1$). Finally, add code to your program so that the distribution's mean and standard deviation are calculated and printed. Run your program and show that $\bar{x}$ and σ have the expected values. Programming tip: The NumPy module includes the functions `mean(list)` and `std(list)`, which calculate the mean and standard deviation of a `list`, respectively.

7. **numerical_integration_numpy.py** It is easy to show analytically that the integral $\int_0^1 3x^2\, dx$ equals 1 exactly. Write a NumPy-based program that calculates the integral $\int_0^1 3x^2\, dx$ numerically as follows: Using `linspace()` and vectorization, create two N-element NumPy arrays x and y by evaluating the quantity $y = 3x^2$ at N values of x, starting at $x = 0$ and ending at $x = 1.000$ in steps of $\Delta x = 0.001$ (think carefully about the proper choice for N). Then use NumPy's numerical integration function `trapz(y,x)` to evaluate $\int_0^1 3x^2\, dx$, and display the result. The function `trapz()` performs numerical integration using the trapezoidal rule.

8. **sinusoid_multiplot_gui.py** In our GUI-equipped sine-wave plotting program, we used the `pyplot` function `subplots()` to create the axes object named `ax` needed to produce the actual graph, which was then placed in the figure container called `fig`. In this problem, we will demonstrate that, by including two extra arguments `nrows` and `ncols` in the function `subplots()`, multiple plots can be placed in `fig`.

Open **sinewave_plot_gui.py** in the Editor, then use **File≫Save As. . .** to create a new program called **sinusoid_multiplot_gui.py** in **YourNamePythonWork**. For our demonstration, we will produce two plots, one of the original sine function and the other of a new cosine function. Modify the sinusoidal-function-creation and window-creation portion of the code as follows:

```
t = np.linspace(start, stop, num)
y = a*np.sin(2*np.pi*f*t)
z = a*np.cos(2*np.pi*f*t)

root = tk.Tk()
root.title('Multiple Sinusoid Plots')
root.geometry('1200x600')
```

Then, include the arguments `nrows=1` and `ncols=2` in `subplots()` to create a single row containing two plots, one at the left, the other at the right. The function `subplots()` now returns `fig, (ax1,ax2)`, where the two axes objects `ax1` and `ax2` are associated with the left and right plots, respectively. Modify the following portion of the code as shown, which configures each plot (on the first line, we are following the Python convention of wrapping arguments of a function, as described in Chapter 2):

```
fig, (ax1,ax2) = plt.subplots(nrows=1, ncols=2,
                              layout='constrained')
canvas = bt.FigureCanvasTkAgg(fig, root)
canvas.get_tk_widget().pack()

ax1.plot(t, y)
ax1.set_xlabel('Time (s)')
ax1.set_ylabel('Sine-Wave Displacement (cm)')

ax2.plot(t, z)
ax2.set_xlabel('Time (s)')
ax2.set_ylabel('Cosine-Wave Displacement (cm)')
```

No other changes to the code are needed. Save your work, then run the program. It will produce the GUI window shown next, if the coding is correct:

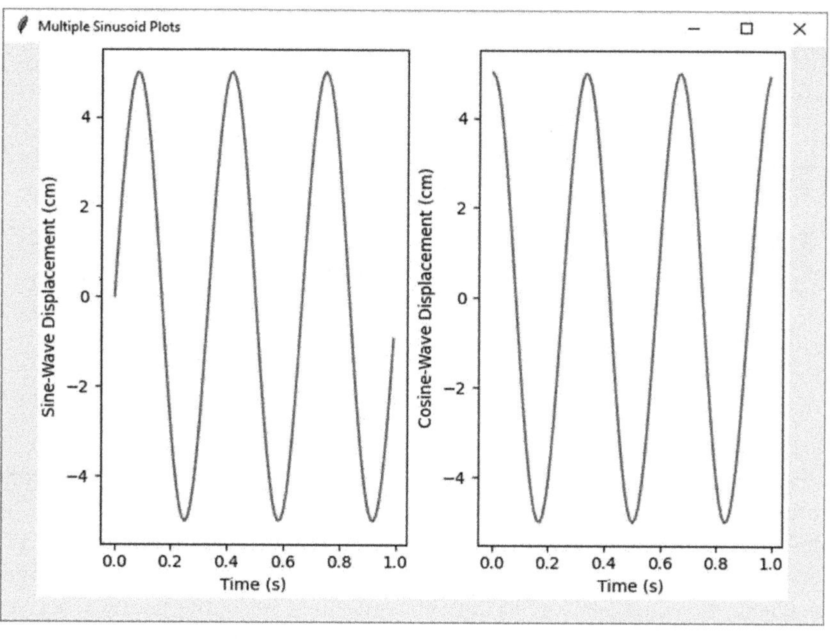
Multiple Sinusoid Plots
Sine-Wave Displacement (cm)
Time (s)
Cosine-Wave Displacement (cm)
Time (s)
0.0
0.2
0.4
0.6
0.8
1.0
4
2
0
−2
−4

CHAPTER 4

Conditional Looping and Real-Time Plots

As mentioned in Chapter 2, many of the Python classes and functions we encounter have longish names and/or quite a few parameters and arguments. To better fit such a class or function on the book's page as well as to render its multiple parameters and arguments more readable, we will follow the Python convention of wrapping parentheses-enclosed arguments and parameters onto multiple lines, with the wrapped elements (typically) aligned with the element next to the opening parenthesis of the class or function. For example, by this convention, the statement

```
lbl = tk.Label(root, text='Enter Integer in Box Below:')
```

may be rewritten as

```
lbl = tk.Label(root,
               text='Enter Integer in Box Below:')
```

4.1 Programming Structures and Real-Time Plots

In a computer program, a looping structure executes a given task repeatedly until a certain condition is met. The Python programming language provides two possible looping structures to control repetitive operations. In Chapter 3, we explored Python's `for` loop in its various guises—the more traditional form of

```
    for item in iterable object:
        execute set of statements
```

as well as the Pythonic versions offered by list comprehension and NumPy vectorization. Now we will focus our attention on the *while loop*, Python's other available looping structure. In contrast to the `for` loop, which executes a specified operation an initially specified number of times (aka *counted looping*), a `while` loop carries out *conditional looping*, where execution is controlled by the Boolean value of a specified condition.

In this chapter, you will write sine-wave plotting programs based on a `while` loop, including programs that produce real-time plots when data are generated one sample at a time. In

The Python-Based Laboratory. John Essick, Oxford University Press. © John Essick (2024). DOI: 10.1093/9780191998478.003.0004

addition, we will further explore the methods of *indexing* and *slicing* for extracting a single element and a series of elements, respectively, from an iterable data structure (such as a string, list, and NumPy array) and introduce a new iterable data structure called the *tuple*.

4.2 Python while Loop

The `while`-loop structure is used to control repetitive operations. In Python, this looping structure will repeatedly execute a block of code until a specified Boolean value is no longer TRUE. Thus, the `while` loop is equivalent to the pseudocode

```
while condition is True
    execute set of statements
    set condition
```

However, the actual form of the Python's `while` loop is as follows. Note that the line beginning with (lowercase) `while` ends in a colon:

```
while <logical test>:
    <statements>
```

Here, if *<logical test>* produces a `True` value, the `while` loop carries out an iteration in which *<statements>* are executed. The execution of the *<statements>* operations may or may not change the resulting value of *<logical test>* on the `while` loop's next iteration. If the value remains `True`, the loop iterates again; if the value is `False`, the loop's job is complete and program execution moves down to whatever code follows the `while` loop.

In the Shell, test drive the `while` loop as follows:

```
>>> i = 1
>>> while i < 6:
>>>     print(i)
>>>     i = i + 1
...
...
    1
    2
    3
    4
    5
>>>
```

Here, we start the sequence of commands by assigning the object named `i` with the integer value of 1. The `while` loop is then instructed to keep running as long as the condition `i < 6` is met. On the first iteration, `i` equals 1. Since the logical test 1 < 6 is `True`, Python prints 1 and then increments `i` by 1. On the second iteration, `i` equals 2, and again the logical test 2 < 6 is `True`, so Python prints 2 and then increments `i` by 1. Similar operations occur in the third, fourth, and fifth iterations. However, at the start of the sixth iteration, `i` equals 6, and so the logical test 6 < 6 is `False`. Thus, the `while` loop stops its execution.

The statement `i = i + 1` commonly appears in looping structure statements. It may appear mysterious at first because it is not valid mathematically; i.e., no value for `i` satisfies that mathematical formula. The resolution to this puzzle is that the equal-sign symbol in this statement does not mean the left side and right side of the relation are equal in a mathematical sense. Instead, the equal-sign symbol denotes the *assignment* operator. In an assignment statement, an operation is given on the right side of the assignment operator (i.e., the "=") and the result of that operation is assigned to the object on the left side of the assignment operator. Thus, the statement `i = i + 1` is deciphered as follows: The right side of the relation says to take the current value of `i` and add 1 to it. Then, assign the result of this operation to the object on the left side of the relation. Since that left-side object is `i`, the statement has simply changed the value of `i` by an increment of 1 in comparison to its original value. This increment-by-one statement `i = i + 1` is so commonly used that Python allows it to be written in the abbreviated form `i += 1`. Similarly, `i = i - 1` can be written `i -= 1`. In fact, this abbreviated statement is valid for any incremental value. That is, `x = x + y` is a valid statement and can be written as `x += y` as long as `x` and `y` are identical Python data types (integer, float, string, or Boolean) or even data structures (e.g., list).

4.3 Sine-Wave Plot Using while Loop

In Chapter 3, we used a `for`-loop-based program called **sinewave_plot_for_loop.py** to plot three cycles of the sine function. Let's try accomplishing the same goal of plotting three cycles of a sine wave, but this time using a `while` loop rather than a `for` loop to generate the data array.

In the Editor, open **sinewave_plot_for_loop.py**, then use **File>>Save As**... to create a new file named **sinewave_plot_while_loop.py** in **YourNamePythonWork**. Replace the `for`-loop-based code with the `while`-loop-based code as follows:

```
import math
import matplotlib.pyplot as plt

a = 5
f = 3

num = 100
delta_t = 0.01

t = []
y = []

i = 0
while i < num:
    t_i = i*delta_t
    y_i = a*math.sin(2.0*math.pi*f*t_i)

    t.append(t_i)
    y.append(y_i)

    i += 1

plt.plot(t, y)
plt.show()
```

Save your work, then run your program to demonstrate it works as expected.

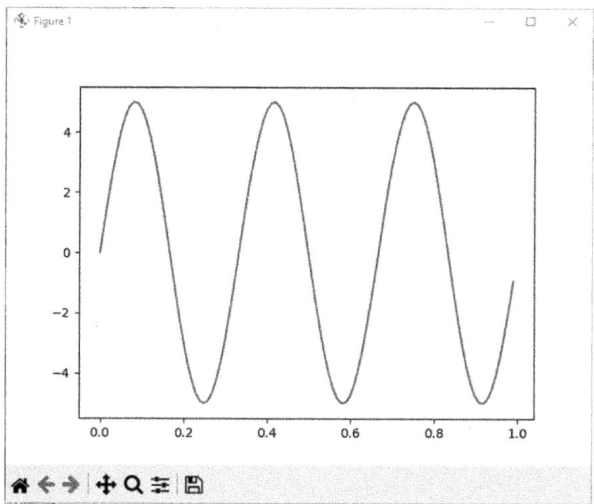

4.4 Real-Time Plotting Using while Loop

Assume you are planning a computer-controlled "data-logging" experiment in which a single temperature reading ("sample") is taken from a temperature sensor every 0.1 second and the experiment is going to be run over an extended, but only fuzzily defined, time period (say, "about" an hour). A `while`-loop-based Python program is ideally suited to control such an experiment, where the time interval for each `while`-loop iteration is set at 0.1 second and a data sample is taken at the start of each iteration. You will also want to store the data in one of the data-structure options offered by Python; let's say you decide to use a list (a NumPy array would be another good option). The **sinewave_plot_while_loop.py** program you just wrote can be easily modified for this experiment—just replace the line that calculates the sine function with code that acquires the one-at-time data sample from the temperature sensor. One downside of this approach, though, is that the program as written only plots the full data set at the end of the data run, i.e., after the one-hour session.

Often, experimenters prefer to view data samples at the moments they are acquired, requiring a plot that behaves like a laboratory *strip chart*, producing a real-time plot as each new data point is generated. With a key change to **sinewave_plot_while_loop.py**, we can produce a program that includes real-time plotting.

In **sinewave_plot_while_loop.py**, a single sample of the sine function is carried out with each `while`-loop iteration, thus simulating the single data sample taken in our assumed experiment. In the present program, the samples from all iterations are stored in a list, which is then plotted at the conclusion of the experimental run. The key change we need to make is a simple one: Move the plotting statements to within the `while` loop, so that they update the plot with each iteration.

With **sinewave_plot_while_loop.py** open in the Editor, use **File>>Save As...** to create and save a file named **sinewave_plot_realtime_loop.py** in **YourNamePythonWork**. Move the `plt.plot(t,y,marker='o')` statement to within the `while`-loop-based code as shown below. The argument `marker='o'` (single quotation marks surrounding a lowercase letter o) is included so that the actual data samples are displayed as filled circles. In addition, two other plot-related statements are needed: `plt.cla()`, which clears the plot from the previous iteration so that the current-iteration plot can be displayed, and `plt.pause(0.1)`, a `pyplot` function that pauses the execution (and thus the `while`-loop iteration) for 0.1 second after the current-iteration plot is displayed:

```
import math
import matplotlib.pyplot as plt

a = 5
f = 3

num = 100
delta_t = 0.01

t = []
y = []

i = 0
while i < num:
    t_i = i*delta_t
    y_i = a*math.sin(2.0*math.pi*f*t_i)

    t.append(t_i)
    y.append(y_i)

    plt.cla()
    plt.plot(t, y, marker='o')
    plt.pause(0.1)

    i += 1

plt.show()
```

Save your work. Then run the program and behold real-time plotting.

While entertaining, this real-time plotting program has a vulnerability—the time it takes your computer to redraw a graph in each `while`-loop iteration. Since the program executes 100 iterations and each iteration is programmed to take 0.1 second, the total runtime of the program is expected to be 10 seconds. Try running your program and time (e.g., using a stopwatch or smart phone) the total execution runtime from when the plotting starts to when it completes. On my desktop system, which has a fast CPU and good GPU, the program took about 11 seconds, 10% longer than expected. When I increased the iteration time to 1.0 second, by using `plt.pause(1.0)`, the total runtime was 101 seconds, which is within 1% of the expected 100 seconds. On the other hand, using `plt.pause(0.05)`, the total runtime was about 6.3 seconds, about 25% longer than the expected 5 seconds. I also repeated this runtime test on a laptop and found almost identical performance. Depending on the speed of your computer, you will find something similar—the program performs (almost) as expected, as long as the `while`-loop iteration time is made long enough (at least 1 second on my computer).

Here is the takeaway message from these runtime tests: If you use this while-loop approach as a data logger and calculate the time-axis values of your plot under the assumption that the time interval for each while-loop iteration is exactly the value programmed into plt.pause(interval), your time-axis will not be precisely calibrated. One option to circumvent this problem is to call the function datetime.now() from the Python Standard Library's datetime module at the moment that each data sample is acquired. Depending on your system, datetime.now() can potentially timestamp the time-of-acquisition of each data sample to millisecond accuracy. As we'll find in the following sections, there are other options as well.

4.5 Real-Time Plotting Using Animation

Matplotlib facilitates the common need for real-time plotting with an assortment of classes in its Animation module. The particular class we will use for our application is called FuncAnimation() that, when called, produces an object we will name ani. This object creates a sequence of frames (like an animated movie), where the content of each frame is a graph that is updated by a given function named animate. To use FuncAnimation(), one must first use the pyplot command fig,ax=plt.subplots() to create a graph with its associated figure (named fig) and axes (named ax) objects. Then, the FuncAnimation() class is designed to produce the object ani, which repeatedly calls the function animate() in order to provide updates to the graph associated with the figure fig. The statements required to create the object named ani by calling the FuncAnimation() class are as follows (note we are using the line-wrapping convention to display this function's six parameters):

```
import matplotlib.animation as animation

ani = animation.FuncAnimation(fig, animate, frames,
                              fargs, interval, repeat)
```

Here, interval is the time (in milliseconds) between the display of one frame and the next (i.e., the update of our graph). If the frame-counting argument frames is set to its default value of None, FuncAnimation() attaches an integer index i to each frame in the animation sequence and this integer is passed to animate() each time it is called by FuncAnimation(), while fargs passes the names of additional variables needed by the function animate(). The parameter repeat determines whether the animation repeats itself after the sequence of frames is completed. By default, repeat=True.

With **sinewave_plot_realtime_loop.py** open in the Editor, use **File>>Save As...** to create and save a file named **sinewave_animation.py** in **YourNamePythonWork**. Modify the program so that FuncAnimation() controls the sine-wave plotting via animation, rather than the while-loop control implemented in **sinewave_plot_realtime_loop.py**. Here, the variable names t and y are defined in the main part of the program and so need to be passed to the function animate() using the parameter fargs. Also, FuncAnimation() passes the index i to animate() tacitly, so animate() is a function of three variables—i, t, and y.

```
import math
import matplotlib.pyplot as plt
import matplotlib.animation as animation

a = 5
f = 3

num = 100
delta_t = 0.01

t = []
y = []

fig, ax = plt.subplots()

def animate(i, t, y):

    t_i = i*delta_t
    y_i = a*math.sin(2.0*math.pi*f*t_i)

    t.append(t_i)
    y.append(y_i)

    plt.cla()
    plt.plot(t, y, marker='o')

    return

ani = animation.FuncAnimation(fig, animate, frames=None,
                              fargs=(t, y), interval=100,
                              repeat=False)

plt.show()
```

Save your work and run the program. You will find the `FuncAnimation()` executes continuously and the assignment of the number of points `num = 100` is inactive. The only way to stop this program is by closing the window. However, the `FuncAnimation()` class does have a `pause()` method, so the program can be paused by adding the following `if` statement after `plt`-related statements:

```
    plt.cla()
    plt.plot(t, y, marker='o')
    if i == num-1:
        ani.pause()
```

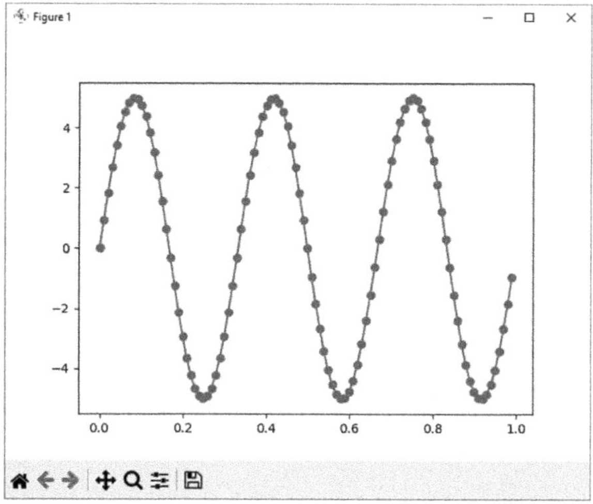

To determine if `FuncAnimation()` had improved performance in comparison to the `while`-loop approach in Section 4.4, I carried out runtime tests with `interval` equal to `50`, `100`, and `1000` ms for which **sinewave_animation.py** is expected to have runtimes of 5, 10, and 100 seconds, respectively. On my desktop system with a fast CPU and GPU, the runtimes were 5.5, 10, and 100 seconds, respectively, while on a laptop, the runtimes were 6.2, 11, and 101 seconds. Thus, the runtime is dependent on available computing resources, but for higher-level machines, the `animation`-based approach does perform well down to `interval` = `100` ms.

4.6 Indexing and Slicing of Iterable Data Structure

As mentioned in Chapter 3, a particular element of an iterable object can be accessed by *indexing*. For example, the third element of the list `mylist = [100, 200, 300, 400]` can be obtained by the command `mylist[2]`, where the square brackets enclose the integer index of the third element. The square-bracketed indexing operation can be applied to any of Python's sequentially ordered data structures, including strings, lists, NumPy arrays, and (the to-be-discussed) tuples. This form of *zero-based (positive) indexing* is common in computer programming and uses a positive integer to indicate the position of an element within an ordered sequence, as measured from the beginning element of the sequence (defined to be the index-0 element).

But what if we're interested in the element at the end of an ordered sequence? Or the penultimate (i.e., next-to-last) element? For this requirement, Python offers *negative indexing*, where the last element of a sequence has an index of −1, the penultimate element has index of −2, and so on. Positive and negative indexing of a 10-element ordered sequence of integers is compared next.

Positive Index	0	1	2	3	4	5	6	7	8	9
Element	100	200	300	400	500	600	700	800	900	1000
Negative Index	−10	−9	−8	−7	−6	−5	−4	−3	−2	−1

Try these negative indexing examples on a 5-element list in the Shell:

```
>>> mylist = [100, 200, 300, 400, 500]
>>> mylist[-1]
    500
>>> mylist[-3]
    300
>>>
```

While indexing allows access to a single element of an ordered sequence, *slicing* permits retrieval of an ordered multiple-element subset of the sequence. The multiple-element subset is termed a *slice*. The syntax of the slice operation is as follows:

```
sequence_name[start:stop:step]
```

Here, `start` is the index of the first element included in the slice, `stop` is (confusingly) the index of the first element not included in the slice, and `step` is the index-increment within the range between `start` and `stop`. In our work, we will almost always take `step` to be its default value of +1, so the slice operation can be abbreviated as

```
sequence_name[start:stop]    (assuming step = +1)
```

Then, the number of elements in the resulting slice equals the difference between `stop` and `start`, hence the reason `stop` is defined as it is. In addition, the default values for `start` and `stop` will correspond with the first and last elements of the ordered sequence, respectively. Thus, if you leave these quantities blank, they will assume these default values. Assuming `step = +1` (and taking the `sequence_name` to be `x`), the following slice commands result in these resulting slices:

Slice Command	Resulting Slice
`x[start:stop]`	Elements with index=`start` to index=`stop-1`
`x[start:]`	Elements with index=`start` through end of sequence
`x[:stop]`	Elements from start of sequence through index=`stop-1`
`x[:]`	Elements of entire sequence

Applying these commands for positive indexing is straightforward. For negative indexing, a few examples may be helpful: `x[-3:]` gives a slice consisting of the last three elements of the ordered sequence `x`, while `x[:-3]` produces a slice composed of all of the elements of `x`, except for the last three elements.

Note that, in the above slices created with `step = +1`, the order of elements in the original sequence is maintained in the slice. Conversely, setting `step = -1` allows one to produce a slice with the order of the elements reversed in comparison to the original sequence. To demonstrate this method, enter the following commands in the Shell:

```
>>> original_list = [1, 2, 3, 4, 5, 6, 7, 8, 9]
>>> forward_slice = original_list[1:4:+1]
>>> print(forward_slice}
    [2, 3, 4]
```

```
>>> reverse_slice = original_list[4:1:-1]
>>> print(reverse_list)
    [5, 4, 3]
```

Use the slice operation to add the following new feature to your real-time plotting program: Only plot the last `num_plot` samples of the acquired data. Modify **sinewave_animation.py** as shown below, where we take `num = 300` and `num_plot = 100` and only plot the most recently acquired `num_plot` samples in the `t` and `y` lists:

```
import math
import matplotlib.pyplot as plt
import matplotlib.animation as animation

a = 5
f = 3

num = 300
num_plot = 100
delta_t = 0.01

t = []
y = []

fig, ax = plt.subplots()

def animate(i, t, y):

    t_i = i*delta_t
    y_i = a*math.sin(2.0*math.pi*f*t_i)

    t.append(t_i)
    y.append(y_i)

    t = t[-num_plot:]
    y = y[-num_plot:]

    plt.cla()
    plt.plot(t, y, marker='o')
    if i == num-1:
        ani.pause()

    return

ani = animation.FuncAnimation(fig, animate, frames=None,
                              fargs=(t, y), interval=100,
                              repeat=False)

plt.show()
```

Save your work and then run the program. You should find that, after the first 100 data samples have been plotted, only the last 100 samples appear on the plot from then on. Also, note that during the first portion of the runtime, when the lists `t` and `y` contain less than 100 elements, `num_plot = 100` produces the commands `t[-100:]` and `y[-100:]`. Thankfully, Python is forgiving and these commands do not return an error message. Instead, in response to these commands, Python produces two slices, each consisting of the then-available elements, i.e., all of the existing less-than-100 elements.

You will find that, when running this program, a warning is written in the Shell. The warning is alerting you that, by default, `FuncAnimation()` caches, i.e., saves in memory, all of the frames produced during runtime. Since this program has the potential to effectively run unbounded (if `num` is chosen to be very large, or the `ani.pause()` statement is removed), a possible memory overload problem is present. To cure this problem, include `save_count=XXX` as an argument in `FuncAnimation()`, where `XXX` is a number of frames. Then, only the last `XXX` frames will be cached as the program runs. For our present program, it might make sense to set `save_count= num_plot` as shown here:

```
ani = animation.FuncAnimation(fig, animate, frames=None,
                              fargs=(t, y), interval=100,
                              repeat=False,
                              save_count=num_plot)
```

With the inclusion of this additional argument, you will find that the warning no longer appears when the program is run.

In the following sections, we will discover that, by placing some constraints on redrawing plots, the performance of a real-time plotting program can be further improved. First, though, we'll take a short detour to learn about another important Python iterable data structure—the *tuple*.

4.7 Tuple Packing and Unpacking

A tuple may be thought of as an *immutable* (i.e., unchangeable) list. A tuple consists of a fixed ordered sequence of elements, separated by commas, and enclosed by parentheses. The syntax for creating a tuple (called *tuple packing*) is as follows:

```
tuple_name = (element_0, element_1, element_2, ...)
```

Like lists, the elements of a tuple can be of any data type; thus, all of the following are valid tuples: `(0, 1, 2)`, `(0.0, 1.1, 2.2)`, `(1,'apple', 8.9)`. Because a tuple is an ordered sequence, its elements are zero-based integer-indexed and so a tuple can be sliced in the same way as lists. However, due to its immutable nature, once created, elements cannot be appended to, or removed from, a tuple.

An *empty tuple* is created with empty parentheses, e.g., `empty_tuple = ()`. To create a tuple containing only one element (termed a *singleton*), the lone element must be given a trailing comma, i.e., `singleton_tuple = (1,)`. The purpose of the comma is to distinguish the parentheses-enclosed object as a tuple as opposed to one of the other types of parentheses-enclosed objects possible in Python programming.

In fact, the defining characteristic of a tuple is that its elements are delimited by commas (as opposed to a list, which consists of elements delimited by commas *and* enclosed in square brackets). Thus, if desired, a tuple's parentheses may be omitted, except in the case of an empty tuple, or when the parentheses are needed to avoid syntactic ambiguity.

In the Shell, enter the following tuple-packing code, which defines a tuple `(1, 2, 3)`, written without the optional parentheses, and then assigns this tuple to the object named `tup`:

```
>>> tup = 1, 2, 3
>>> tup
    (1, 2, 3)
```

The reverse of tuple packing is (you probably guessed it) *tuple unpacking*. Tuple unpacking provides the means for assigning values to multiple variables in one line and is the preferred alternative to using separate assignment statements on separate lines. For example, in our just-completed program **sinewave_animation.py**, the sine-wave amplitude and frequency were initialized with two separate statements on two successive lines:

```
a = 5
f = 3
```

These assignments can be more concisely written using tuple unpacking, as follows:

```
a, f = 5, 3
```

Here, the right-hand side of the assignment statement packs the tuple `(5,3)`. Then, based on the relative position of each element, the assignment statement assigns the values `5` and `3` to the two elements of the tuple on the left-hand side of the statement, whose names are `a` and `f`, respectively.

In addition, the following statement appeared in our just-completed program:

```
fig, ax = plt.subplots()
```

In this statement, `fig` and `ax` are the elements of a tuple. Through tuple unpacking, these variables are being assigned the `figure` and `axes` tuple values produced by the function `plt.subplots()`.

Tuples are useful in the above two examples because the given sequence of items are not to be altered (i.e., they are immutable). In the first example, the tuple object `(5,3)` only exists in order to assign values to the variables `a` and `f`. Once the right-hand tuple `(5,3)` is unpacked into the left-hand tuple `(a,f)`, Python deletes the right-hand tuple from computer memory in a memory-conserving process called *garbage collection*. In the second example, the function `plt.subplots()` returns more than one object and these objects are unpacked into a tuple. This tuple provides a convenient way of organizing and accessing these objects.

4.8 Real-Time Plotting Using Blitting

Okay, back to real-time plotting. This is our goal: In a controlled experiment, where an independent variable x is varied and the resulting response of a dependent variable y is measured, plot y versus x as the data are being acquired so that an experimenter can view the data acquisition in real time. An example of this type of experiment might be applying a range of voltages across a resistor and measuring the resulting current associated with each applied voltage level. Key to our work is the recognition that, in such an experiment, the experimenter will know at the start of the data run the sequence of x-values being employed and will likely have an estimate of the expected range of resulting y-values. Thus, instead of implementing the computationally intensive process of continuously autoscaling and redrawing a real-time plot as the experiment unfolds, the axes and scaling can be fixed as a static background at the outset of the data run and then data points can simply be added to this unchanging background as they are acquired.

The scheme outlined in the previous paragraph is a well-known technique in the field of computer graphics called *blitting*, so we will not have to invent this process from scratch. Even better, `FuncAnimation()` can be configured to implement blitting. As outlined above, blitting speeds up repetitive drawing by rendering all unchanging graphic elements into a static background image at the start of program execution. In our case, the background image consists of a blank plot, whose axes are outfitted with properly scaled tick marks and axis labels. To create this blank plot, our program will call `fig,ax=plt.subplots()` as well as a custom-made function we will name `init()`, which will set the scaling of the axes. Then, for every new animation frame, only the changing element (i.e., the data-point plot) will need to be drawn onto this background.

To update just the data-point plot requires some `pyplot` functionality we have not used previously. Many times before, we have called the function `plt.plot()` in order to plot a curve, but we have not taken advantage of the fact that this function returns a tuple consisting of `line` objects. These objects are produced by a class called `Line2D()`, which is called by `plt.plot()`, and they contain all the configuration information (e.g., marker type) regarding each data plot, including its sequence of data points. To obtain these returned objects, an assignment command is needed as follows: `(line_0,line_1, ...)=plt.plot()`, where the tuple contains multiple line objects, if there are multiple data curves on the single plot. In the past, we simply wrote the command as `plt.plot()` and, since we didn't include the assignment of the `line` objects to the tuple, these objects were garbage collected by Python. In our current program, however, we will need the `line` object associated with our data plot and, since our plot only contains a single data curve, the tuple containing our `line` object is a singleton. Naming this object `line`, the command to access it is as follows:

```
line, = plt.plot(y, t, marker='o')
```

The comma after `line` is required because this object is contained in a single-element tuple. Once we have obtained this object, our plot's data curve can be redrawn using the `set_data()` method associated with the `Line2D()` class. Thus, the data curve associated with the `line` object can be updated by the command: `line.set_data(t,y)`.

Finally, to configure `FuncAnimation()` to implement blitting while producing a sequence of `num` frames, the set of arguments is as follows:

```
ani = animation.FuncAnimation(fig, animate, frames=num,
                              fargs=(t, y), interval=100,
                              init_func=init, repeat=False,
                              blit=True)
```

Here, `init_func=init` chooses the `init()` function for initializing the plot axes. Taking `frames=num` and `repeat=False` will produce an animation, whose number of frames is `num`, with the animation stopped once these num frames are plotted. With frames=num, FuncAnimation() attaches an integer index i to each frame in the animation sequence and this integer is passed to animate() each time it is called by FuncAnimation(), while fargs passes the names of additional variables needed by the function animate().

With **sinewave_animation.py** open in the Editor, use **File>>Save As**... to create and save a file named **sinewave_animation_blit.py** in **YourNamePythonWork**. Modify the program as shown below. In the three lines below the import statements, we are using tuple unpacking to initialize quantities. In the init() function, ax.set_xlim(left,right) and ax.set_ylim(bottom,top) are used to fix scaling of the axes, with the maximum time-value (denoted as right) determined by (num-1)*delta_t; note, labeling of the axes is now included:

```
import math
import matplotlib.pyplot as plt
import matplotlib.animation as animation

a,f = 5, 3
num, delta_t = 100, 0.01
t, y = [], []

fig, ax = plt.subplots()
line, = plt.plot(t, y, marker='o')

def init():

    ax.set_xlim(0, (num-1)*delta_t)
    ax.set_ylim(-a, a)
    ax.set_xlabel('Time (s)')
    ax.set_ylabel('Sine-Wave Displacement (cm)')

    return line,

def animate(i, t,y):

    t_i = i*delta_t
    y_i = a*math.sin(2.0*math.pi*f*t_i)

    t.append(t_i)
    y.append(y_i)

    line.set_data(t, y)

    return line,

ani = animation.FuncAnimation(fig, animate, frames=num,
                              fargs=(t, y), interval=100,
                              init_func=init, repeat=False,
                              blit=True)

plt.show()
```

Save your work, then run the program.

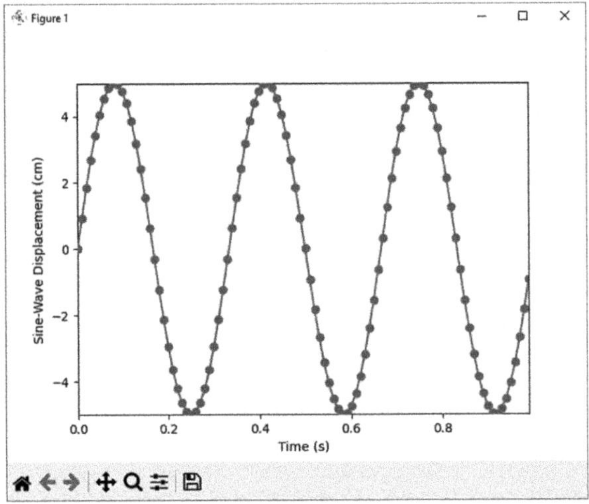

To gauge if blitting achieved any improvements in performance, I again carried out runtime tests with `interval` equal to `50`, `100`, and `1000` ms for which the expected runtimes are 5, 10, and 100 seconds, respectively. Remarkably, on my desktop system with a fast CPU and GPU, the runtimes were 5, 10, and 100 seconds, respectively, indicating that the program was performing well down to `interval = 50` ms. On the other hand, (somewhat surprisingly) blitting degraded the laptop performance at shorter `interval`-values, yielding runtimes of 7.2, 12, and 101 seconds.

4.9 Real-Time Plotting with GUI

By combining the program **sinewave_plot_gui.py** from Chapter 3 with that of the just completed **sinewave_animation_blit.py**, we can create a real-time plotting program with a Tkinter-based graphical user interface.

With **sinewave_animation_blit.py** open in the Editor, use **File>>Save As...** to create and save a file named **sinewave_animation_blit_gui.py** in **YourNamePythonWork**. Modify the program as shown below. The GUI-related commands can be found in **sinewave_plot_gui.py**, so consider using copy-and-paste operations to expedite your coding. The Tkinter button allows a user to pause the animation, if desired:

```
import math
import matplotlib.pyplot as plt
import matplotlib.animation as animation
import matplotlib.backends.backend_tkagg as bt
import tkinter as tk

a, f = 5, 3
num, delta_t = 100, 0.01
t, y = [], []
```

```
root = tk.Tk()
root.title('Real-Time Sine Wave Plot With Blitting')
root.geometry('800x600')

def stop():

    ani.pause()

    return

btn = tk.Button(root, text= 'Stop Plot', command=stop)
btn.pack()

fig, ax = plt.subplots(layout='constrained')
line, = plt.plot(t, y, marker='o')
canvas = bt.FigureCanvasTkAgg(fig, root)
canvas.get_tk_widget().pack()

def init():

    ax.set_xlim(0, (num-1)*delta_t)
    ax.set_ylim(-a, a)
    ax.set_xlabel('Time (s)')
    ax.set_ylabel('Sine-Wave Displacement (cm)')

    return line,

def animate(i,t,y):

    t_i = i*delta_t
    y_i = a * math.sin(2.0*math.pi*f*t_i)

    t.append(t_i)
    y.append(y_i)

    line.set_data(t, y)

    return line,

ani = animation.FuncAnimation(fig, animate, frames=num,
                              fargs=(t,y), interval=100,
                              init_func=init, repeat=False,
                              blit=True)

canvas.draw()

root.mainloop()
```

Save your work and run your program with various choices of `interval`, including `50`, `100`, and `1000` ms. Are the graphical operations of Matplotlib and Tkinter noticeably impacting the performance for any of those choices of `interval`? On my desktop system, equipping the program with a GUI did degrade performance a bit, yielding runtimes of 5.6, 10.5, and 100 seconds for `interval` equal to `50`, `100`, and `1000` ms, respectively.

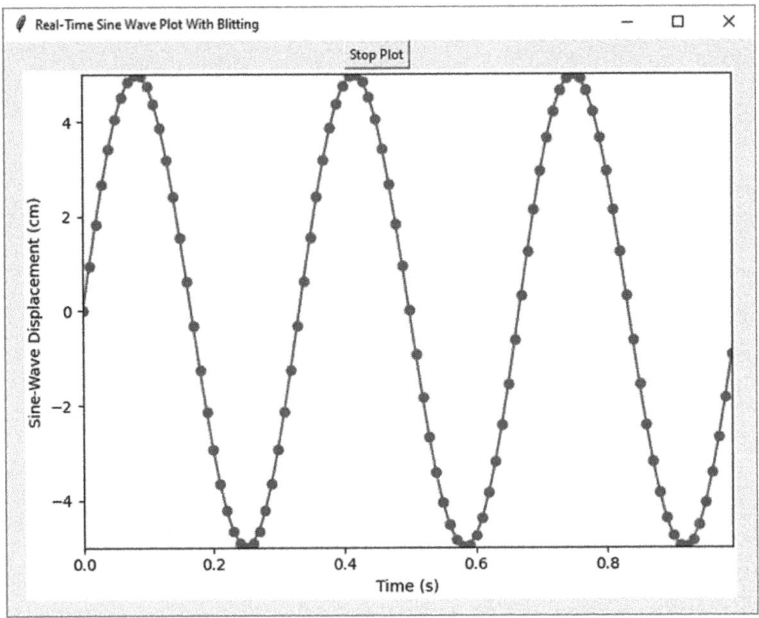

In summary, we've been able to write real-time plotting programs, which function as autoscaling strip charts. However, the computation overhead of rendering the plots can tax the capabilities of a given computing system's CPU and GPU. By basing real-time plotting programs on the `FuncAnimation()` class from Matplotlib's `Animation` module, excellent performance is possible up to a system-dependent maximum frame update rate. On my desktop system, this maximum frame update rate is about 1/(0.1 s) = 10 Hz. Finally, embedding the real-time plot in a GUI does reduce performance slightly.

Do It Yourself

odd_integer_search.py The odd integers are given by $2i+1$, where $i = 0,\ 1,\ 2, \ldots$ Write a `while`-loop-based program that answers the following question: What is the smallest odd integer that is divisible by 3 and also, when cubed, yields a value greater than 4000? Print the answer in the Shell. Programming tip: Given two operands `x` and `y`, the AND operation is performed by the Python command `x and y`. Also, x^y is written as `x**y`. For this problem, you may want to write the `while` statement as follows: `while condition:`, where the quantity `condition` is initialized appropriately.

Use It!

Terminate Loop with break Command This exercise will demonstrate how to exit a repetitive loop through use of the `break` command. In Chapter 3's Problem 6 and Do It Yourself project, we learned about the `random()` function in the Python Standard Library's `random` module that returns a random floating-point number x in the range $0 \leq x < 1.0$ when called. The following function named `create_random_integer()` uses `random()` to produce a random integer in the range from 0 to 9. Here, the `int()` function truncates the float in the range from 0.000... to 9.999... to produce an integer, e.g., 3.867.... becomes 3. (Note: the `random` module includes the function `randint()`, which can also be used to create a random integer.)

```
import random

def create_random_integer():

    random_float = random.random()
    random_int = int(10*random_float)

    return random_int
```

Create a program named **while_loop_break.py** and save it in **YourNamePythonWork**. Include the above definition of `create_random_integer()` at its top, then add the following code to the program:

```
integer_to_match = create_random_integer()

iterations = 1
while True:
    random_integer = create_random_integer()
    if random_integer == integer_to_match:
        break
    else:
        iterations += 1

print('Number of Iterations to Match = ',iterations)
```

Save your work. This program will first create a random `integer_to_match` in the range from 0 to 9 using `create_random_integer()`, then implement a `while` loop to count the number of iterations that `create_random_integer()` must be called in order to produce the same integer as `integer_to_match`. When a match is produced, execution of the program jumps out of the `while` loop via the `break` command. That is, `break` terminates the loop when it is encountered and program execution proceeds to the code immediately below the loop (in this case, the `print()` statement). Without the use of the `break` command, the alternative more-cumbersome coding is needed:

```
integer_to_match = create_random_integer()

iterations = 1
condition = True
while condition:
    random_integer = create_random_integer()
```

```
    if random_integer == integer_to_match:
        condition = False
    else:
        iterations += 1
```

A related command that can be handy in loop structures is `continue`. Where `break` terminates the execution of the loop, `continue` simply terminates the current iteration of the loop and skips ahead to the next iteration. Both `break` and `continue` are available for use in `while` loops as well as `for` loops.

Problems

1. **iterate_to_ten.py** Write a `while`-loop-based program that performs 10 iterations numbered 1 through 10. Make each iteration last one second and print the message **Iteration Number = XXX**, where XXX is the iteration number. Programming tip: The *less than or equal to* operator is <=. The execution of a Python program can be delayed by `delay` seconds using the `sleep(delay)` command in the `time` module. The `time` module is in the Python Standard Library, but must be imported to a program using the command `import time`.
2. **single_sine_cycle.py** Create a `while`-loop-based program that stops automatically after its `while` loop has iterated exactly 1000 times. As the `while` loop performs these 1000 iterations, evaluate one cycle of a sine wave (where the final data point is at the equivalent point on the sine wave as the initial point) and, after the loop completes its execution, use Matplotlib to produce the plot shown below. Programming tip: Python's *less than or equal to* operator is <=.

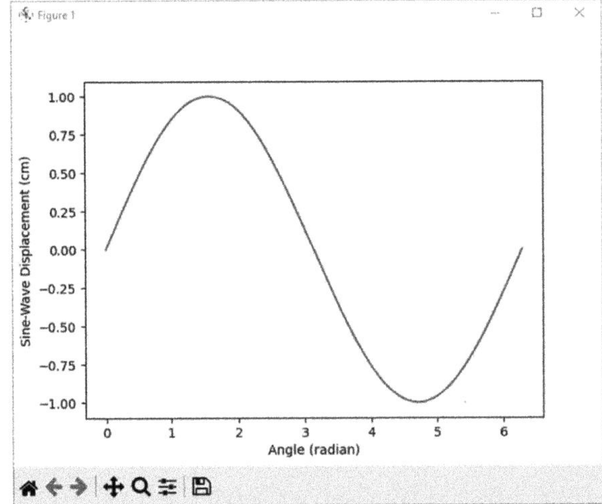

3. **palindromic_word_detector.py** Write a program that identifies single-word palindromes (i.e., words that read the same backward as forward). In the program, use `input(prompt)` to present the user with prompt **Enter Single Word to Test**: in the Shell. After the word is input, change it to all lowercase to allow proper nouns (with their first letter capitalized) to be tested. Then, use a test of your own design to determine if the lowercase word is a

palindrome. If the word passes the test, print **Word is Palindrome** in the Shell; if the word fails the test, print **Word is Not Palindrome**.

When completed, test your program with the following inputs: *radar*, *rotator*, *Otto*, *Hannah*. Programming tip: The string method `string.lower()` changes the string object `string` to all lowercase.

4. **while_for_time_comparison.py** Write a program that compares the time it takes a `while` loop and a `for` loop to create a sequence of integers from 0 to 999. To carry out this timing operation, use the `timeit(stmt)` function in the Python Standard Library's `timeit` module, which executes a specified code `stmt` (in our case, a function) one million times and return the total execution time. Thus, the time value reported is numerically equal to the number of microseconds per single execution of the function. Code the following program, where `while_loop_list()` and `for_loop_list()` create the 1000-integer sequence as lists:

```
import timeit as ti

num = 1000

def while_loop_list():

    while_list = []
    i = 0
    while i<num:
        while_list.append(i)
        i += 1

    return

def for_loop_list():

    for_list = []
    for i in range(num):
        for_list.append(i)

    return

while_time = ti.timeit(stmt=while_loop_list)
while_time = round(while_time, 1)
print('while loop time (microsecond) = ', while_time)

for_time = ti.timeit(stmt=for_loop_list)
for_time = round(for_time, 1)
print('for loop time (microsecond) = ', for_time)
```

Save your work, then run the program. Since `timeit(stmt)` executes each function one million times, the program will take a noticeable time to run.

As a percentage of the `for` loop execution time, how much longer does it take the `while` loop to build the list than the `for` loop? The use of the iterator `range(num)` in the `for` loop, which "under the hood" is written in C, accounts for more optimized execution of the `for` loop in this list-building example.

5. **CPU Hogging** Explore the impact of `while`-loop execution on central processing unit (CPU) usage. In Windows, open the **Task Manager** by right-clicking an empty area on the taskbar at the bottom of your monitor or, in macOS, open **Applications>>Activity Monitor**. You will find that the Task Manager/Activity Monitor displays CPU usage (expressed as a percentage of your computer's maximum value).

   Open **sinewave_plot_realtime_loop.py** and run this program with `plt.pause(0.2)` to produce one `while`-loop iteration every 200 ms. What is the approximate CPU usage by the Python application when the program is running? Next, run the program with `plt.pause(0.1)` to produce one `while`-loop iteration every 100 ms. By halving the iteration time (now it is 100 ms versus the original 200 ms), what is the approximate increase in CPU usage by Python over the original usage? Finally, run the program with `plt.pause(0.05)` to produce one `while`-loop iteration every 50 ms. The iteration time is now one-fourth that of our original run. What is the approximate increase in CPU usage by Python now in comparison to the original (200-ms iteration) usage? Takeaway: To avoid CPU hogging, carefully consider required loop iteration time. For example, if a loop is simply updating numbers on a GUI display, probably a small number of loop iterations per second is sufficient.

6. **iterations_until_integer_equals_five.py** (a) Begin your program by using the information in this chapter's Use It! project to define a function named `iterations_to_five()`, which carries out the following task: Within each iteration of a `while` loop, randomly generate an integer in the range from 0 to 9, then check to see if this integer equals 5. Iterate this process until the random integer does equal 5. Count the number of `while`-loop iteration required to achieve the match and return this integer quantity under the name `iterations` (but don't have your function print it).

   Knowing that the integers are created randomly (i.e., equal probability for producing one of the ten integers 0 through 9 with each `while`-loop iteration), what do you expect the value of `iterations` to be, on average? Let's add some code to our program to find out. First, at the top of your program, include the import statement `import numpy as np`. Then, below the definition of `iterations_to_five()`, place code that calls `iterations_to_five()` within a `for` loop so that, when executed, the `for` loop will repeat the integer-matching task N times (we'll call each of these times a *trial*) and store the resulting N values of `iterations` in a list named `trials_list`. Finally, use the NumPy function `np.mean()` to calculate the average of the N elements in `trials_list` and display this average in the Shell with the message **Mean Number of Iterations = XXX**, where XXX is the value calculated by `np.mean()`. Save your work, then run the program with the number of trials $N = 100{,}000$. Is the mean value for `iterations` that you obtain equal to the value that you expected?

   (b) Plot a histogram of the data in `trials_list`, where the histogram displays the number of trials (out of N total trials) when `iterations` equals 1, 2, 3,. . . . To create the histogram, use the `hist()` function in the `pyplot` module by including import `matplotlib.pyplot as plt` at the top of your program, then using the following code:

```
plt.hist(trials_list, bins=200)
plt.xlabel('Number of Iterations to Produce Integer 5')
plt.ylabel('Number of Trials')
plt.show()
```

(c) For a given `while`-loop iteration within the `iterations_to_five()` function, there is equal probability for producing one of the ten integers 0 through 9 and so, on that iteration, the probability of producing 5 is $\frac{1}{10}$ and the probability of producing one of the other nine integers is $\frac{9}{10}$. Thus, for example, if 5 is produced on the third `while`-loop iteration, (i.e., another integer is produced on the first and second iteration, while 5 is produced on the third iteration), the probability of this chain of (independent) events is $\left(\frac{9}{10}\right)\left(\frac{9}{10}\right)\left(\frac{1}{10}\right) = \left(\frac{1}{10}\right)\left(\frac{9}{10}\right)^2$. Applying this reasoning, the probability P that 5 is obtained after a given number of iterations (called `iterations`) is $P(\text{iterations}) = \left(\frac{1}{10}\right)\left(\frac{9}{10}\right)^{\text{iterations}-1}$. Thus, if our program runs `iterations_to_five()` N times ("trials"), the number of trials associated with each possible value of `iterations` is $N \times P(\text{iterations})$.

Create a sequence of `iterations` values from 1 to 100. This sequence can be contained in a 100-element list generated by `iterations_list = list(range(1,101))`. Then, add `for`-loop-based code that uses the above statistical analysis to predict the number of trials for each value of `iterations` from 1 to 100. Place these 100 predicted values in a list named `predict_list`. Finally, add the following code, which places the predicted curve on the same plot as the histogram:

```
plt.hist(trials_list, bins=200)
plt.xlabel('Number of Iterations to Produce Integer 5')
plt.ylabel('Number of Trials')
plt.plot(iterations_list, predict_list)
plt.show()
```

If all goes well, the following plot will be produced:

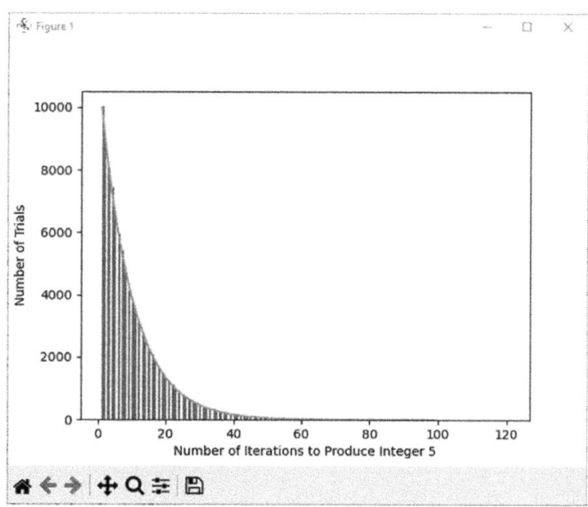

7. **single_point_animation.py** Three (x,y) positions (1, 3), (5, 5), and (9, 7) are represented by the two lists `x=[1,5,9]` and `y=[3,5,7]`. Create an animation-based program in which a single point starts at (1, 3), then after 500 ms moves to (5, 5), and then after 500 ms to (9, 7), where it stays for 500 ms. The process then repeats.

   To write your program, use **sinewave_animation.py** as your guide, replacing the sinewave creating code with the two `x` and `y` lists given above. The correct configuration of `FuncAnimation()` for this program is as follows:

```
ani = animation.FuncAnimation(fig, animate, frames=3,
                              fargs=(x, y), interval=500,
                              repeat=True)
```

Use the commands `plt.xlim(0,10)` and `plt.ylim(0,10)` to fix the limits of each axis to be 0 and 10.

CHAPTER 5

GUI-Equipped Waveform Simulator

In this chapter, you will create several modules (i.e., files containing Python code, each having with **.py** as its extension) that will be used here as well as throughout the remainder of the book. These "utility" modules will be stored in a folder named **Utilities**, which is a subfolder of the **YourNamePythonWork** folder, i.e., a file path we represent in shorthand by **YourNamePythonWork>>Utilities**. You may wish to create the **Utilities** folder now.

As explained at the end of Chapter 2, our graphical user interface (GUI) is designed to mimic the front panel of a laboratory instrument, so an *entry box* and an *output label* will be called a *control* and *indicator*, respectively.

5.1 Digitized Waveform Simulator

The goal for this chapter is to generalize our previous work and write a digitized data simulation program that can produce periodic waveforms of various functional shapes. We'll call this program **waveform_simulator.py** and it is our most ambitious project yet. In carrying out this project, we'll glean experience in Python's modular programming methods and the programming components we develop (in the form of Python modules comprised of functions) will be of use to us throughout later chapters.

To construct **waveform_simulator.py**, we'll first write three Python modules, each of which carries out one of the three tasks we require in the overall program—creation of the desired waveform data, creation of a GUI with interactive controls, and creation of a GUI-embedded graph to display the waveform. Once these three modules are finished, we will import them into our top-level ("main") module, which will coordinate the various operations to produce the desired waveform simulation. The completed GUI for **waveform_simulator.py** will look as shown here:

The Python-Based Laboratory. John Essick, Oxford University Press. © John Essick (2024). DOI: 10.1093/9780191998478.003.0005

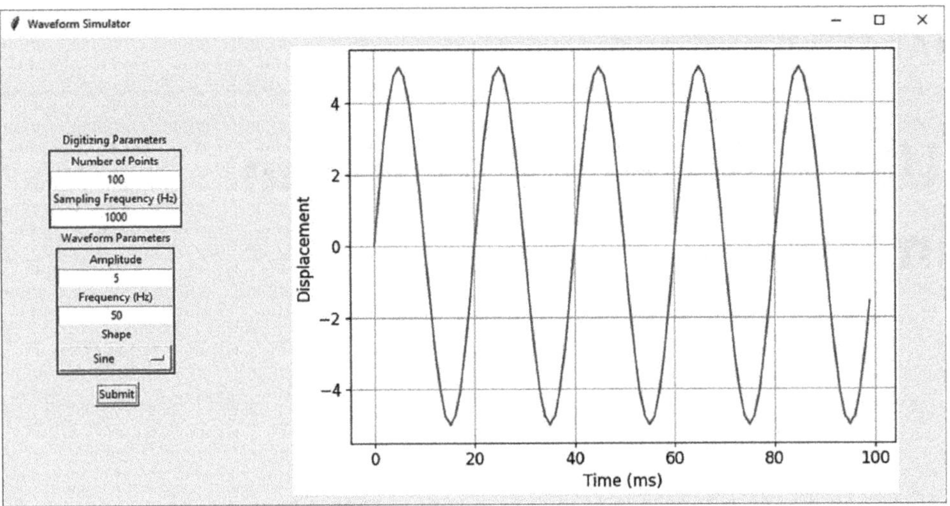

After completing **waveform_simulator.py**, we will be able to use it to study the properties of digitized data of the type produced by a data-acquisition device in a real experiment. In a typical laboratory experiment, a time-varying analog (continuous) signal $y(t)$ is input to an instrument, which is programmed to measure ("sample") this signal at N equally spaced times t_i. If Δt is the time interval between one sample and the next, then $t_i = i\,\Delta t$, where $i = 0, 1, 2, \ldots, N-1$. Since there will be one data sample taken per time interval Δt, the *sampling frequency* is defined to be

$$f_s = \frac{1}{\Delta t} \tag{5.1}$$

In previous our work, we have built data simulation programs, always assuming that a physical system under investigation is producing a pure sine-wave signal with a time-varying *displacement* given by $y(t) = A\sin(2\pi ft)$, where f and A are the sine wave's frequency and amplitude, respectively. Furthermore, we assumed that this analog signal is input to a digitizing instrument, which samples the signal at N equally spaced times at a sampling rate $f_s = 1/\Delta t$ and then outputs the result. Defining y_i to be the sample taken at time t_i, the resulting data set will be given by the two relations

$$t_i = i\,\Delta t \tag{5.2}$$

$$y_i = A\sin(2\,\pi\,ft_i), \tag{5.3}$$

where $i = 0, 1, 2, \ldots, N-1$ and $\Delta t = 1/f_s$. Note that the first and last data samples are taken at times $t_0 = 0$ and $t_{N-1} = (N-1)\,\Delta t$, respectively. In **waveform_simulator.py**, we will generalize this approach by allowing $y(t)$ to take on periodic shapes other than the sine function, including cosine, DC level, square, sawtooth, triangle, and user-defined forms.

5.2 Installing SciPy From PyPI

In the past few chapters, we met and learned how to use the Matplotlib library for plotting and the NumPy library for array-based mathematical operations. Now, at last, we meet *SciPy* (pronounced "sigh-pie"), the third member in this triad of modules that make Python so useful for the work of scientists and engineers. Scientific computing frequently involves processing large sequences of data, often facilitated by matrix mathematics. The central importance of NumPy is its `ndarray`, a data structure that was created to provide a fast, vectorization method for carrying out array-based operations. In addition to `ndarray` functionality, the NumPy library provides a limited collection of routines that implement the `ndarray` for performing useful tasks (e.g., linear algebra, Fourier transform, random number generation, and data file operations). The SciPy library of Python modules is "built on top" of NumPy in that its routines are implemented using the `ndarray`. However, the range of routines offered by SciPy is much broader than that of NumPy and includes modules for the tasks commonly encountered in science and engineering such as least-squares curve fitting, root finding, statistical analysis, numerical integration, image processing, and signal processing. As with the NumPy library, many of SciPy's underlying algorithms are executed in compiled C code, so they run much faster than the same algorithm written using standard Python lists or tuples.

SciPy is free, open-source software, available for download from PyPI. To install (lowercase) `scipy` on your Windows computer using `pip`, start a terminal session (see Section 1.2), then type the following command (for macOS, replace `pip` with `pip3`.):

```
pip install scipy
```

To find the directory into which `scipy` has been installed, enter the following command in your terminal session (for macOS, change `pip` to `pip3`):

```
pip show scipy
```

5.3 SciPy Waveform Generation Functions

With SciPy now installed on your computer, this comprehensive computing toolkit is available for use in a wealth of scientific and engineering projects. For our present work in building the **waveform_simulator.py** program, we'll call upon SciPy's waveform generation functions. In later chapters, we'll explore SciPy's curve-fitting and fast Fourier transform routines.

SciPy's waveform generation functions are in its `signal` module. For example, when called, the *sawtooth waveform generator* function returns an array `y` of values following a periodic sawtooth or triangle shape, where one period consists of a rising ramp, then a falling ramp. This function has the form

```
y = scipy.signal.sawtooth(2πft, width)
```

Here, `f` is the waveform frequency, `t` is an array of time values, and `width` is the fraction of a period that the waveform is rising. When `width = 1`, the waveform is a sawtooth, while `width = 0.5` produces a triangle waveform.

Open IDLE and test-drive the sawtooth function in the Shell to produce a 3-Hz sawtooth waveform:

```
>>> import numpy as np
>>> import scipy.signal as sig
>>> import matplotlib.pyplot as plt
```

```
>>> t = np.linspace(0., 1., 100)
>>> y = sig.sawtooth(2*np.pi*3*t, 1.0)
>>> plt.plot(t, y)
>>> plt.show()
```

The resulting plot will be as shown here:

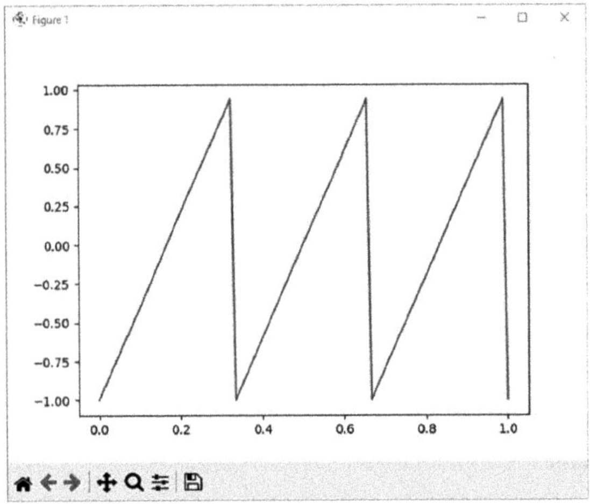

Close the plot window to restore a Python prompt (>>>), then continue your interactive session with the commands

```
>>> y = sig.sawtooth(2*np.pi*3*t, 0.5)
>>> plt.plot(t, y)
>>> plt.show()
```

A plot of a 3-Hz triangle waveform will result:

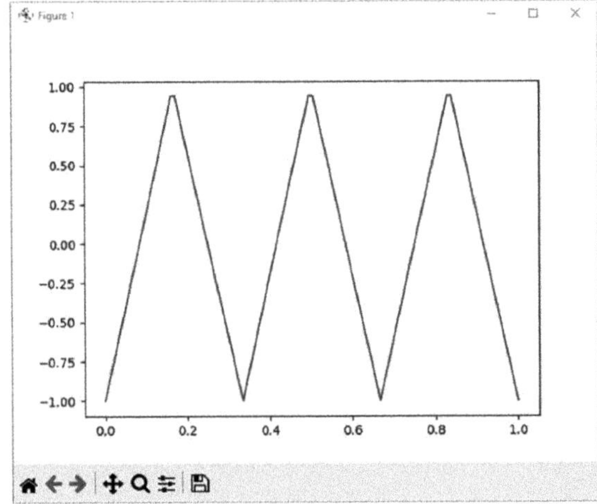

5.4 Waveform Creator Module

Let's begin by writing the heart of our waveform simulator program as a module named **waveform_creator.py**. This module will contain two functions, which together can be used to describe a digitized waveform. The first of our module's functions is designed to calculate a `num`-element NumPy array `t` of evenly spaced time values, ranging from times `start` to `stop`. This function encodes the information given by Eqs. (5.1) and (5.2). The second function then samples a waveform with a particular shape the times `t` and places these sampled values in a `num`-element NumPy array named `y`. This function generalizes Eq. (5.3) to include waveforms with shapes beyond the sine function.

In the Editor, use **File>>New File** to create the file named **waveform_creator.py** and store it in the **Utilities** folder, which is a subfolder of **YourNamePythonWork**, a file path we will represent by the shorthand **YourNamePythonWork>>Utilities**. For our work ahead, it is important that the **Utilities** folder is located in the **YourNamePythonWork** folder in order for Python to find the "utility" modules we will write and import into our top-level programs. If you haven't created the **Utilities** subfolder yet, create it now and store the module **waveform_creator.py** there.

We'll be using commands from `numpy` and `scipy.signal` in this module, so start it with the `import` statements

```
import numpy as np
import scipy.signal as sig
```

Follow the two import statements with the definition of a function named `create_time_data()`, which has the two parameters `num` and `samp_freq` for the number of samples and the sampling frequency, respectively, and uses the NumPy function `linspace()` to create the time array `t`:

```
def create_time_data(num, samp_freq):

    delta_t = 1./samp_freq
    start, stop = 0, (num-1)*delta_t

    t = np.linspace(start, stop, num)

    return t
```

Below the `create_time_data()` function, write the following definition of another function named `create_waveform_data()`, which uses an *if . . . elif* structure to create a waveform with a shape chosen from seven possibilities. This function has four parameters, the first of which is the time array `t` produced by `create_time_data()`, while the remaining three parameters are related to waveform properties—`a`, `f`, `shape`—the waveform's amplitude, frequency, and shape, respectively. The Python function `len()` is used to determine the array `t`'s number of elements ("length"), which equals the number of samples `num`. The waveform called `DC Level` is a `num`-element array whose elements are all have the value `a`; the NumPy function `ones(num)` creates a `num`-element NumPy array whose elements are all equal to 1. From SciPy's `signal` module, `square(`$2\pi ft$`)` is used to create a square waveform of frequency `f` and `sawtooth(`$2\pi ft$`, width)` is utilized to make both a sawtooth and triangle waveform of frequency `f`. Finally, an option to create a `User-Defined` waveform is provided, with a default value of $5\sin[2\pi(50)t] + 5\sin[2\pi(100)t]$, i.e., the sum of two equal-amplitude sine waves, one with frequency 50 Hz, the other with frequency 100 Hz:

```
def create_waveform_data(t, a, f, shape):

    num = len(t)

    if shape == 'Sine':
        y = a*np.sin(2*np.pi*f*t)
    elif shape == 'Cosine':
        y = a*np.cos(2*np.pi*f*t)
    elif shape == 'DC Level':
        y = a*np.ones(num)
    elif shape == 'Square':
        y = a*sig.square(2*np.pi*f*t)
    elif shape == 'Sawtooth':
        y = a*sig.sawtooth(2*np.pi*f*t, 1.0)
    elif shape == 'Triangle':
        y = a*sig.sawtooth(2*np.pi*f*t, 0.5)
    elif shape == 'User-Defined':
        y = (5*np.sin(2*np.pi*50*t)
             + 5*np.sin(2*np.pi*100*t))

    return y
```

Save your work. We'd now like to run this module to make sure it is programmed correctly. One option for assessing the soundness of this code is to write a "bug-finder" program called **debugger.py**, which serves as a main program that calls the functions in **waveform_creator.py** to see whether they perform properly. If both **waveform_creator.py** and **debugger.py** reside in **YourNamePythonWork>>Utilities** (i.e., the two programs are in the same folder), the code for **debugger.py** is the following:

```
import matplotlib.pyplot as plt
import waveform_creator as wc

num, samp_freq = 100, 100
a, f = 5, 3
shape = 'Sine'

t = wc.create_time_data(num, samp_freq)
y = wc.create_waveform_data(t, a, f, shape)

plt.plot(t, y)
plt.show()
```

You can try this debugging approach by writing **debugger.py** and running it. However, Python provides a better way.

5.5 Running Module as "Main" Program

A well-written Python program is hierarchical in nature. It consists of a top-level ("main") module, which calls code within lower-level modules. In this modular approach to programming, the top-level module is the program that the Python Interpreter is instructed to run, while the lower-level modules are those imported into the top-level module.

When the command is given to the Python Interpreter to run a program, the Interpreter defines a special internal attribute called __name__ for each module. This quantity includes two underscores on either side of it and is an example of a *dunder attribute*, where *dunder* stands for double underscore. The Interpreter assigns the value of __name__ as follows: For the top-level module, its __name__ is set to the value '__main__' because it's the "main" program; for each module that is imported into the top-level module, its __name__ is set to its module's name. Thus, the value of __name__ can be used to check whether the current program is being run on its own as a top-level module or whether it is being used as a lower-level imported module. For example, when the bug-finder program given at the end of Section 5.4 is run, **debugger.py** and **waveform_creator.py** will be assigned __name__ == '__main__' and __name__ == 'waveform_creator', respectively.

In writing a modular Python program, each module is written to carry out a particular component task required for the successful completion of the greater program's overall mission. As we saw in Section 5.4, when developing a module such as **waveform_creator.py**, it would simplify the programmer's inevitable job of program debugging to be able to execute that module as a stand-alone main program (as opposed to having to import it into program like **debugger.py**, which has been created for the sole purpose of debugging the imported module). And, by a programming trick involving __name__, that simplification is possible. The trick lies in including the dunder attribute __name__ in this *if* statement structure:

```
if __name__ == '__main__':
    <Statements to Execute>
```

By placing this block of code at the end of a module (which we will call a *test block*), <Statements to Execute> will execute if the module is being run as a main program, as opposed to being imported into another module. Thus, by including this *if* statement at the end of **waveform_creator.py**, it is no longer necessary to write the separate bug-finder **debugger.py** program. The debugging code that was previously in **debugger.py** can now instead be the <Statements to Execute>.

Complete **waveform_creator.py** as shown next. Note that the import waveform_creator as wc statement used in **debugger.py** is no longer needed.

```
import numpy as np
import scipy.signal as sig

def create_time_data(num, samp_freq):

    delta_t = 1./samp_freq
    start, stop = 0, (num-1)*delta_t

    t = np.linspace(start, stop, num)

    return t
```

```
def create_waveform_data(t, a, f, shape):

    num = len(t)

    if shape == 'Sine':
        y = a*np.sin(2*np.pi*f*t)
    elif shape == 'Cosine':
        y = a*np.cos(2*np.pi*f*t)
    elif shape == 'DC Level':
        y = a*np.ones(num)
    elif shape == 'Square':
        y = a*sig.square(2*np.pi*f*t)
    elif shape == 'Sawtooth':
        y = a*sig.sawtooth(2*np.pi*f*t, 1.0)
    elif shape == 'Triangle':
        y = a*sig.sawtooth(2*np.pi*f*t, 0.5)
    elif shape == 'User-Defined':
        y = (5*np.sin(2*np.pi*50*t)
             + 5*np.sin(2*np.pi*100*t))

    return y

if __name__ == '__main__':

    import matplotlib.pyplot as plt

    num, samp_freq = 100, 100
    a, f = 5, 3
    shape = 'Sine'

    t = create_time_data(num, samp_freq)
    y = create_waveform_data(t, a, f, shape)

    plt.plot(t, y)
    plt.show()
```

Save your work. Run **waveform_creator.py** and demonstrate that the test block causes the code to execute as a stand-alone top-level program, producing three cycles of a 3-Hz sine-wave waveform that is plotted by Matplotlib. Change `shape` to some of its other possible values (e.g., `'Square'`, `'Sawtooth'`) and show that the expected plots result. Then, close **waveform_creator.py**.

5.6 GUI Creator Module

Next, let's write a module called **gui_creator.py** that contains all of the functions needed to create a Tkinter-based graphical user interface. In the Editor, use **File>>New File** and **File>>Save** to create and then save a program named **gui_creator.py** in **YourNamePythonWork>>Utilities**.

We'll be building our GUI from Tkinter widgets, so start the program with the `import` statement

```
import tkinter as tk
```

Our first function in this module, called `create_window()`, will generate an operating-system window named `root` by calling the `Tk()` class, i.e., using the assignment statement `root=tk.Tk()`. The `root` window will be the container for all of our other Tkinter widgets. The `Tk()` class has several methods and, among these, we will use `title()` and `geometry()` to control of the window's `title` and `size`, respectively. Below the `import` statement, start coding the `create_window()` function as follows:

```
def create_window(title, size='1000x500'):

    root = tk.Tk()
    root.title(title)
    root.geometry(size)

    return root
```

With this coding, the window's background color will be the default color for your computing system (often, light gray). If, instead, you prefer white, simply include the command the `root.config(bg='white')`. There are many other color options available, e.g., `'red'`, `'green'`, `'cyan'`. As written, the `create_window()` function allows a programmer to choose the arguments `title` and `size`. By assigning a value to the `size` argument in the function's definition, that value become its default value, i.e., the argument's value if that argument does not appear when the function is called.

To complete setting up the window, we'll divide it into sectors using the Tkinter frame widget we experimented with in Section 2.7. Here, we will configure three frames that will shortly be used as the containers for controls (widgets that facilitate input of various parameters needed for waveform simulation) and display (a Matplotlib graph of the produced waveform). As we learned in Chapter 2, a Tkinter frame is a rectangular region inside of a specified container (e.g., the `root` window or another frame). To create a borderless frame object named `frame`, the command is as follows:

```
    frame = tk.Frame(container)
```

Once created, the `frame` object is placed within the specified `container` using the command `frame.pack(side,padx,pady)`. In Chapter 2, we discussed how the argument `side` determines the placement of a widget within its container via the options `tk.TOP` (default), `tk.BOTTOM`, `tk.LEFT`, and `tk.RIGHT`. The arguments `padx` and `pady` specify the amount of horizontal and vertical padding, respectively, to leave outside of each side of the borderless frame, where the padding is measured in pixels. The following code divides a generic `container` (which we will always choose to be the `root` window when this function is called) into two frames—`left_frame` and `right_frame`—on the left and right sides of the `container`, respectively. The option `padx=50` ensures that these frames are indented from the left and right edges of the `container`. A frame named `control_box`, which will contain the program's controls, is then placed within the `left_frame`; i.e., `left_frame` is the container for control_box. In **gui_creator.py**, below the code that defines create_window(), enter the following definition for the create_frames(container) function:

```
def create_frames(container):

    left_frame = tk.Frame(container)
    left_frame.pack(side=tk.LEFT, padx=50)

    right_frame = tk.Frame(container)
    right_frame.pack(side=tk.RIGHT, padx=50)

    control_box = tk.Frame(left_frame)
    control_box.pack(side=tk.TOP, pady=10)

    return control_box, right_frame
```

To build our GUI, this function only needs to return the frame objects control_box and right_frame. Save your work.

At the bottom of your **gui_creator.py** program, input the test block

```
if __name__ == '__main__':

    root = create_window(title='Waveform Simulator')
    control_box, right_frame = create_frames(root)
```

Save your work, then run the module. An operating window of the specified size and titled **Waveform Simulator** should appear. The empty frames are also present in the window, but that empty space needs to be filled with something useful. It's time to put some controls in control_box.

Our function create_time_data(num,samp_freq) in the **waveform_creator.py** module calculates the sequence of times at which a signal is digitized in a particular experiment, where num is the number of times in the time sequence and samp_freq is the sampling frequency. Thus, num and samp_freq are the *digitizing parameters* for such an experiment. Let's place the label **Digitizing Parameters** within the control_box frame. Then, in a new frame below that label, we'll place two Tkinter text entry widgets labeled **Number of Samples** and **Sampling Frequency**, which will allow a user to interactively input values for num and samp_freq, respectively. In Chapter 2, we found that a label named lbl can be placed within a container by the following command: lbl=tk.Label(container,text), where text is the message displayed in the label. We also learned that a control called ent is created by ent=tk.Entry(container,textvariable,justify), where textvariable specifies the string variable in which the text entered in the control is stored and justify gives the justification of the text within the entry box. A string variable is created by the statement var=tk.StringVar().

Immediately below the code that defines the create_frames() function, code the following definition for the function digitizing_control(). This function creates the **Digitizing Parameters** control described above and returns the string variables var_num and var_samp_freq, which hold the entered values for num and samp_freq, respectively. For the frame object, bg='blue' and bd=2 sets the frame's background color to blue and places a blue border around the frame that is two pixels thick, respectively. The argument fill=tk.X in the four pack() functions causes the objects being placed within the frame to completely fill the frame in the horizontal (X) direction. To understand the result of fill=tk.X, try removing it from the pack() functions. You'll find that the widgets are shortened and uncover some of the frame's background blue color.

```
def digitizing_control(container):

    lbl_1 = tk.Label(container,
                     text='Digitizing Parameters')
    lbl_1.pack()

    frame = tk.Frame(container, bg='blue', bd=2)
    frame.pack()

    lbl_2 = tk.Label(frame, text='Number of Points')
    lbl_2.pack(fill=tk.X)

    var_num = tk.StringVar()
    ent_1 = tk.Entry(frame, textvariable=var_num,
                     justify=tk.CENTER)
    ent_1.pack(fill=tk.X)

    lbl_3 = tk.Label(frame, text='Sampling Frequency (Hz)')
    lbl_3.pack(fill=tk.X)

    var_samp_freq = tk.StringVar()
    ent_2 = tk.Entry(frame, textvariable=var_samp_freq,
                     justify=tk.CENTER)
    ent_2.pack(fill=tk.X)

    return var_num, var_samp_freq
```

Save your work. At the bottom of your **gui_creator.py** program, add the third statement to the test block.

```
if __name__ == '__main__':

    root = create_window(title='Waveform Simulator')
    control_box, right_frame = create_frames(root)
    var_num, var_samp_freq = digitizing_control(control_box)
```

Save your work and run the module. A blue-bordered box labeled **Digitizing Parameters** and containing two controls labeled **Number of Points** and **Sampling Frequency (Hz)** should now appear on the left side of the **Waveform Simulator** window.

Note that the variable names `frame`, `lbl_1`, `lbl_2`, `lbl_3`, `ent_1`, and `ent_2` are only used internally within the definition of the `digitizing_control()` function. Thus, these same variable names can be reused within the definition of another function, as long as they are only used internally there as well. Stated in technical terms, variables only used within a particular function are called *local variables* and the collection of those local variables is contained in the function's *namespace*. The term *scope* then specifies the region in your program where the objects in a namespace are accessible. Applying these terms to the `digitizing_control()` function, the local variables `frame`, `lbl_1`, `lbl_2`, `lbl_3`, `ent_1`, and `ent_2` are contained in `digitizing_control()`'s namespace and, since these variables are only accessible within `digitizing_control()`, the scope of this namespace is simply the code within `digitizing_control()`. We'll have more to say about namespace and scope at the end of the chapter.

Next, we need controls related to our function `create_waveform_data(t,a,f,shape)` in the **waveform_creator.py** module. Given the time array `t`, this function calculates the digitized samples of a waveform with amplitude, frequency, and functional form given by `a`, `f`, and `shape`, respectively. Thus, `a`, `f`, and `shape` are the *waveform parameters* for our simulation. Below the **Digitizing Parameters** control, let's place the label **Waveform Parameters** within the `control_box` frame. Then, in a new frame below that label, we'll place two Tkinter entry widgets labeled **Amplitude** and **Frequency (Hz)**, which will allow a user to interactively input numeric values for `a` and `f`, respectively. Finally, below those two entry widgets, we'll place an option menu widget, which offers a dropdown menu of waveform-shape options. To create an option menu object named `opm`, the `OptionMenu()` class is implemented as follows:

```
opm = tk.OptionMenu(contrainer, textvariable, *options)
```

Here, `textvariable` specifies the variable in which the selected dropdown menu item is to be stored and `options` is a Python list that contains the desired dropdown menu items (the asterisk alerts Python that this argument is a multielement item).

Immediately below the code that defines the `digitizing_control()`, input the following definition for the function `waveform_control()`. This function creates the **Waveform Parameters** control described above and returns the string variables `var_a`, `var_f`, and `var_shape`, which hold the entered values for `a`, `f`, and `shape`, respectively.

```
def waveform_control(container):

    lbl_1 = tk.Label(container, text='Waveform Parameters')
    lbl_1.pack()

    frame = tk.Frame(container, bg='blue', bd=2)
    frame.pack()

    lbl_2 = tk.Label(frame, text='Amplitude')
    lbl_2.pack(fill=tk.X)

    var_a = tk.StringVar()
    ent_1 = tk.Entry(frame, textvariable=var_a,
                     justify=tk.CENTER)
    ent_1.pack(fill=tk.X)

    lbl_3 = tk.Label(frame, text='Frequency (Hz)')
    lbl_3.pack(fill=tk.X)

    var_f = tk.StringVar()
    ent_2 = tk.Entry(frame, textvariable=var_f,
                     justify=tk.CENTER)
    ent_2.pack(fill=tk.X)

    lbl_4 = tk.Label(frame, text='Shape')
    lbl_4.pack(fill=tk.X)

    var_shape = tk.StringVar()
    options = ['Sine', 'Cosine', 'DC Level', 'Square',
               'Sawtooth', 'Triangle', 'User-Defined']
    opm = tk.OptionMenu(frame, var_shape, *options)
    opm.pack(fill=tk.X)

    return var_a, var_f, var_shape
```

Save your work. At the bottom of your **gui_creator.py** program, add the fourth statement to the test block.

```
if __name__ == '__main__':

    root = create_window(title='Waveform Simulator')
    control_box, right_frame = create_frames(root)
    var_num, var_samp_freq = digitizing_control(control_box)
    var_a, var_f, var_shape = waveform_control(control_box)
```

Save your work, then run the module. Below the labeled **Digitizing Parameters** control, a blue-bordered box labeled **Waveform Parameters** should appear, with two controls labeled **Amplitude** and **Frequency (Hz)** as well as an option menu box labeled **Shape** with seven selection options in its dropdown menu.

The last item for the `control_box` is a button labeled **Submit**. When pressed, this button will call a to-be-written function, which will read the values that a user has entered for the five input parameters. In Chapter 2, we acquired skill with the Tkinter button widget. As a reminder, the button object named `btn` is created by invoking the `Button()` class via the following statement:

```
btn = tk.Button(container, text, command)
```

For this statement, `text` is the text label placed within the button and `command` specifies an action to take (e.g., name of a function to execute) when the button is clicked by the mouse.

Right below the code that defines `waveform_control()`, enter the following definition for the function `action_button(container,label,action)`. This function creates a frame, then places a button inside it. The button is identified by its text `label` and executes the specified `action` when pressed. Due to the option `pady=8`, the button will be separated top-and-bottom from neighboring GUI objects by eight pixels. In our finalized **waveform_simulator.py** program, we'll program `action` to be the retrieval of the values entered into the five interactive inputs.

```
def action_button(container, label, action):

    frame = tk.Frame(container)
    frame.pack(pady=8)

    btn = tk.Button(frame, text=label, command=action)
    btn.pack()

    return btn
```

Save your work. At the bottom of your **gui_creator.py** program, add the fifth statement to the test block. For now, we set `action=None`, which obviously means no action is triggered by a button press.

```
if __name__ == '__main__':

    root = create_window(title='Waveform Simulator')
    control_box, right_frame = create_frames(root)
    var_num, var_samp_freq = digitizing_control(control_box)
    var_a, var_f, var_shape = waveform_control(control_box)
    btn = action_button(control_box, label='Submit',
                        action=None)
```

Save your work and run the module. Below the labeled **Waveform Parameters** controls, there will be a button labeled **Submit**. Our GUI control panel is now complete!

Finally, we just need to add two short functions to retrieve the entered values, one function named `get_dig_entries()` to fetch the two **Digitizing Parameters** entries and `get_wav_entries()` to deliver the three **Waveform Parameter** entries. In Chapter 2, we found that when using the `StringVar()` class to create a string named `var` (which is then used to store the text a user has entered in a control), the value of `var` can be accessed via the `get()` method of `StringVar()`, i.e., through the dot-notation command `var.get()`.

Right below the code that defines `action_button()`, implement the `get()` method in writing the following definitions named `get_dig_entries()` and `get_wav_entries()`. Since the entries are retrieved as string data types, the operators `int()` and `float()` are used to convert quantities to their correct data type (i.e., integer or float):

```
def get_dig_entries(var_num, var_samp_freq):

    num = int(var_num.get())
    samp_freq = float(var_samp_freq.get())

    return num, samp_freq
```

```
def get_wav_entries(var_a, var_f, var_shape):

    shape = var_shape.get()
    a = float(var_a.get())
    f = float(var_f.get())

    return a, f, shape
```

To test this code, immediately below it, define the following function named `print_entries()`:

```
def print_entries():

    num, samp_freq = get_dig_entries(var_num,
                                     var_samp_freq)
    a, f, shape = get_wav_entries(var_a, var_f, var_shape)
    print(num, samp_freq)
    print(a, f, shape)

    return
```

Save your work. At the bottom of your **gui_creator.py** program, program the button to execute `print_entries` when pressed by setting `action=print_entries`, and start the event-handling loop by including the `root.mainloop()` statement:

```
if __name__ == '__main__':

    root = create_window(title='Waveform Simulator')
    control_box, right_frame = create_frames(root)
    var_num, var_samp_freq = digitizing_control(control_box)
    var_a, var_f, var_shape = waveform_control(control_box)
    btn = action_button(control_box,label='Submit',
                        action=print_entries)
    root.mainloop()
```

Save your work and run the module. When the GUI window with the control panel appears, enter values into the five inputs. After these values are entered, press the **Submit** button. If all goes well, your five entries will be printed in the Shell.

There is a bug in our `get_wav_entries()` function, which is not readily apparent, but will pop up in future use, if not corrected. The bug relates to when a user chooses `shape` to be either `'DC Level'` or `'User-Defined'`. In the former case, the user will enter a value at the **Amplitude** control in order to select the desired constant DC level, but may leave the **Frequency (Hz)** control blank, as it is not relevant for a DC level waveform. However, when running the program, the `get_wav_entries()` function is called, which contains the `float(var_f.get())` command. Since the **Frequency (Hz)** control is blank, `var_f.get()` returns an empty string, i.e., `''` (two quotation marks without a space in between). Unfortunately, the `float()` function is not designed to handle an empty string as its argument and so Python will raise the following exception: `ValueError: could not convert string to` `float: ''`. Test out this scenario yourself by running the program with **Number of Points**, **Sampling Frequency (Hz)**, **Amplitude**, and **Shape** set to *100*, *1000*, *5*, and *DC Level*, respectively, while leaving the **Frequency (Hz)** control blank (i.e., no entered value). When you press the **Submit** button, you will see the `ValueError` exception raised in the Shell. A similar problem relates to when `shape` is chosen to be `'User-Defined'`. Since all of the waveform parameters are programmed into the user-defined function, a user may leave the **Amplitude** and **Frequency (Hz)** controls blank, again raising the `ValueError` exception when the **Submit** button is pressed. To fix this bug, we can simply assign placeholder values (we'll choose `0`) to the problematic quantities when either `'DC Level'` or `'User-Defined'` is chosen as the waveform shape by a user. Modify `get_wav_entries()` as shown here, in order to fix the bug:

```
def get_wav_entries(var_a, var_f, var_shape):

    shape = var_shape.get()
    if shape == 'User-Defined':
        a = 0
        f = 0
    elif shape == 'DC Level':
        a = float(var_a.get())
        f = 0
    else:
        a = float(var_a.get())
        f = float(var_f.get())

    return a, f, shape
```

Save your work, then run the program under the *DC Level* scenario outlined above to verify that the bug has been fixed. Also, try choosing the *User-Defined* shape and show no exception is raised if the **Amplitude** and **Frequency (Hz)** controls are both left blank.

The complete **gui_creator.py** program is as follows:

```
import tkinter as tk

def create_window(title, size='1000x500'):

    root = tk.Tk()
    root.title(title)
    root.geometry(size)

    return root


def create_frames(container):

    left_frame = tk.Frame(container)
    left_frame.pack(side=tk.LEFT, padx=50)

    right_frame = tk.Frame(container)
    right_frame.pack(side=tk.RIGHT, padx=50)
```

```
    control_box = tk.Frame(left_frame)
    control_box.pack(side=tk.TOP, pady=10)

    return control_box, right_frame


def digitizing_control(container):

    lbl_1 = tk.Label(container,
                     text='Digitizing Parameters')
    lbl_1.pack()

    frame = tk.Frame(container, bg='blue', bd=2)
    frame.pack()

    lbl_2 = tk.Label(frame, text='Number of Points')
    lbl_2.pack(fill=tk.X)

    var_num = tk.StringVar()
    ent_1 = tk.Entry(frame, textvariable=var_num,
                     justify=tk.CENTER)
    ent_1.pack(fill=tk.X)

    lbl_3 = tk.Label(frame, text='Sampling Frequency (Hz)')
    lbl_3.pack(fill=tk.X)

    var_samp_freq = tk.StringVar()
    ent_2 = tk.Entry(frame, textvariable=var_samp_freq,
                     justify=tk.CENTER)
    ent_2.pack(fill=tk.X)

    return var_num, var_samp_freq


def waveform_control(container):

    lbl_1 = tk.Label(container, text='Waveform Parameters')
    lbl_1.pack()

    frame = tk.Frame(container, bg='blue', bd=2)
    frame.pack()

    lbl_2 = tk.Label(frame, text='Amplitude')
    lbl_2.pack(fill=tk.X)

    var_a = tk.StringVar()
    ent_1 = tk.Entry(frame, textvariable=var_a,
                     justify=tk.CENTER)
    ent_1.pack(fill=tk.X)
```

```
    lbl_3 = tk.Label(frame, text='Frequency (Hz)')
    lbl_3.pack(fill=tk.X)

    var_f = tk.StringVar()
    ent_2 = tk.Entry(frame, textvariable=var_f,
                     justify=tk.CENTER)
    ent_2.pack(fill=tk.X)

    lbl_4 = tk.Label(frame, text='Shape')
    lbl_4.pack(fill=tk.X)

    var_shape = tk.StringVar()
    options = ['Sine', 'Cosine', 'DC Level', 'Square',
               'Sawtooth', 'Triangle', 'User-Defined']
    opm = tk.OptionMenu(frame, var_shape, *options)
    opm.pack(fill=tk.X)

    return var_a, var_f, var_shape


def action_button(container, label, action):

    frame = tk.Frame(container)
    frame.pack(pady=8)

    btn = tk.Button(frame, text=label, command=action)
    btn.pack()

    return btn


def get_dig_entries(var_num, var_samp_freq):

    num = int(var_num.get())
    samp_freq = float(var_samp_freq.get())

    return num,samp_freq


def get_wav_entries(var_a, var_f, var_shape):

    shape = var_shape.get()
    if shape == 'User-Defined':
        a = 0
        f = 0
    elif shape == 'DC Level':
        a = float(var_a.get())
        f = 0
```

```
    else:
        a = float(var_a.get())
        f = float(var_f.get())

    return a, f, shape


def print_entries():

    num, samp_freq = get_dig_entries(var_num,
                                     var_samp_freq)
    a, f, shape = get_wav_entries(var_a, var_f, var_shape)
    print(num, samp_freq)
    print(a, f, shape)

    return


if __name__ == '__main__':

    root = create_window(title='Waveform Simulator')
    control_box, right_frame = create_frames(root)
    var_num, var_samp_freq = digitizing_control(control_box)
    var_a, var_f, var_shape = waveform_control(control_box)
    btn = action_button(control_box,label='Submit',
                        action=print_entries)
    root.mainloop()
```

5.7 Plot Creator Module

Our final module, named **plot_creator.py**, will produce the Matplotlib graph to be placed within the `right_fame` of the Tkinter-produced **Waveform Simulator** window. In the Editor, use **File>>New File** to create a file named **plot_creator.py**, then save it in **YourNamePython-Work>>Utilities.**

In Chapters 3 and 4, we learned that to place a Matplotlib graph in a Tkinter-produced container, our program needs to start with two `import` statements:

```
import matplotlib.pyplot as plt
import matplotlib.backends.backend_tkagg as bt
```

Below these import statements, enter the following definition for the `create_plot(container)` function. This function creates a Matplotlib plot with its associated figure and axes objects, which are given the names `fig` and `ax`, respectively. The `layout='constrained'` option assures that the graph and its labeling will fit (i.e., without chipping) in the rectangular `fig` area. Then, `bt.FigureCanvasTkAgg(fig, container)`, along with `canvas.get_tk_widget().pack()`, use `fig` to create a plot object named `canvas`, which is placed in the designated Tkinter-related `container`:

```
def create_plot(container):

    fig, ax = plt.subplots(layout='constrained')

    canvas = bt.FigureCanvasTkAgg(fig, container)
    canvas.get_tk_widget().pack()

    return ax, canvas
```

Save your work. In a few moments, we'll write our top-level program **waveform_simulator.py**, which will call `create_plot(container)` with the assignment `container=right_frame`.

Finally, below your definition of the `create_plot()` function, add this definition for the function `update_plot()`. This function provides the option to clear any existing plot (via the `clear_flag` argument), then performs a *y* versus *x* plot of the `x` and `y` data arrays, which are passed to it as arguments. The choice of marker style is controlled by the `marker` argument. Other statements produce axis labeling and *tick labels* (i.e., the numbers that scale each axis) of a chosen `fontsize`; you may wish to optimize the value of `text_size` for your particular computing system.

```
def update_plot(ax, canvas, x, y, label_x, label_y,
                marker='-', clear_flag=True):

    text_size = 12

    if clear_flag == True:
        ax.clear()

    ax.tick_params(axis='both', labelsize=text_size)
    ax.set_xlabel(label_x, fontsize=text_size)
    ax.set_ylabel(label_y, fontsize=text_size)
    ax.grid()

    ax.plot(x, y, marker)
    canvas.draw()

    return
```

Save your work. By default, `clear_flag=True`; i.e., the existing plot will be cleared before the current plot is rendered. This option is the one that we will use most often. But, in future work, we will place two curves on the same plot, which will require setting `clear_flag=False` before plotting the second curve (so the first curve is not erased). The default argument for `marker` is `'-'` (dash inside the two quotation marks), which produces a solid-line curve. An option we will frequently choose is `'o-'` (lowercase letter o followed by dash, both enclosed by two quotation marks), which produces filled circles at each data point with a solid line interpolated between neighboring points. If you wish to simply use the default values for `marker` and `clear_flag`, these arguments do not have to be written when calling the function. That is, if you opt for `marker='-'` and `clear_flag=True`, the call for this function in your top-level program can be written simply as follows:

```
    update_plot(ax, canvas, x, y, label_x, label_y)
```

This module is complete. As a quick evaluation of your coding, append the following test block at the bottom of the program:

```
if __name__ == '__main__':

    import tkinter as tk

    label_x, label_y = 'x-axis', 'y-axis'
    x, y = [], []
    root = tk.Tk()

    ax, canvas = create_plot(root)
    update_plot(ax, canvas, x, y, label_x, label_y)
```

Save your work, then run this module. It should produce an operating system window that holds a blank graph with axis labeling.

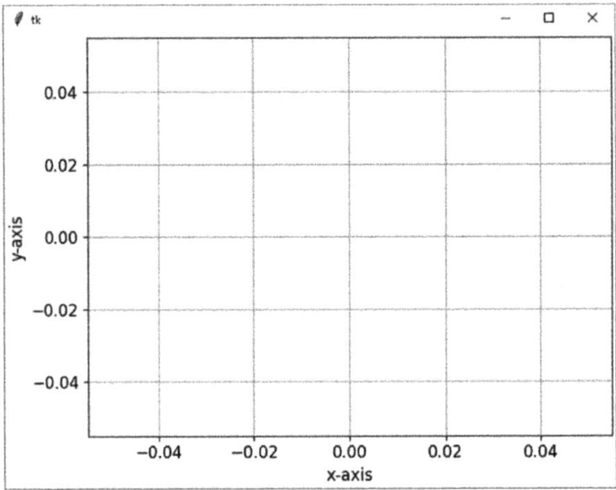

The complete **plot_creator.py** program is as follows:

```
import matplotlib.pyplot as plt
import matplotlib.backends.backend_tkagg as bt


def create_plot(container):

    fig, ax = plt.subplots(layout='constrained')

    canvas = bt.FigureCanvasTkAgg(fig, container)
    canvas.get_tk_widget().pack()

    return ax, canvas
```

```
def update_plot(ax, canvas,x, y, label_x, label_y,
                marker='-', clear_flag=True):

    text_size = 12

    if clear_flag == True:
        ax.clear()

    ax.tick_params(axis='both', labelsize=text_size)
    ax.set_xlabel(label_x, fontsize=text_size)
    ax.set_ylabel(label_y, fontsize=text_size)
    ax.grid()

    ax.plot(x, y, marker)
    canvas.draw()

    return


if __name__ == '__main__':

    import tkinter as tk

    label_x, label_y = 'x-axis', 'y-axis'
    x, y = [], []
    root = tk.Tk()

    ax, canvas = create_plot(root)
    update_plot(ax, canvas, x, y, label_x, label_y)
```

Finally, just a quick note on plot markers. Previously, we plotted data using the high-level `pyplot.plot()` command, which always produces a solid-line curve. If, in addition, one wishes a filled circle symbol to appear on that line at the location of each data point, then the argument `marker='o'` is included. In **plot_creator.py**, we are instead using the lower-level `ax.plot()` command to plot data. For this command, the argument `marker='o'` simply produces a filled circle at each data point (i.e., a scatter plot). If, in addition, one wishes a solid line to be interpolated between these filled circles, the argument for `ax.plot()` is `marker='o-'`.

5.8 Waveform Simulator Program

Now that the hard work of writing our three modules is finished, we're ready to write the top-level module that calls them. We'll name this main program **waveform_simulator.py.**

In the Editor, use **File>>New File** to create a main module named **waveform_simulator.py**, then save it in **YourNamePythonWork**. The three lower-level modules—**waveform_creator.py**, **gui_creator.py**, and **plot_creator.py**—should be in **YourNamePythonWork>>Utilities**.

Begin your program by importing the three modules, each with abbreviated aliases. Because these modules are in a subfolder of the folder that holds this program, the dot notation `subdirectory_name.module_name` points Python to the correct location of the module:

```
import Utilities.waveform_creator as wc
import Utilities.gui_creator as gc
import Utilities.plot_creator as pc
```

Write the body of this module as shown. Most of these statements will be familiar from the test-block coding in our modules. After defining labels, functions from **gui_creator.py** are called to create a window with the `control_box` and `right_frame` frames, and then to place the **Digitizing Parameters** and **Waveform Parameters** controls in the `control_box` frame. Next, a blank plot is placed in the `right_frame` frame, using functions from the **plot_creator.py** module. Finally, the button-creation function in **gui_creator.py** is called to place a **Submit** button at the bottom of `control_box`. When pressed, the button is programmed to execute a to-be-written function named `calculate_plot_waveform()`. Finally, an event-handling loop is activated via the `root.mainloop()` command. Remember to place the module-alias prefix in front of each imported function; you may wish to optimize the `size` of the `root` window for your particular computing system.

```
import Utilities.waveform_creator as wc
import Utilities.gui_creator as gc
import Utilities.plot_creator as pc


label_x, label_y = 'Time (ms)', 'Displacement'

root = gc.create_window(title='Waveform Simulator',
                        size='1000x500')
control_box, right_frame = gc.create_frames(root)

var_num, var_samp_freq = gc.digitizing_control(control_box)
var_a, var_f, var_shape = gc.waveform_control(control_box)

ax, canvas = pc.create_plot(right_frame)
pc.update_plot(ax, canvas, [], [], label_x, label_y)

btn = gc.action_button(control_box, label='Submit',
                       action=calculate_plot_waveform)

root.mainloop()
```

Save your work. Our last step is to write the `calculate_plot_waveform()` function. When the **Submit** button is pressed, this function calculates the requested waveform and then plots it. The code for the function is shown next and it again includes statements familiar from our test blocks. First, using functions from **gui_creator.py**, the entered values are fetched from the **Digitizing Parameters** and **Waveform Parameters** controls. Second, **waveform_creator.py** functions are used to calculate the waveform's `t` and `y` data arrays. Third, a function from

plot_creator.py is used to update the existing GUI graph. The `t` array is multiplied by 1000 in order to convert from the time unit of seconds to milliseconds.

```
def calculate_plot_waveform():

    num, samp_freq = gc.get_dig_entries(var_num,
                                        var_samp_freq)
    a, f, shape = gc.get_wav_entries(var_a, var_f, var_shape)

    t = wc.create_time_data(num, samp_freq)
    y = wc.create_waveform_data(t, a, f, shape)

    pc.update_plot(ax, canvas, 1000*t, y, label_x, label_y,
                   marker='-')

    return
```

It may seem natural to complete our program by placing this definition for `calculate_plot_waveform()` below the previously written code as shown here:

```
import Utilities.waveform_creator as wc
import Utilities.gui_creator as gc
import Utilities.plot_creator as pc

label_x, label_y = 'Time (ms)', 'Displacement'

root = gc.create_window(title='Waveform Simulator',
                        size='1000x500')
control_box, right_frame = gc.create_frames(root)

var_num, var_samp_freq = gc.digitizing_control(control_box)
var_a, var_f, var_shape = gc.waveform_control(control_box)

ax, canvas = pc.create_plot(right_frame)
pc.update_plot(ax, canvas, [], [], label_x, label_y)

btn = gc.action_button(control_box, label='Submit',
                       action=calculate_plot_waveform)

root.mainloop()

def calculate_plot_waveform():

    num, samp_freq = gc.get_dig_entries(var_num,
                                        var_samp_freq)
    a, f, shape = gc.get_wav_entries(var_a, var_f, var_shape)

    t = wc.create_time_data(num, samp_freq)
    y = wc.create_waveform_data(t, a, f, shape)
```

```
    pc.update_plot(ax, canvas, 1000*t, y, label_x, label_y,
                   marker='-')

    return
```

However, when this program is run with this code structure, a Traceback exception will be raised stating `NameError: name 'calculate_plot_waveform'is not defined.` This error statement results because, as the Python Interpreter reads through the program, it encounters the name `calculate_plot_waveform` in the `action_button` statement before this object is defined and so the Interpreter is at a loss as to what to do. Thus, the correct coding places the definition of the `calculate_plot_waveform` at a position within the program before it is called as shown below. Complete this correct coding scheme for **waveform_simulator.py**.

```
import Utilities.waveform_creator as wc
import Utilities.gui_creator as gc
import Utilities.plot_creator as pc

def calculate_plot_waveform():

    num, samp_freq = gc.get_dig_entries(var_num,
                                        var_samp_freq)
    a, f, shape = gc.get_wav_entries(var_a, var_f,
                                     var_shape)

    t = wc.create_time_data(num, samp_freq)
    y = wc.create_waveform_data(t, a, f, shape)

    pc.update_plot(ax, canvas, 1000*t, y, label_x, label_y,
                   marker='-')

    return

label_x, label_y = 'Time (ms)', 'Displacement'

root = gc.create_window(title='Waveform Simulator',
                        size='1000x500')
control_box, right_frame = gc.create_frames(root)

var_num, var_samp_freq = gc.digitizing_control(control_box)
var_a, var_f, var_shape = gc.waveform_control(control_box)

ax, canvas = pc.create_plot(right_frame)
pc.update_plot(ax, canvas, [], [], label_x, label_y)

btn = gc.action_button(control_box, label='Submit',
                       action=calculate_plot_waveform)

root.mainloop()
```

Save your work, then run **waveform_simulator.py**. The GUI you have built will appear with the control panel awaiting your entries, along with a blank plot. Enter *100* and *1000* for **Number of Samples** and **Sampling Frequency (Hz)**, and *5*, *50*, and *Sine* for **Amplitude**, **Frequency (Hz)**, and **Shape**, respectively. These choices for the parameters simulate sampling a 50-Hz sine wave for 100 ms (100 samples at a rate of 1000 samples per second takes 0.1 s) and so should result in outputting five cycles of the 50-Hz sine wave. Click on the **Submit** button, and if all goes well, you will see the five cycles.

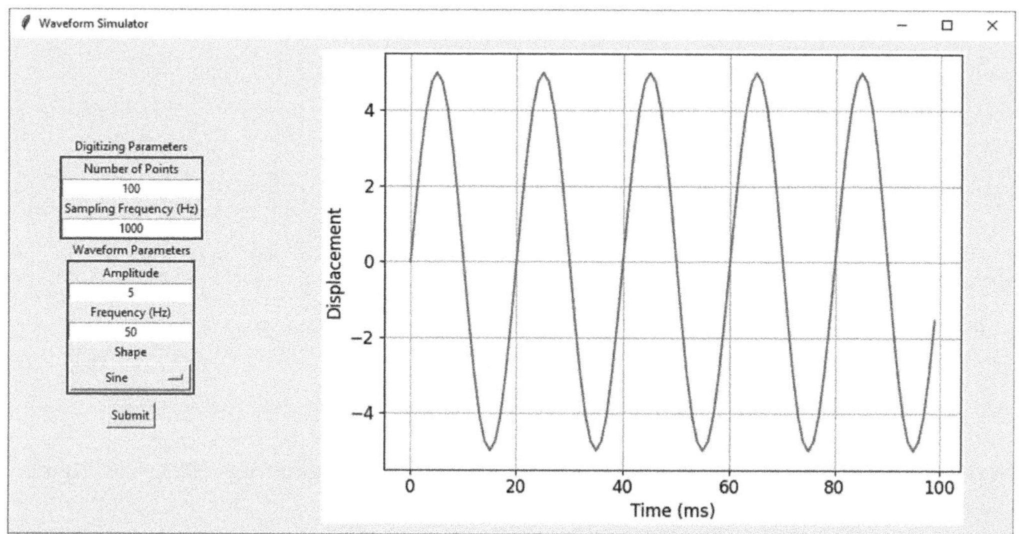

The `root.mainloop()` command will maintain program execution in a continuous loop, waiting for new clicks of the **Submit** button. Try pressing **Submit** after selecting other **Shape** options. The program will run until you close the **Waveform Simulator** window. Close this window.

Let's explore the properties of digitized data a bit more. In order to better view each digitized sample, with **waveform_simulator.py** open in the Editor, change the plot marker to a filled circle, i.e., change `marker='-'` to `marker='o-'` in the `update_plot()` function inside `calculate_plot_waveform()` as follows:

```
pc.update_plot(ax, canvas, 1000*t,y, label_x, label_y,
                    marker='o-')
```

Save your work. Now, instead of simply an interpolated solid-line plot of the data samples, the individual data samples will be indicated by filled circles with a solid line still interpolated between neighboring samples.

Run **waveform_simulator.py**. This time create a 100-Hz sine wave by entering *100* and *1000* for **Number of Samples** and **Sampling Frequency (Hz)**, and *5*, *100*, and *Sine* for **Amplitude**, **Frequency (Hz)**, and **Shape** respectively, then press **Submit**. Given the **Time**-axis is calibrated in milliseconds, with this choice of parameters, neighboring data samples are separated by a time interval of 1 ms (at a rate of 1000 samples per second, the time interval between neighboring samples is 0.001 s), and we are simulating an experiment that acquires 100 data samples over the time span from t = 0.000 to t = 99 ms. Thus, for a 100-Hz sine-wave signal whose period is 10 ms (= 0.01 s), we observe 10 of its cycles on the plot and note that each cycle is described by a succession of 10 samples (there are two closely spaced samples near the crest as well as near the trough of the sine wave). At this level of sampling, the sine wave is resolved, but the representation is somewhat crude.

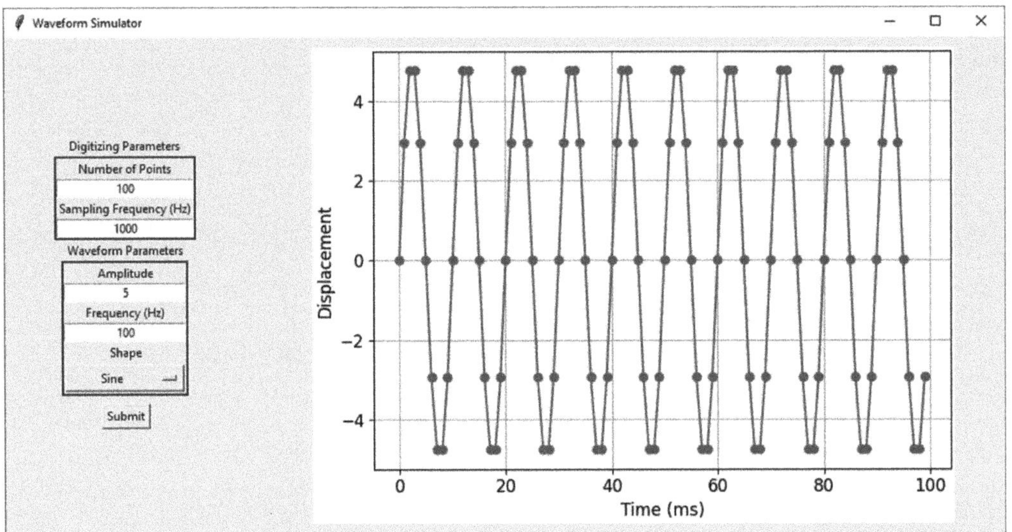

Next, without changing **Number of Samples**, **Amplitude**, **Frequency (Hz)**, and **Shape**, change **Sampling Frequency** to *10000*, and press **Submit**. With this choice of parameters, neighboring data samples are separated by a time interval of 0.1 ms (= 0.0001 s), so now the 100 samples are acquired over the time span from t = 0.0000 to t = 9.9 ms. Thus, for the 100-Hz sine-wave signal with a period of 10 ms, we observe only one of its cycles on the plot, where this single cycle is well resolved because it is described by a succession of 100 samples.

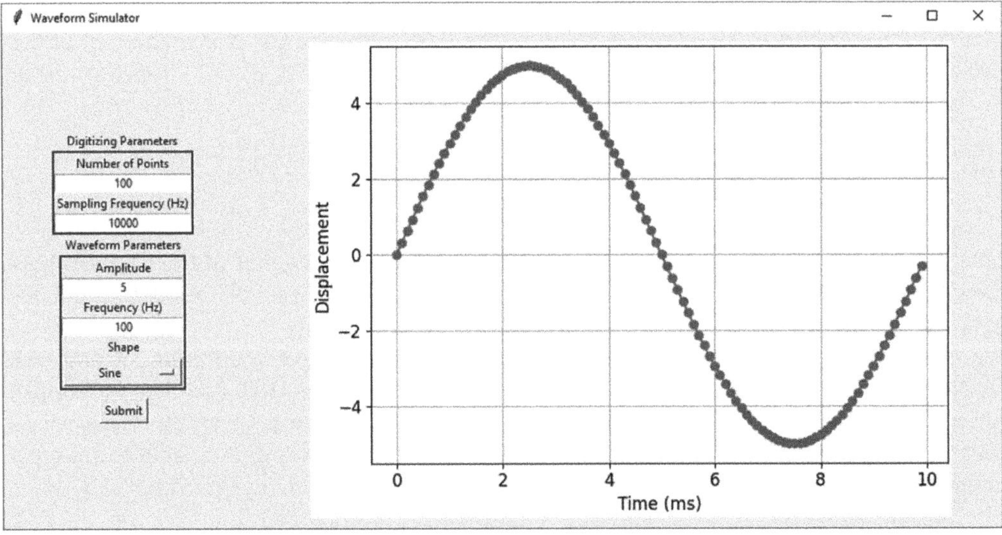

Finally, run your program with **Shape** set to its other available options and enjoy the results.

5.9 Namespace and Scope

To conclude this chapter, let's circle back to our discussion of namespace and scope. Understanding of these concepts is essential in writing well ordered and properly functioning Python code. Stated concisely, Python uses namespaces to manage the region of accessibility ("scope") of objects within your program, thus preventing naming conflicts and enabling you to organize your program efficiently.

When execution of a Python program ("module") is initiated, several namespaces are created, where each namespace is a set of identifiers (such as variable names, function names, and class names) for objects within the module. First, each module has a *built-in namespace*, which consists of all the programming functions and objects built-in to the Python language, e.g., `print()`, `int()`, `True`, `False`. Objects within the built-in namespace are accessible everywhere within a program and so the scope of the built-in namespace is the entire program.

Second, a *global namespace* is created, which comprises identifiers for all objects within the core body of the module, i.e., all code within the program, except that within function definitions. Similar to the built-in namespace, all objects within the global namespace are accessible everywhere within the module and so the scope of the global namespace is the entire module. In addition, if a given module is a top-level ("main") program with another lower-level module imported into it (via an `import` statement), a global namespace is created for the imported module. However, the scope of the main module's global namespace is only the main module and the scope of the imported module's global namespace is only the imported module. That is, objects in the main and imported modules are independent of each other.

Third, a *local namespace* is created for each function within a module. Objects identified in a function's local namespace are not accessible outside of that function and so the scope of this local namespace is simply the region within the function.

Here is the importance of each namespace and its associated scope: During program execution, when the Python Interpreter encounters an object identifier, it determines which object is being referenced by progressing through the **local>>global>>built-in** hierarchy. That is, first the local namespace is checked and, if the identifier is found there, the object associated with that identifier is used. If the identifier is not in the local namespace, then the global namespace is checked and, if found there, that associated object is used. Finally, if the identifier is not found in the global namespace, the built-in namespace is checked and, if found there, that associated object is used. If the identifier is not found in the built-in namespace, then a `NameError` exception is thrown.

Let's see how these concepts are used in the **waveform_simulator.py**. program. As a reminder, the code for this program is as follows:

```
import Utilities.waveform_creator as wc
import Utilities.gui_creator as gc
import Utilities.plot_creator as pc

def calculate_plot_waveform():

    num, samp_freq = gc.get_dig_entries(var_num,
                                        var_samp_freq)
    a, f, shape = gc.get_wav_entries(var_a, var_f,
                                     var_shape)
```

```
    t = wc.create_time_data(num, samp_freq)
    y = wc.create_waveform_data(t, a, f, shape)

    pc.update_plot(ax, canvas, 1000*t, y, label_x, label_y,
                   marker='-')

    return


label_x, label_y = 'Time (ms)', 'Displacement'

root = gc.create_window(title='Waveform Simulator',
                        size='1000x500')
control_box, right_frame = gc.create_frames(root)

var_num, var_samp_freq = gc.digitizing_control(control_box)
var_a, var_f, var_shape = gc.waveform_control(control_box)

ax, canvas = pc.create_plot(right_frame)
pc.update_plot(ax, canvas, [], [], label_x, label_y)

btn = gc.action_button(control_box, label='Submit',
                       action=calculate_plot_waveform)

root.mainloop()
```

When execution of the **waveform_simulator.py** module is initiated, in addition to a built-in namespace, a global namespace is created that comprises identifiers for each object outside of the definition for `calculate_plot_waveform()`. Thus, this global namespace includes the variable names `label_x`, `label_y`, `var_num`, `var_samp_freq`, `ax`, `canvas`, and so on. In addition, a local namespace is created for `calculate_plot_waveform()`, which contains the variable names `num`, `samp_freq`, `a`, `f`, `shape`, and so forth. Finally, for each imported module such as **plot_creator.py**, an independent global namespace is created, along with local namespaces for each function within that module.

Then, for example, as **waveform_simulator.py** executes and program execution moves into `calculate_plot_waveform()`, the function `get_dig_entries()` is called, which requires the objects `var_num` and `var_samp_freq` as its arguments. To find these objects, the Python Interpreter first checks the local namespace of the function `calculate_plot_waveform()` but does not find them there. So, the Interpreter next checks the global namespace and does find these objects there (i.e., in the region outside the function definition) and so those objects are used as the arguments of `get_dig_entries()`. Soon after, program execution within `calculate_plot_waveform()` moves to the location where the function `create_time_data()` is called, which requires the objects `num` and `samp_freq` as its arguments. To find these objects, the Python Interpreter first checks the local namespace of the function `calculate_plot_waveform()` and finds these objects there and so those objects are used as the arguments of `create_time_data()`. Throughout the entire program execution, the Python Interpreter resolves the identity of each referenced objects in this manner, i.e., by progressing through the **local>>global>>built-in** hierarchy.

Finally, remembering that imported modules have their own unique namespaces, the function `update_plot()` within the **plot_creator.py** module does not have direct access to the global namespace of **waveform_simulator.py**, which is where needed quantities such as `ax`, `canvas`, `label_x`, and `label_y` reside. Thus, these quantities must be passed between the **waveform_simulator.py** and **plot_creator.py** modules as arguments of the `update_plot()` function.

For completeness, I'll mention that, for example, `calculate_plot_waveform()` and `get_dig_entries()` are an example of *nested functions*, where the latter is enclosed within the former. However, the definitions of these two functions are found is different modules and so the above-described rules of namespace and scope apply.

Alternately, the definitions of two functions can be nested within the same module. An example of such a module is as follows, where the definition of `inner()` is enclosed within the definition of `outer()`, and the region within `outer()`, save the definition of `inner()`, establishes the *enclosing namespace*:

```
a = 1
b = 2

def outer():

    b = 3

    def inner():
        print(a, b)
        return

    inner()

    return

outer()
```

For this situation, the namespace hierarchy is modified as follows: **local>>enclosing>> global>>built-in**, often summarized as *LEGB*.

Try saving this code in a file named **nested_functions_single module.py** in **YourNamePythonWork** and then run the program. You will find that the output in the Shell is as follows:

```
    1 3
>>>
```

Can you explain why?

Do It Yourself

sum_of_sines.py Fourier analysis tells us that a square wave $y(x)$ with a frequency and amplitude of 1 is given by the following sum of sine waves with odd integer frequencies:

$$y(t) = \frac{4}{\pi}\sum_n \frac{\sin(2\pi nt)}{n} \quad n = 1, 3, 5, 7, ..., \infty, \tag{5.4}$$

where n is the frequency of the sine wave in each term.

Write a GUI-equipped program that approximates the above sum by evaluating only its first N terms (i.e., if $N = 3$, then only the $n = 1$, $n = 3$, and $n = 5$ terms are included in the sum). To observe three cycles of the square wave with high resolution on the time axis, evaluate the sum over the range from $t = 0$ to $t = 3$ s with $\Delta t = 0.001$ s.

The GUI for your program should appear as shown next, with a single control labeled **Number of Terms** to input the desired value of N and a plot of y versus t.

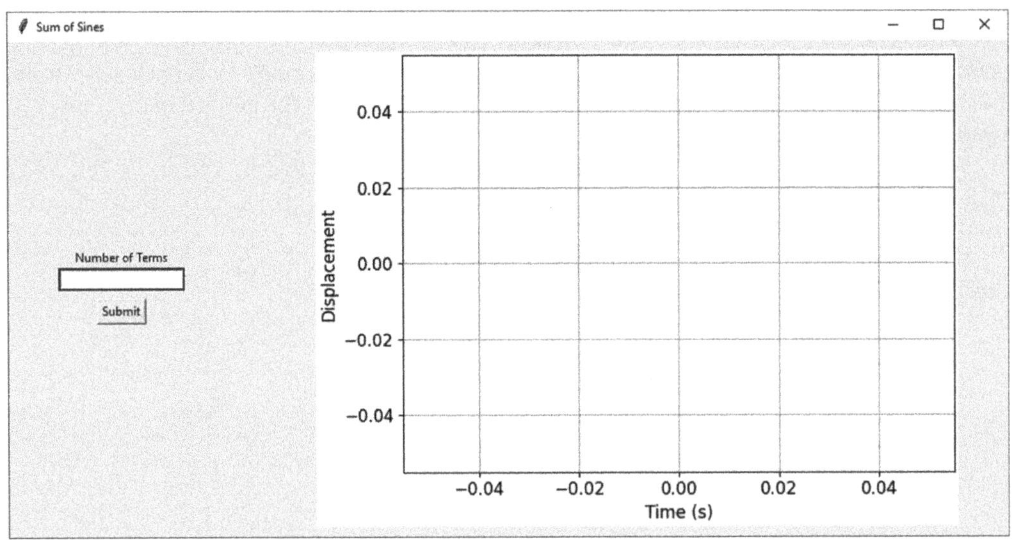

To write your program, begin with **waveform_simulator.py** open in the Editor, then use **File>>Save As...** to create **sum_of_sines.py** and modify the code as needed. To aid in the modification process, define the following new functions:

- For creating a single blue-border control with descriptive `label`, open **gui_creator.py** and add these two definitions:

```
def create_control(container, label, char_width=20):

    lbl = tk.Label(container, text=label)
    lbl.pack()

    frame = tk.Frame(container, bg='blue', bd=2)
    frame.pack()

    var_ent = tk.StringVar()
    ent = tk.Entry(frame, textvariable=var_ent,
                   justify=tk.CENTER, width=char_width)
    ent.pack(fill=tk.X)

    return var_ent
```

```
def get_control_string(var_ent):

    control_string = var_ent.get()

    return control_string
```

- For summing the first N terms in the summation given by Eq. (5.4), open **waveform_creator.py** and add the following for-loop-based function. Here, `np.zeros(num)` creates a `num`-element NumPy array, where every element has a value of 0. Remember all of the positive odd integers are given by $n = 2i + 1$, where $i = 0, 1, 2, 3, \ldots$ and also that `y += x` is shorthand for `y = y + x`:

```
def sum_sines(t, num_terms):

    num = len(t)
    y = np.zeros(num)
    largest_odd = 2*num_terms + 1

    for odd in range(1, largest_odd, 2):
        y += (1/odd)*np.sin(2*np.pi*odd*t)
    y = (4/np.pi)*y

    return y
```

Once your program is complete, run it with a fairly small value of **Number of Terms**. As shown in the next illustration, near the transitions between $y = +1$ and $y = -1$, you will observe "overshoot" peaks (also known as *ringing* artifacts), an effect called the *Gibbs phenomenon*. As **Number of Terms** is increased, you should be able to minimize this effect. For what value of **Number of Terms** does the Gibbs phenomenon become negligible (e.g., overshoot peak is only 5% of square wave's amplitude)? Finally, show that a square wave is well approximated (i.e., overshoot is minimized) when N is large. Note that with our choice of Δt, the maximum allowed value for **Number of Terms** is 500. Why?

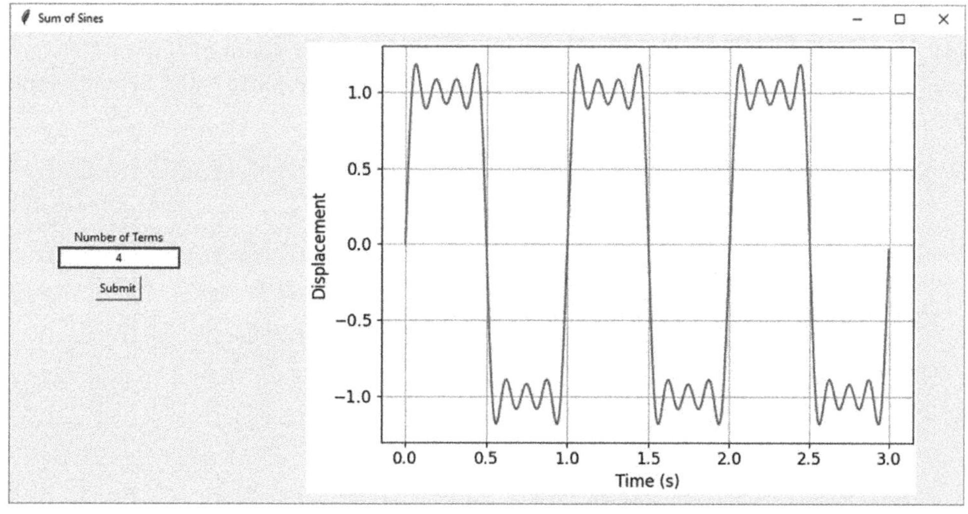

Use It!

timer.py Let's build a GUI-equipped timer that, when a **Start** button is pressed, counts one-second time intervals until a **Stop** button is clicked.

For convenience, we'll use the same button, making it toggle between **Start** and **Stop** each time it is pressed. Begin your **timer.py** program with the code shown below, which produces the toggling button. The button `btn` is created by the `Button()` class (within the `action_button` function). Here, we use a method of the `Button()` class called `cget(option)`, which reports the current value of a given `option` of the `btn`, in this case, its `text` label. The *if-elif* statement toggles the value of the button's `text` label.

```
import Utilities.gui_creator as gc

def toggle():

    if btn.cget('text') == 'Start':
        btn.config(text='Stop')
    elif btn.cget('text') == 'Stop':
        btn.config(text='Start')

    return

root = gc.create_window('Timer', '150x100')

btn = gc.action_button(root, label='Start', action=toggle)

root.mainloop()
```

Save your work. Run **timer.py** to verify that the button's text label toggles between **Start** and **Stop** when pressed.

Next, we need timing code that counts time in seconds and an indicator to display these counts on our GUI. To create the indicator, open **gui_creator.py** and add the following function named `create_indicator()`. This function creates a blue-bordered frame within the specified `container` and, within that frame, places a descriptive `label_1` above a second `label_2`. `label_2` will display a string-formatted version of our count of elapsed time and `label_1` is captioning text. If left unspecified, `label_2` will initialize to its default empty string.

```
def create_indicator(container, label_1,
                     label_2='', char_width=20):

    lbl_1 = tk.Label(container, text=label_1)
    lbl_1.pack()

    frame = tk.Frame(container, bg='blue', bd=2)
    frame.pack()
```

```
    lbl_2 = tk.Label(frame, text=label_2, width=char_width)
    lbl_2.pack(fill=tk.X)

    return lbl_2
```

Save your work and close **gui_creator.py.**

Add the following code to **timer.py**, which places the blue-bordered indicator on the GUI, labeled as **Elapsed Time (s)**. Also, the function `time_display()` uses a `while` loop to count one-second delays produced by `time.sleep(1.)`. The execution of a Python program can be delayed by `delay` seconds using the `sleep(delay)` command in the `time` module. The `time` module is in the Python Standard Library, but must be imported to a program using the command `import time`.

```
import Utilities.gui_creator as gc
import time

def time_display():

    display = 0
    while btn.cget('text') == 'Stop':
        time.sleep(1.)
        display += 1
        lbl.config(text=str(display))

    return

def toggle():

    if btn.cget('text') == 'Start':
        btn.config(text='Stop')
    elif btn.cget('text') == 'Stop':
        btn.config(text='Start')

    time_display()

    return


root = gc.create_window(title='Timer', size='150x100')

lbl = gc.create_indicator(root, 'Elapsed Time (s)')

btn = gc.action_button(root, label='Start', action=toggle)

root.mainloop()
```

Save your work. Try running **timer.py**. As written, the function `time_display()` will be called when the **Start** button is pressed. However, you will find that the program execution hangs up and you may have to abort Python to escape the program.

The problem you are observing is common in data-acquisition programs where a computer is being asked to maintain a GUI at the same time it is supposed to be carrying out a time-intensive data-acquisition task, in this case, counting seconds by involving the `sleep()` command. In this present case, the GUI-related operation `root.mainloop()` is running an infinite loop, constantly checking for events like button presses, while the `while` loop is trying to acquire data. Because of how we programmed these two operations, they are trying to run on the same execution path within the computer (termed a *thread*) and so end up "blocking" each other's execution.

The solution to this blocking problem is to program the GUI- and data-related operations to execute on separate threads. Python's `threading` module (included in the Standard Python Library) makes this solution easy to program. Modify **timer.py** as shown below. Here, GUI-related operations execute on the original thread of the execution, while data-related operation in the `time_display()` function execute on another thread named `t` created by `threading.Thread()`. Remember to start the new thread with `t.start()` and to add `import threading` at the top of your program.

```
import Utilities.gui_creator as gc
import time
import threading

def time_display():

    display = 0
    while btn.cget('text') == 'Stop':
        time.sleep(1.)
        display += 1
        lbl.config(text=str(display))

    return


def toggle():

    if btn.cget('text') == 'Start':
        btn.config(text='Stop')
    elif btn.cget('text') == 'Stop':
        btn.config(text='Start')

    t = threading.Thread(target=time_display)
    t.start()

    return

root = gc.create_window(title='Timer', size='150x100')

lbl = gc.create_indicator(root, 'Elapsed Time (s)')

btn = gc.action_button(root, label='Start', action=toggle)

root.mainloop()
```

Save your work and run **timer.py**. When counting seconds, its GUI will appear as shown here:

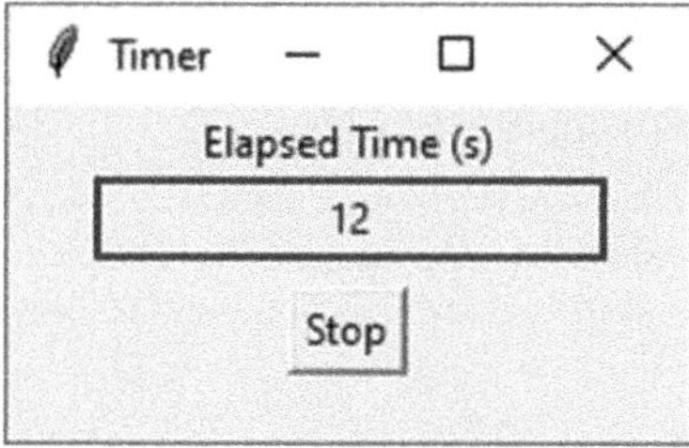

You might try comparing its timing with a stopwatch (e.g., on your smart phone) to see how accurate this program marks time.

Problems

1. **triangle_wave.py** A unity-amplitude 1-Hz triangle wave $y(t)$ can be approximated by the following sum of four sine waves:

$$y \approx \frac{8}{\pi^2}\left[sin\left(2\pi t\right) - \frac{1}{9}\sin\left(6\pi t\right) + \frac{1}{25}\sin\left(10\pi t\right) - \frac{1}{49}\sin\left(14\pi t\right)\right].$$

   Using the **User-Defined** option of **waveform_simulator.py**, evaluate this expression to create a plot of y versus t in the range $0 \leq t \leq 3$ s and show that a good approximation to a triangle wave results.
2. **spiral.py** In polar coordinates (r, θ), a particular spiral is defined by the relation $r = \theta$, where $0 \leq \theta \leq 6\pi$. Build a GUI-equipped program that transforms the description of this spiral to Cartesian coordinates (x, y), and then displays the spiral as a y versus x plot. For this program, use functions from **gui_creator.py** and **plot_creator.py** to embed the plot in an operating-system window as shown below. The transformation equations from polar to Cartesian coordinates are $x = r\cos\theta$ and $y = r sin\theta$.

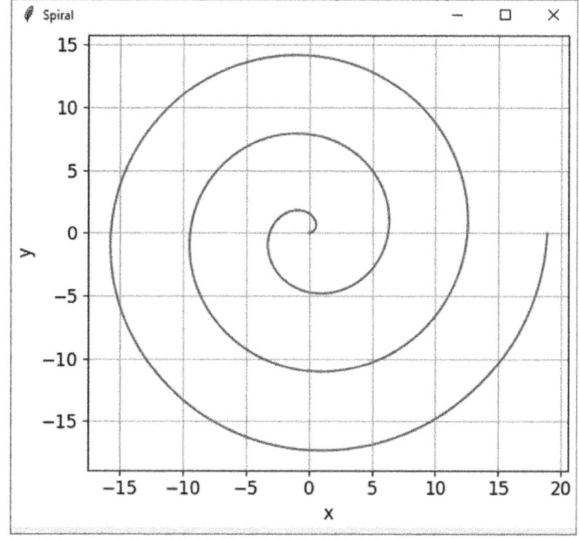

3. **metronome.py** Write a GUI-equipped program that functions as a 60-beats-per-minute metronome, where beats are indicated by the on–off blinking of a label widget. For each blink, make the on-time and off-time equal to one-half second. The GUI for this program should appear as shown here (when the label is on):

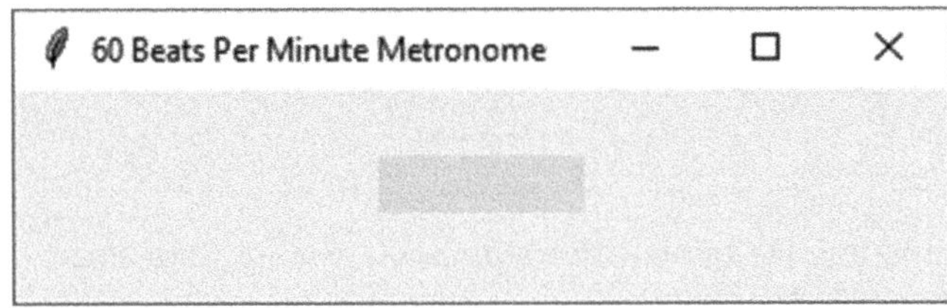

Construct the program's GUI by placing a label widget on a `root` window (if desired use a frame to provide padding). Programming tip: The label's foreground (`fg`) and background (`bg`) colors can be set by the command `lbl.config(bg='cyan',fg='white')`. Defining `'cyan'` and `'white'` as the on and off states, respectively, the label can be made to alternate between these two states every 0.5 seconds using this `while`-loop-based code:

```
while True:
    time.sleep(0.5)
    bg = lbl.cget('bg')
    fg = lbl.cget('fg')
    lbl.config(bg=fg, fg=bg)
```

Place this `while` loop within the definition of a function called `alternate_states()`. In your program, after the code that creates the GUI, spawn a thread that executes `alternate_states()`. If all goes well, the label will be cyan for one-half second, followed by being white for one-half second, and then repeat that pattern until you close the `root` window. Thus, the label "blinks on–off" once a second or 60 times per minute.

4. **root_finder.py** SciPy's `optimize` module provides several methods for obtaining the *roots* of a function $f(x)$, i.e., the values of x for which $f(x) = 0$. For this problem, after the `import scipy.optimize as opt` statement, use `opt.brentq(f,a,b)`, which implements the popular Brent's method, to find the two roots of the second-order polynomial $x^2 + x - 1 = 0$.

   The parameters for `opt.brentq(f,a,b)` include the function $f(x)$ as well as two values `a` and `b`, which define a *bracketing interval* $[a, b]$. In determining each root of $f(x)$, the root must lie within the interval $[a, b]$ and $f(x)$ must switch signs in going from $f(a)$ to $f(b)$, i.e., $\text{sgn}[f(a)] = -\text{sgn}[f(b)]$.

   To find the roots of $f(x) = x^2 + x - 1$, first write plotting code as follows in order to determine values for `a` and `b`:

```
import numpy as np
import matplotlib.pyplot as plt

def func(x):
    y = x**2 + x - 1
    return y
```

```
a, b = ....<make choice of a and b here>
x = np.linspace(a, b, 1000)
y = func(x)

plt.plot(x, y)
plt.show()
```

Once values for a and b are determined for a particular root, add the command `root = opt.brentq(func,a,b)` to your program to find the root.

To check that your two results for `root` are correct, compare these results with those found using the quadratic formula.

5. **numerical_integration_scipy.py** In contrast to NumPy's `trapz()` function, which implements the simplest numerical integration algorithm (called the trapezoidal rule), SciPy's `integrate` module provides access to several higher-precision integration methods. In this problem, we test drive `scipy.integrate.quad(func,a,b)`, the workhorse of SciPy's integration methods. The input parameters for `quad()` are `func`, the name of the function to integrate, as well as `a` and `b`, the lower and upper limits of the integral, respectively.

   It is easy to show analytically that the integral $\int_0^1 3x^2\,dx$ equals 1 exactly. In your program, first define a function named `func(x)` that evaluates $y = 3x^2$. Then, below the definition of `func(x)`, evaluate the integral using the command

```
value, abserr = integ.quad(func, 0, 1)
```

   Here, `value` is the determined value of the integral and `abserr` is an estimate of the absolute error of `value`. Print both of these quantities in the Shell. Programming tip: To print a float named `value` to XXX decimal places, use the *f-string* formatting `print(f'{value:.XXXf}')`.

6. **projectile_range.py** Quite commonly, a projectile lands at a level that is different than that from which it is launched. As shown below, consider a projectile launched with speed v (m/s) at angle θ above the horizontal at a height H above the ground and define range R to be the horizontal distance it travels during its flight (Fig. 5.1).

   For this situation, it can be shown that

$$R = \frac{v^2}{2g}\left[\sin(2\theta) + \sqrt{\sin^2(2\theta) + \frac{8gH}{v^2}\cos^2\theta}\right],$$

   where θ is in the range from 0° to 90°.

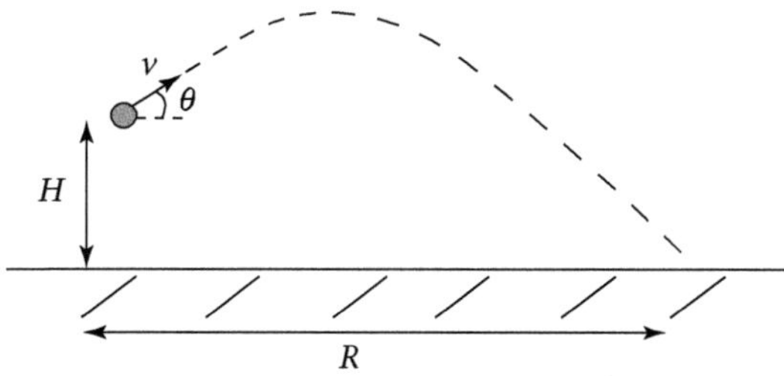

Fig. 5.1 Problem 6.

(a) Write a GUI-equipped program that, given v and H as input via controls labeled **Launch Speed (m/s)** and **Launch Height (m)**, respectively, calculates R for a sequence of 91 equally spaced angles in the range from $\theta = 0°$ to $\theta = 90°$ and then displays a plot of R versus θ on an embedded Matplotlib graph. Also, from the R versus θ data, determine the launch angle that produces the largest range and display that result in an indicator labeled **Angle for Maximum Range (deg)**. The design for the program's GUI is shown next.

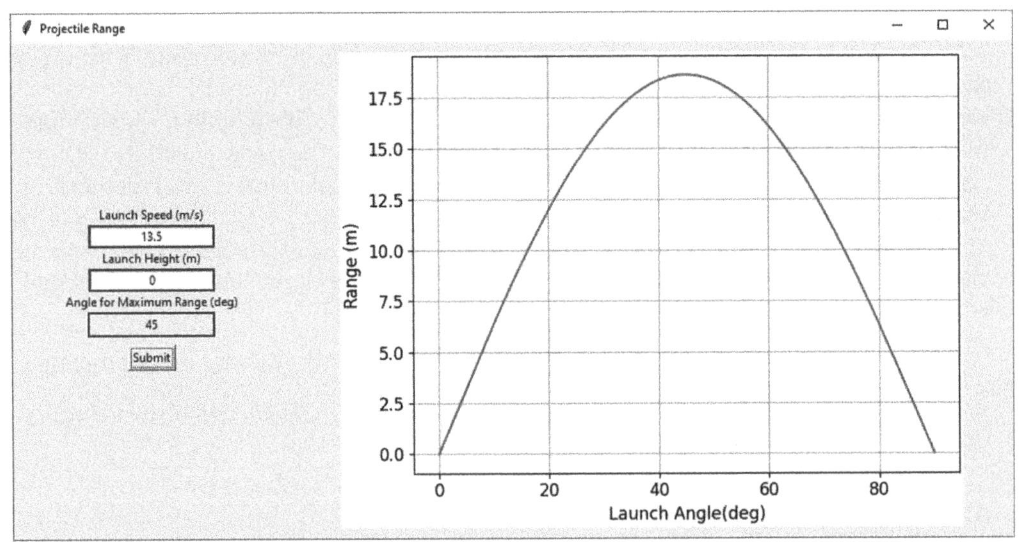

(b) At a track-and-field meet, a shot putter launches a shot (heavy metal ball) at speed of 13.5 m/s. Run your program with this launch speed, assuming the shot lands at the same level from which it was launched. Show that your program verifies the well-known fact that the maximum range results if the launch angle equals 45°.

(c) In reality, a shot putter launches the shot from a height close to (but, above) their head. With $v = 13.5$ m/s and $H = 2.1$ m, run your program and then determine the launch angle that produces the maximum range.

(d) Finally, if the shot putter launched the shot with $v = 13.5$ m/s from a cliff of height $H = 100$ m, run your program and then determine the launch angle that produces the maximum range.

Programming tip: In creating the GUI, use the control- and indicator-related functions described in the Do It Yourself and Use It! projects in this chapter. Create the R and θ data in Python lists named `ranges` and `thetas`, respectively, where `thetas = list(range(0,91))`. In **waveform_creator.py**, create the `ranges` list by using a function called `projectile(thetas,v,h)`, which returns `ranges` and `max_angle`. The index of the maximum range value can be found using the NumPy command `max_index = np.argmax(ranges)`. Once `max_index` is known, the corresponding launch angle can be found by `thetas[max_index]`.

7. **falling_in_air.py** In the presence of air, a falling object's acceleration a (= change in velocity per unit time) can be described by the relation

$$a = g - \alpha v^2, \tag{5.5}$$

where $g = 9.8\text{m/s}^2$ is the acceleration caused by gravity in a vacuum, v is the object's instantaneous speed in meters per second, and α is a constant with units of m^{-1}. The velocity-squared term describes the effect of air resistance, and it becomes more and more significant as the object attains greater speeds. When the object achieves terminal velocity v_T, the object no longer accelerates because, at that speed, $g = \alpha {v_T}^2$.

Assume an elevated object is released from rest at time $t = 0$ and that the subsequent time is divided into small intervals, each of duration Δt. Then the object's fall can be sampled at N times given by $t_n = n\,\Delta t$, where $n = 0, 1, 2, \ldots, N - 1$. Let v_n be the object's instantaneous speed at time t_n. Then, if the object's speed v_{n-1} at time t_{n-1} is known and Δt is small, Eq. (5.5) can be used to determine v_n as follows:

$$v_n = v_{n-1} + a\,\Delta t = v_{n-1} + \left(g - \alpha\, {v_{n-1}}^2\right)\,\Delta t \tag{5.6}$$

(a) Write a GUI-equipped program that implements Eq. (5.6) to determine a falling object's speed v_n at the N times t_n and then plots v versus t on a Matplotlib graph. Take $\Delta t = 0.001$ s. The GUI of your program should appear as follows.

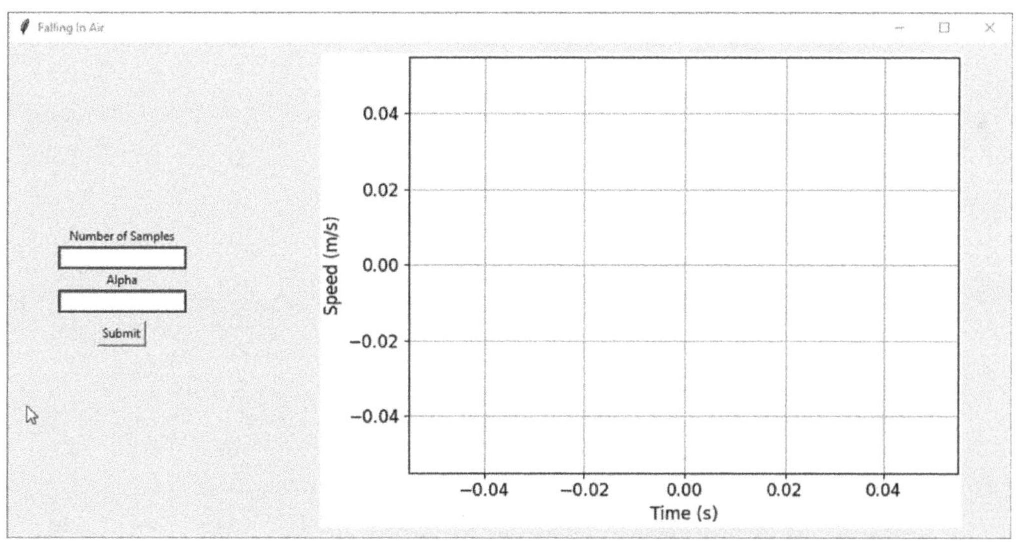

(b) For a 1.0 kg steel ball, a human skydiver, and a raindrop, the numeric value of α is about 0.00092, 0.0035, and 0.11, respectively. Run your program with each value of α. Determine the resulting terminal velocity v_T for each object. After being released, about how long does it take each object to attain v_T?

Programming tip: In creating the GUI, use the control-related functions described in the Do It Yourself project in this chapter. Create the t data as a NumPy array using the `create_time_data()` function in **waveform_creator.py**. In **waveform_creator.py**, add a function called `falling_in_air()`, which creates the v data as a NumPy array using a `for` loop as shown here:

```
v = np.zeros(num)
for i in range(2,num):
    v[i] = code for Eq. (5.6) here
```

CHAPTER 6

Introduction to Data-Acquisition Device Features

In this chapter, you will be introduced to the capabilities of a multifunction data-acquisition (DAQ) device connected to your computer. Here, it is assumed that you have such a DAQ device available, along with a DC voltage source (e.g., battery, power supply, function generator), function generator, voltmeter, and required cabling.

Prior to starting this chapter, please download the latest version of the software available from your DAQ device's manufacturer that provides access to your DAQ device's functionality. For NI DAQ devices, download **NI-DAQmx** (Windows or Linux) at https://www.ni.com. For MCC devices, download **DAQami** (Windows) at digilent.com.

6.1 Data-Acquisition Hardware

In this chapter, you will be introduced to some of the powerful capabilities of computer-controlled data acquisition and generation. These capabilities enable an engineer or scientific investigator to reach out through a personal computer to monitor and create events in the outside world. For this introduction, you will control the basic operations of a multifunction data-acquisition (DAQ) device using its handy interactive utility program. We'll be highlighting DAQ devices from two popular manufacturers—NI (formerly, National Instruments) and Measurement Computing Corporation (MCC, for short)—but the important issues we will explore in acquiring and generating digitized data apply to multifunction DAQ devices from all manufacturers.

NI and MCC, headquartered in Austin, Texas and Norton, Massachusetts, respectively, offer a wide variety of computer-controlled multifunction data-acquisition devices. Each of the NI devices can be controlled by the interactive utility program *NI Measurement & Automation Explorer*, nicknamed *MAX*, which is included as a component of the NI device driver *NI-DAQmx*, while MCC devices are controlled by an intuitive drag-and-drop interface program named *DAQami*. NI-DAQmx (which is available for Windows and Linux only) can be downloaded for free at https://www.ni.com, while DAQami (available for Windows only) is downloaded at no cost at https://digilent.com. MAX and DAQami provide a quick and easy method for becoming familiar with the features of your DAQ device. For this chapter, I'll assume that you have access to a DAQ device from either NI or MCC. If your DAQ device is from another manufacturer, it's hoped that your device has an associated interactive software program similar to MAX or DAQami so that you can follow along with the exercises.

The Python-Based Laboratory. John Essick, Oxford University Press. © John Essick (2024). DOI: 10.1093/9780191998478.003.0006

In DAQspeak, the operations that data-acquisition devices perform include *analog input (AI)*, *analog output (AO)*, *digital input/output (DIO)*, and *counter input/output*. Respectively, these terms are shorthand for analog-to-digital conversion (acquiring a digitized representation of an incoming analog voltage); digital-to-analog conversion (generating an outgoing analog signal from a sequence of digital numbers); digital line control (setting and reading the line's HIGH/LOW state); and digital waveform operations (event timing, pulse counting, pulse train generation). Some of the available DAQ devices are quite specialized (for instance, only designed to process digital signals), while others are multifunctional, performing some or all of the above-mentioned operations. There are high-speed devices, which digitize incoming signals at rates over 10^9 samples per second (S/s), and more garden-variety systems with maximum sampling frequencies on the order of 10^5 or 10^6 S/s. Digital resolution varies from 8 to 24 bits, and each device connects to a PC through one of about ten possible ways (including the USB and PCI Express interfaces).

As written, this chapter supposes you are working with one of two popular low-cost multifunction DAQ devices—the NI USB-6002 or the MCC USB-202. These representative devices will be used to illustrate the generic features of data-acquisition products. In fact, most NI and MCC DAQ devices are controlled by the same software that we will be using and so, if you have access to any NI or MCC multifunction DAQ device, you should be able to directly carry out the exercises. For those with a DAQ device from another manufacturer, check whether your device has an associated interactive software program similar to MAX or DAQami. If so, with a little work on your part, it's hoped that you can adapt the presentation and follow along with the exercises. The exercises are designed so that they can be completed using a DAQ device with similar (or superior) specifications to those of either the NI USB-6002 or MCC USB-202 unit.

A brief description of each of our two representative devices follows.

NI USB-6002: This low-cost multifunction DAQ device provides basic data-acquisition functionality for applications such as simple data logging, portable measurements, and academic laboratory experiments. It is a small, stand-alone unit that connects to your computer via a USB port. The 6002 has eight analog input channels that can perform 16-bit analog-to-digital conversion operations at rates up to 50 kS/s. There are also two 16-bit hardware-timed analog output channels, which can update output voltages at a maximum rate of 5 kS/s and produce well-resolved analog signals with frequencies up to a few hundred hertz. Additionally, the device has thirteen digital I/O lines and a 32-bit counter (with no hardware-timing measurement capabilities). The USB-6002 is on the higher-specification end of NI's USB-6000/6001/6002/6003 product line, which has replaced the popular (but now legacy) USB-6008 and USB-6009 DAQ devices.

MCC USB-202: Like the NI product, this multifunction DAQ device provides basic functionality for experiments and is a small, stand-alone unit that connects to your computer via a USB port. At about one-fourth the cost of the NI USB-6002, this MCC unit is quite affordable; however, as is to be expected, this affordability coincides with a reduction in some specifications. The USB-202 has eight analog input channels that can perform 12-bit analog-to-digital conversion operations at rates up to 100 kS/s. There are also two 12-bit software-timed analog output channels, which can update output voltages at a maximum rate of 250 S/s via software commands, but cannot produce analog waveforms with accurate frequencies. Additionally, the device has eight digital I/O lines and a 32-bit counter (with no hardware-timing measurement capabilities). The USB-202 is one of the three units in MCC's USB-201/202/205 product line.

Going forward, I will simply call the NI USB-6002 and MCC USB-202 units “our NI device” and “our MCC device,” respectively.

A conduit carrying signals to and from the outside world must somehow be connected to a DAQ device. For example, among its various capabilities, a multifunction DAQ device may have, say, eight analog input channels labeled AI 0 through AI 7, each standing ready to digitize an incoming analog signal. So how does one get the meaning-laden analog signals from an experiment to the DAQ unit’s channel inputs? In one wiring scheme for NI products, an experimenter connects wires from their experiment to appropriate terminals on an *I/O terminal block*. An I/O terminal block has screws or BNC terminals to which the experimental wires can be attached. A cable is then connected from the I/O terminal block to a connector at the end of the DAQ device. This connector has (depending on the particular device) 50, 68, or 100 pins, where each pin is associated with a particular function of the DAQ device. Alternately, some DAQ devices are configured with an on-board terminal block or prototyping breadboard to which wires from an experiment are directly attached. Our representative NI and MCC devices are outfitted in this scheme. Regardless of connection method, the function of each DAQ-device pin can be determined from the DAQ device’s pinout diagram. The pinout diagrams for our NI and MCC DAQ devices are reproduced in Fig. 6.1.

6.2 Interactive Utility Programs: MAX and DAQami

Everyone: To carry out the exercises in this chapter, ideally you have a NI or MCC multifunction DAQ device connected to your computer and its associated software—NI-DAQmx or DAQami, respectively—correctly installed. If you haven’t downloaded either software yet, now is the time to do so. As long as your DAQ device is a NI or MCC product, you should be able to complete these exercises, even if your device isn’t the NI USB-6002 or MCC USB-202 model.

NI device users: The interactive utility program *NI Measurement & Automation Explorer*, nicknamed *MAX*, is included as a component of the NI-DAQmx download. First, use MAX to verify that the NI hardware and software are properly configured on your system. Find MAX on your computer (one place to look is in a folder named **National Instruments**) and open it. After MAX opens, double-click on **Devices and Interfaces** under the **My System** heading at the left. This action will command MAX to determine all of the DAQ devices present within your computing system.

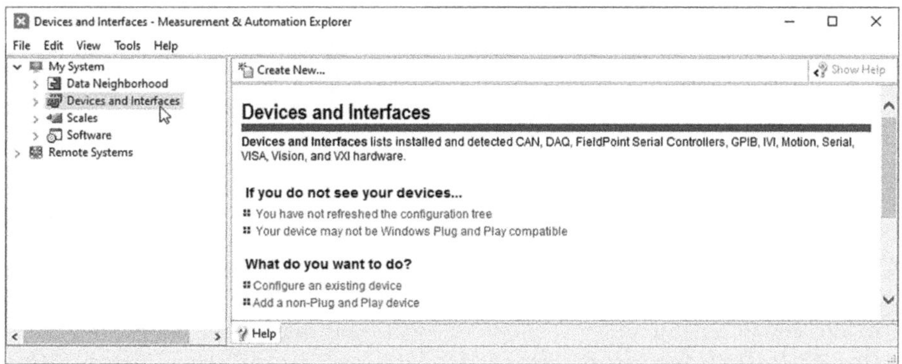

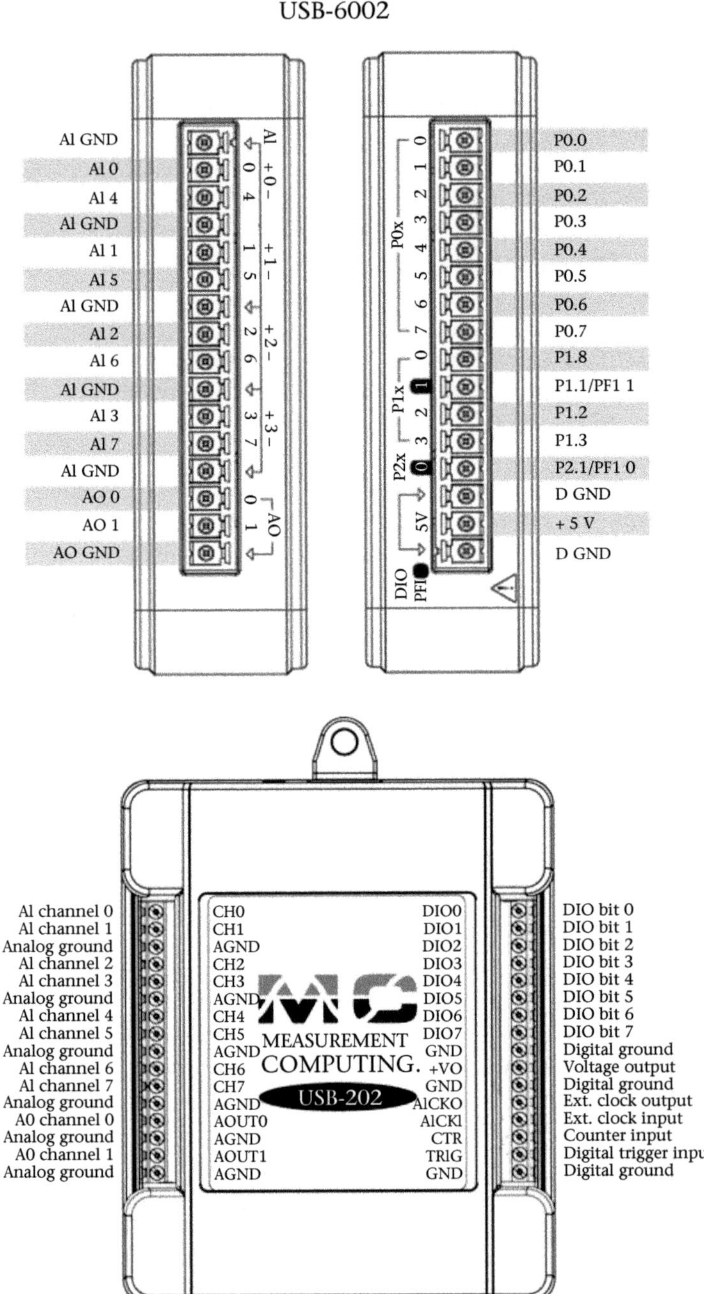

Fig. 6.1 Pinout diagrams for DAQ devices: the terminal configuration for the NI USB-6002 and MCC USB-202 units.

MAX will list the findings of its device survey in hierarchical tree fashion as shown next. For the system used below, we see that a *NI USB-6002* device was found to be connected to the computer (for older versions of MAX, you may have to double-click on a folder called **NI-DAQmx Devices** to view the DAQ devices that have been found). Note that the driver software has (automatically) given this device the shorthand name *Dev1*. The DAQ device on your system may have a different shorthand name. If you go through this procedure and a device that you believe is connected to your computer does not appear on the **Devices and Interfaces** list, try selecting **View≫Refresh** (or press <*F5*> on your keyboard) to command MAX to perform its device survey again. If the device in question doesn't appear, then something is wrong with the device's connection that must be remedied.

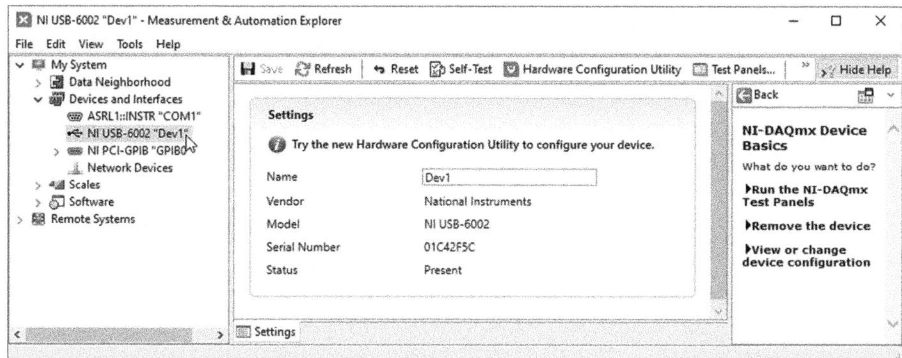

Next, right-click on the name of your DAQ device and select **Test Panels**... (or press the **Test Panels**... button in the toolbar). An interactive window named **Test Panels** will appear, which we will use to test the various functions of your DAQ device.

MCC device users: Let's use DAQami to verify that the MCC hardware and software are properly configured on your system. First, find DAQami on your computer (likely in a folder named **Measurement Computing**) and open it. After DAQami opens, a **Getting Started** window will appear. Click on the **New Configuration** option.

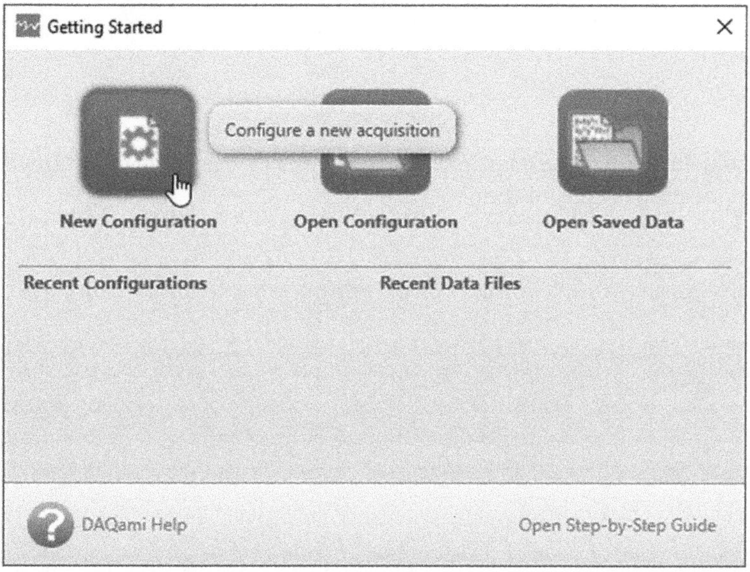

In the **New Configuration Options** window that follows, select **Manual**. A DAQami window will open. Click on the **Available Devices** button in the upper left-hand corner.

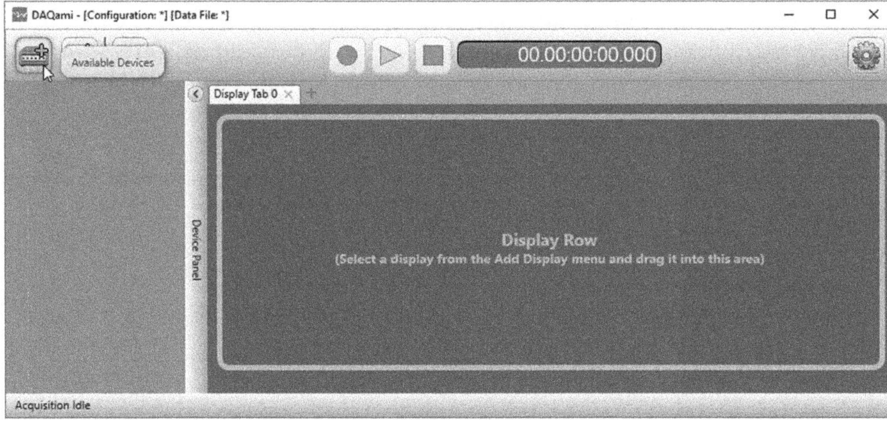

A dropdown menu will display all of the detected devices connected to your computer (if your device doesn't appear in the listing, there is something wrong with the device's connection that must be remedied). Place the mouse cursor over the name of your DAQ device.

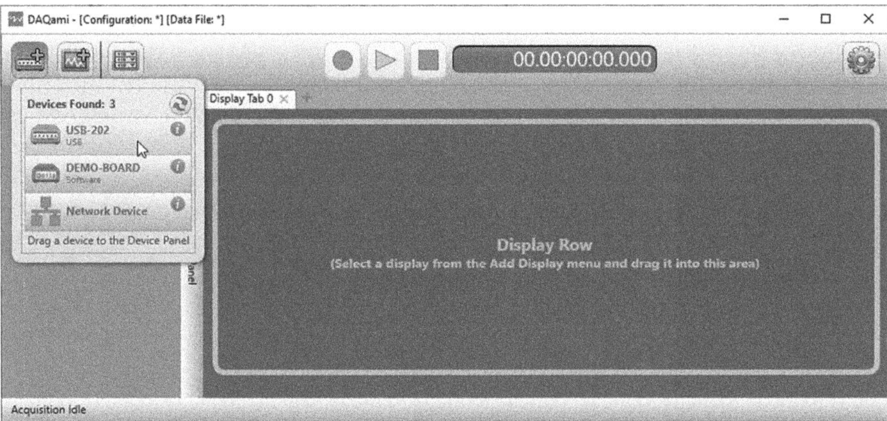

Press and hold down the mouse button as you drag the device name to the **Device Panel** on the left-hand side of the DAQami window.

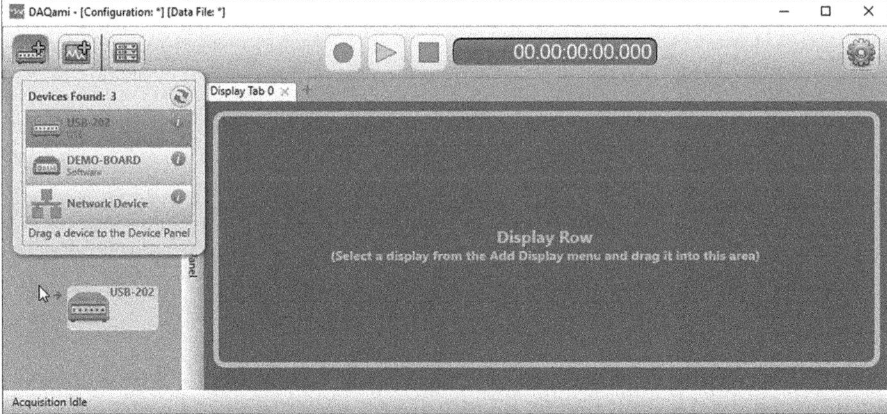

When you release the mouse button, DAQami will be configured to control your selected DAQ device.

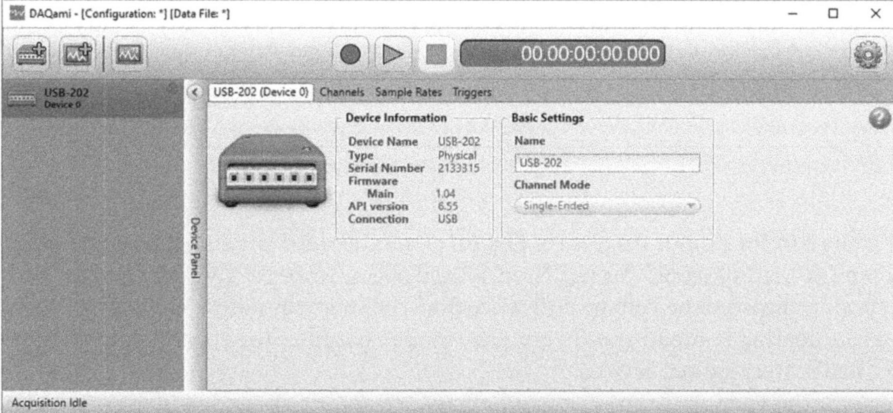

Your DAQami window will end up in the *Configuration Panel* as shown above, with a toolbar containing menu items **Channels**, **Sample Rate**, and **Triggers**. The **Panel Button** (the rightmost of the three buttons at the left-hand top of the window) can be used to toggle between this **Configuration Panel** and the **Display Panel** (where acquired data are displayed).

Everyone: For starters, we'll use the interactive program appropriate for your device (i.e., MAX or DAQami) to explore the analog input operation. However, one needs to understand several important concepts to properly implement this operation, so we'll take a few minutes to obtain this needed grounding in Sections 6.3–6.5.

6.3 Analog Input Modes

To implement analog input (AI) operations, multifunction DAQ devices typically have eight or sixteen AI channels. From the pinout diagrams in Fig. 6.1, we see that both of our highlighted devices have eight AI channels. Through software settings, the inputs on the NI device can be configured to operate in two distinct AI modes—*single-ended* and *differential*. The AI inputs on the less-expensive MCC device only accommodate single-ended operation.

In the single-ended mode, each available analog input pin is an AI channel, with all of these channels referenced to the same common ground. If your DAQ device is connected to a computer powered by building power, this common ground is the building ground and is present at the *analog ground* pins, i.e., AI GND pins on our NI device and AGND pins for our MCC device. If your DAQ device is powered by a laptop under battery power, the analog ground is established by the laptop battery. Within the DAQ device, all AI ground pins are internally connected to the same analog ground.

As an aside, the single-ended scheme described here is termed the *referenced singled-ended* (*RSE*) mode to differentiate it from an alternate mode available on higher-end DAQ devices called the *nonreferenced singled-ended* (*NRSE*) mode. In the NRSE mode, users supply their own ground level at an AI SENSE pin. The NRSE mode is not available on our NI and MCC devices.

In the single-ended input mode, when measuring a voltage difference produced by a source, one of the two lead wires from the voltage source is attached to the pin of an analog input

Table 6.1 NI USB-6002 Channel Assignments for Differential Input Mode

Channel name	Positive input pin	Negative input pin
ai0	AI 0	AI 4
ai1	AI 1	AI 5
ai2	AI 2	AI 6
ai3	AI 3	AI 7

Note: Pin labels are in uppercase, channel name in lowercase.

channel shown in the pinout diagram of Fig. 6.1; the other lead wire is attached to an analog ground pin (AI GND or AGND for our NI or MCC devices, respectively). The DAQ device will then accurately measure the voltage difference from the source as long as the source is *floating*. Examples of floating voltage sources are transformer outputs, thermocouples, some power supplies, and battery-power devices.

However, if the voltage source is *ground-referenced*, the single-ended input measurement mode may yield inaccurate voltage readings due to the presence of *ground loops*. Here's an explanation of the problem: Voltage sources such as function generators are ground-referenced; that is, they are powered by building power and so one of the wire leads emanating from the source is connected to building ground. In the single-ended input measurement mode, this grounded wire lead must be connected to the DAQ device's analog ground pin. If analog ground for the DAQ device is also building ground (e.g., the DAQ device is connected to a desktop computer powered by building power), connecting a ground-referenced source to a DAQ device in this manner establishes two grounding points within the measurement circuitry, and these two ground levels may differ from each other (typically, this voltage difference is on the order of 1 to 100 mV). In addition, internal building wiring effectively produces conducting loops in which voltages can be induced through Faraday's law, the phenomenon of electromagnetic induction. These effects combine to introduce noise into the measured signal. Thus, the single-ended input mode can be problematic when measuring a small-amplitude signal from a ground-referenced source as ground-loop-generated noise may overwhelm the signal you are trying to detect.

In differential input mode, available analog input pins are paired to form independent AI channels, each of which is sensitive only to the voltage difference between paired pins. On our NI device, differential channel *ai0* is pin AI 0 paired with pin AI 4, differential channel *ai1* is AI 1 paired with AI 5, and so on (note that pin labels are uppercase, while differential channel names are lowercase). In this *differential amplifier* configuration, noise pickup by the AI channels is suppressed. Thus, if your device has differential AI capability (our NI device does, our MCC device does not) and it is not a problem in your situation to halve the number of available measurement channels, the differential input mode is the preferred modus operandi. The pin assignments for the differential input mode AI channels for the NI device are given in Table 6.1.

6.4 Range and Resolution

When measuring analog input signals with a DAQ device, there are several issues to keep in mind. First, a DAQ device can only measure voltages that fall within an allowed *range*, which extends from a minimum voltage V_{min} to a maximum voltage V_{max}. If the magnitudes of V_{max} and V_{min} are the same, the range is written as $\pm V_{\text{max}}$. A DAQ device might have a fixed range (for both our NI and MCC devices, the range is fixed at ± 10 V) or offer the choice of several software-selected ranges (some higher-end DAQ device have seven possible ranges starting at ± 0.1 V and ending at ± 10 V). Check the datasheet for your particular DAQ device to find the available ranges. Next, the DAQ device's analog-to-digital converter has a built-in *resolution* of n-bits, which means that this digitizer represents the analog voltage level being sampled with a binary number of n digits. This binary resolution places a limit on the resulting *voltage resolution* (i.e., the smallest detectable voltage difference ΔV). An n-bit digitizer divides the measurable voltage span $V_{span} \equiv V_{\max} - V_{\min}$ (e.g., $V_{span} = 20$ V when the range is ± 10 V) into 2^n divisions, and the resulting voltage resolution of the AI signal being sampled is given by

$$\Delta V = \frac{V_{span}}{2^n}. \tag{6.1}$$

Given the value of $V_{span} = 20$ V, the voltage resolution provided by a 12-bit (MCC) and 16-bit (NI) DAQ device is 5 and 0.3 mV, respectively.

6.5 Sampling Frequency and Aliasing Effect

Another important issue peculiar to digitized data is related to the sampling frequency f_s, where f_s is the rate at which analog-to-digital conversions take place. The sampling frequency places an upper limit on the range of frequencies allowed within the original analog signal, if the digitizing process is to result in a faithful representation of the input. This upper threshold is called the *Nyquist frequency* $f_{nyquist}$, which we will show equals one-half of the sampling frequency, that is, $f_{nyquist} = f_s/2$. The physical significance of the Nyquist frequency is this: For a given sampling rate f_s, an input sinusoidal wave of frequency $f_{nyquist}$ will be sampled in the minimal manner necessary to represent the sinusoid's peaks and valleys—just twice per cycle. If the input frequency exceeds this threshold, the sampling rate becomes insufficient and the analog-to-digital conversion process becomes inaccurate in a way described next.

Fairly often, the bandwidth limitation placed on a "to-be-digitized" analog input signal does not cause you—the experimentalist—much of a problem. For example, if dealing with audio-produced electrical signals, you might know on physical grounds that the frequency content of the analog input is bracketed by 20 and 20,000 Hz. Then, with a sampling frequency of at least 40 kHz, this input can be properly acquired. The maximum sampling frequency for our NI and MCC device is 50 and 100 kS/s, respectively, so both units meet the sampling-frequency requirement of this audio application. In other applications, an analog signal may have passed through an amplifier that behaves as a lowpass filter because of its finite bandwidth response. In this case, the sampling frequency must be twice the maximum frequency passed by the amplifier. If no natural frequency bracketing exists in your experiment, then you must impose a high-frequency cutoff by placing a lowpass filter in your data-gathering circuitry. Given the available sampling frequency f_s, the filter's components then are chosen so that frequencies higher than $f_s/2$ are not passed.

What happens if a signal with frequency higher than the Nyquist limit accidentally is input into your digitizer? Something much worse than you might expect. In an effect called *aliasing*, that signal with a too-high frequency appears falsely as a signal with a lower-than-Nyquist frequency when processed by the digitizer. This phenomenon, which is peculiar to discrete sampling, is illustrated in Fig. 6.2.

Quantitatively, assume that a sine-wave signal with (high) frequency f is input into a digitizer and sampled at N discrete times $t_i = i\Delta t$, where $i = 0, 1, 2, ..., N-1$ and Δt is the time

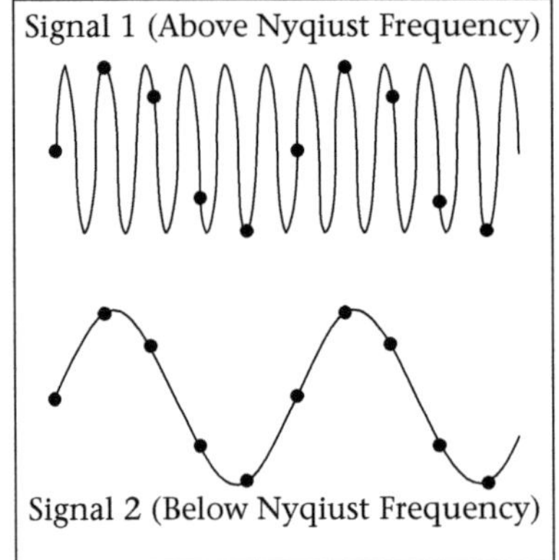

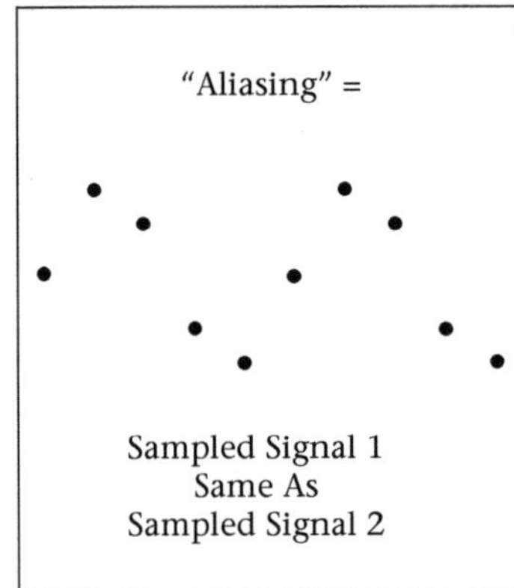

Fig. 6.2 Aliasing effect when signal with a frequency greater than the Nyquist frequency is digitally sampled.

increment between neighboring data samples. Then the sampling frequency and Nyquist frequencies are $f_s = 1/\Delta t$ and $f_{nyquist} = f_s/2$, respectively. If $f > f_{nyquist}$, it can be shown that at every digitizing time t_i, the displacement of the input signal $\sin(2\pi f t_i)$ is equal to plus or minus the displacement of a sine wave with the "alias frequency" f_{alias}, where $0 \leq f_{alias} \leq f_{nyquist}$. By applying the condition $\sin(2\pi f t_i) = \pm \sin(2\pi f_{alias} t_i)$ at every t_i (see Problem 1), one finds that

$$f_{alias} = |f - nf_s| \quad \text{(aliasing condition)}, \tag{6.2}$$

where $n = 1, 2, 3....$

Demonstrate the aliasing effect by running **waveform_simulator.py** and programming it to produce a 100-Hz sine wave of amplitude 5; i.e., set **Amplitude**, **Frequency (Hz)**, and **Shape** equal to *5*, *100*, and *Sine*, respectively. Then, enter *100* and *1000* for **Number of Samples** and **Sampling Frequency (Hz)**, respectively. Press **Submit** to demonstrate what a digitized sinusoid with frequency less than the Nyquist frequency (in this case, $f_{nyquist}$ = 500 Hz) looks like. Note that 10 cycles occur over the course of 100 ms, so the frequency of the wave is 10 cycles/0.1 s = 100 Hz, as expected. Now, with the other **Digitizing Parameters** and **Waveform Parameters** left as is, change the sine wave's frequency to *1100* Hz, and press **Submit**. You will find that, when digitized, this "higher-than-Nyquist" frequency wrongly appears as a 100-Hz sine wave with 10 cycles over a time span of 100 ms, consistent with the prediction of Eq. (6.2) with $n = 1$; that is, $f_{alias} = |1100 \text{ Hz} - (1)\ 1000 \text{ Hz}| = 100$ Hz. Predict what will happen if you program **waveform_simulator.py** with some other "higher-than-Nyquist" frequencies such as 800 and 2100 Hz, and then see if your expectations are realized when you press **Submit**.

6.6 Analog Input Operation Using MAX and DAQami

Everyone: We are now ready to program your DAQ device to digitize an analog voltage difference applied at its input. We'll start by digitizing a DC (constant) voltage that, for example, is sourced from a power supply, battery, or function generator (using its DC option). To demonstrate the use of both analog input modes, we'll have NI device users perform this AI operation in the differential mode using differential channel *ai0*, while MCC device users will use their AI channel 0, but in the single-ended mode (which is the only option available on our MCC device).

To execute the desired AI operation, first secure the connections from the DC voltage source to the correct pins on your DAQ device. Choose a DC voltage source that is within the allowed range (e.g., −10 V to +10 V) of your DAQ device, then connect its two wires to the appropriate pins. For our NI device's differential channel *ai0*, the high- and low-voltage wires from your source should be connected to AI 0 and AI 4, respectively; for our MCC device's AI channel 0, the high- and low-voltage wires from your source should be connected to CH0 and AGND, respectively.

NI device users: With MAX open (if you haven't done so already), right-click on the name of your DAQ device in **Devices and Interfaces** under the **My System** heading at the left and select **Test Panels**... (or press the **Test Panels**... button in the toolbar). In the interactive **Test Panels** window that appears, select the **Analog Input** tab, and then enter the settings **Mode≫On Demand** and **Input Configuration≫Differential** as shown below. The values of **Min Input Limit** and **Max Input Limit** are appropriate for the range of ±10 V. In the **Channel Name** box, the text *Dev1/ai0* selects AI differential channel *ai0* on the DAQ device with the shorthand name *Dev1* (on your **Test Panel**, use the shorthand name of your particular DAQ device, which may be different than *Dev1*).

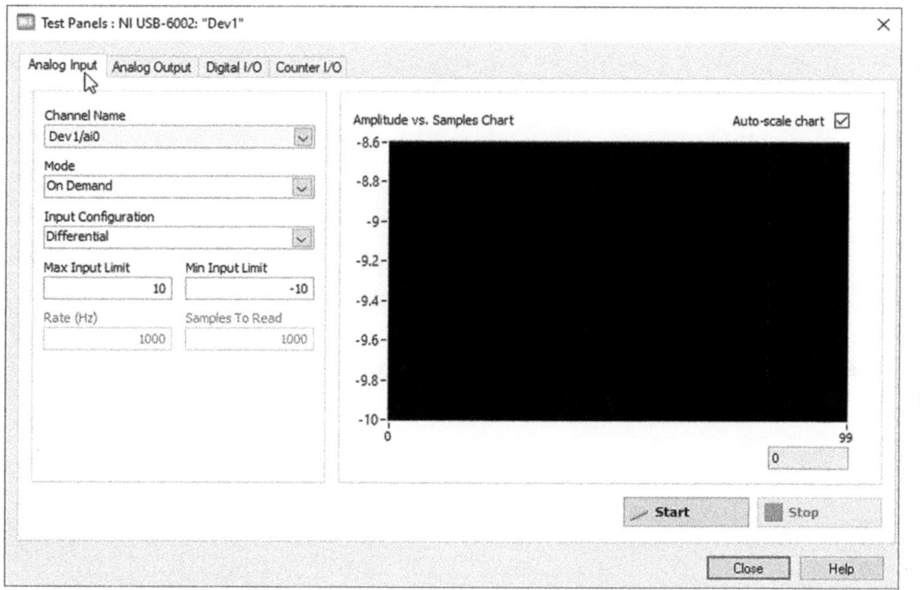

MCC device users: In the DAQami window, use the **Panel Button** (the rightmost of the three buttons at the left-hand top of the window) to select the **Configuration Panel**. In the **Configuration Panel**, click on the **Channels** menu item, then switch on channel **AI0** by clicking on the **Active** button.

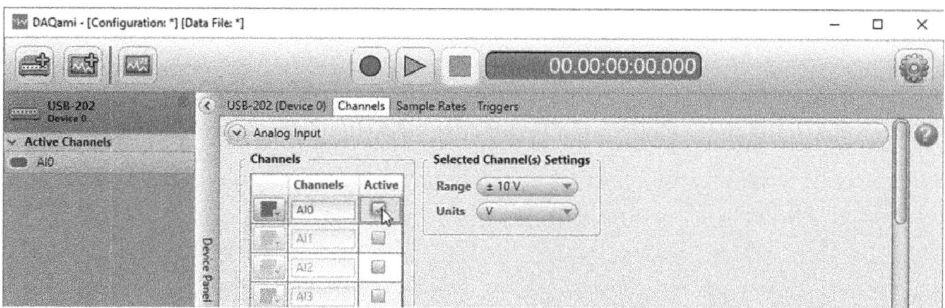

Note that channel AI0 now appears in the Device Panel's **Active Channels** listing. Next, click on the **Displays** button in the upper left-hand corner.

You will be toggled to the **Display Panel**, where a dropdown menu lists the possible display modes. Place the mouse cursor over **Scalar**, a display option that shows the numeric value of an acquired voltage sample.

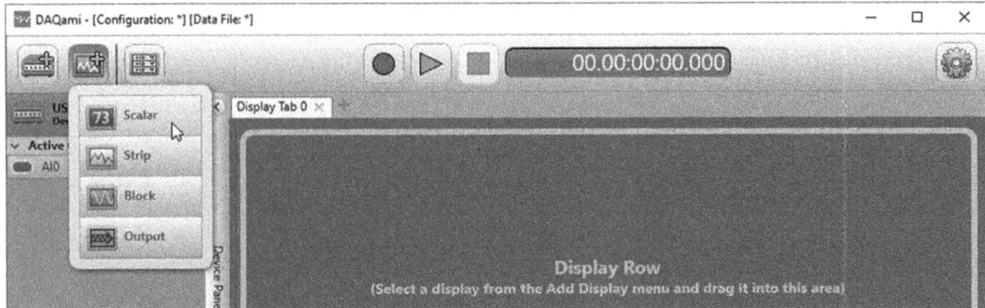

Press the mouse button down and drag the **Scalar** icon into the **Display Row** area.

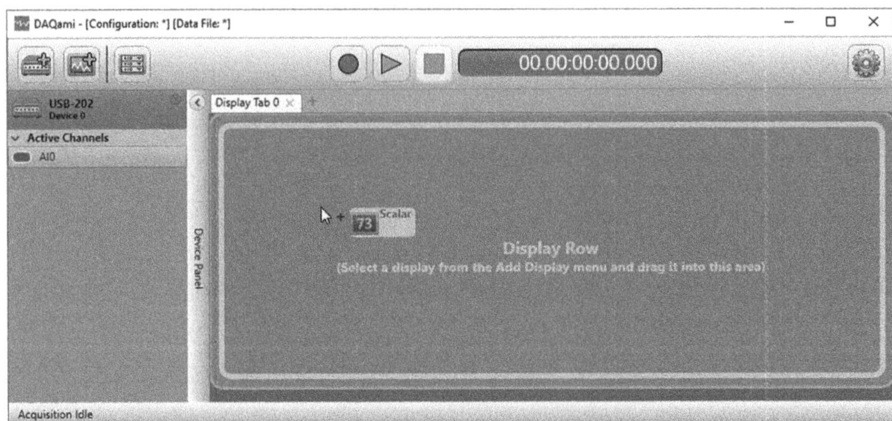

When you release the mouse button, a **Scalar** display will appear in the Display Panel.

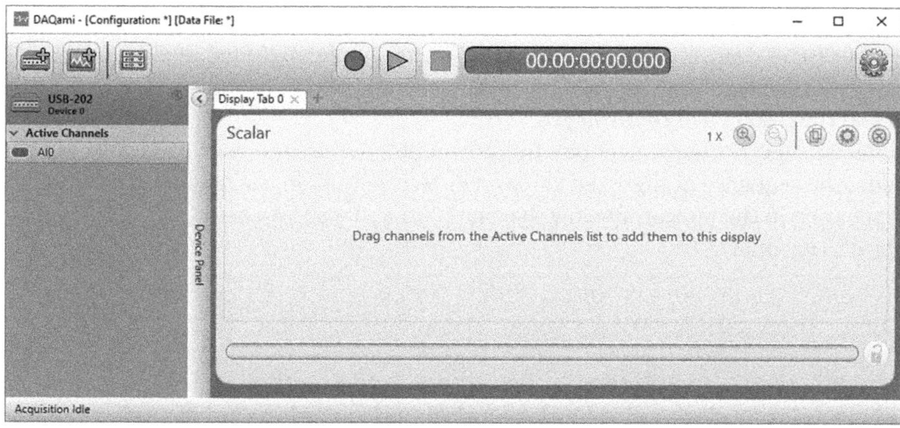

Finally, click on the **AI0** icon in the **Active Channels** listing and drag it to the **Display Row**.

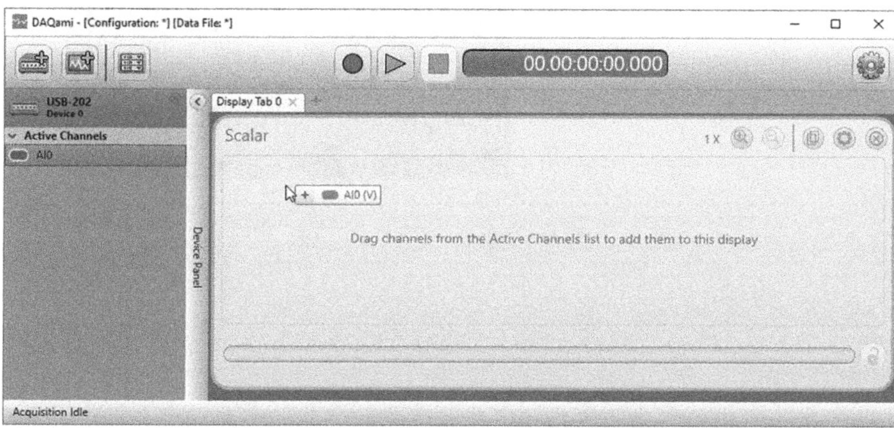

When you release the mouse button, the **Scalar** display of acquired voltage readings from channel AI0 will be ready to do its work.

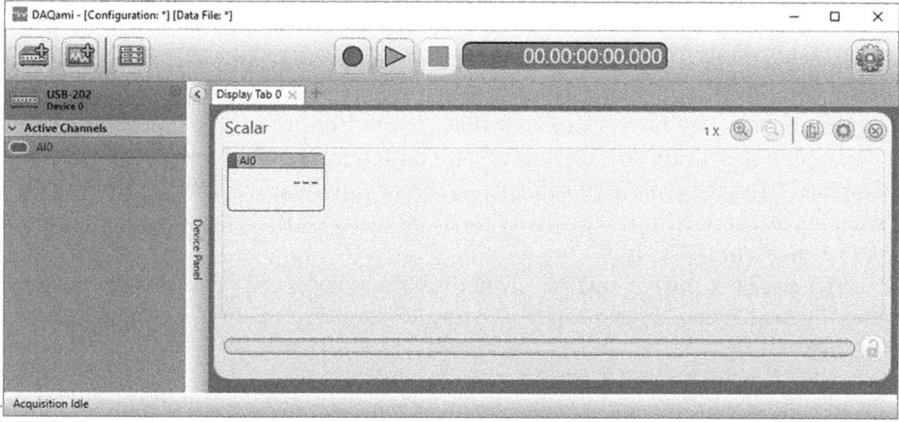

Everyone: Press the green-arrow **Start/Run** button. In the **On Demand** mode, we are using for both MAX and DAQami, the DAQ device will acquire one digitized reading of the applied voltage. Until you press the **Stop** button, this measurement will be repeated several times per second and the results will be displayed.

NI device users: If the connections from the voltage source to your DAQ device are properly made, the sequence of digitized values will be reported on a strip chart with the current value appearing in the indicator below the chart. For a plot shown below, a 2-V DC source was connected to channel *ai0*.

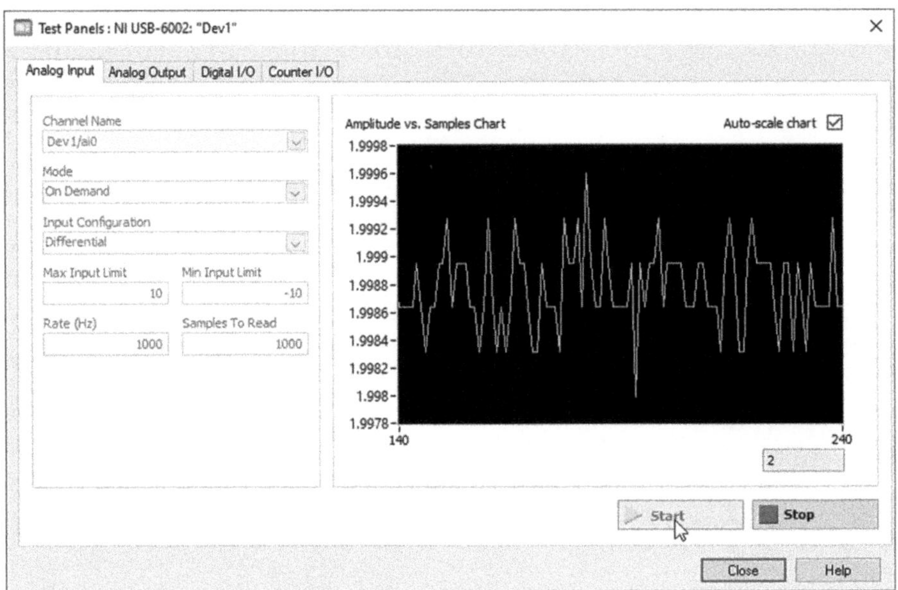

Small fluctuations in the measured voltage are expected because of experimental noise, as shown above. Note that these fluctuations occur at only certain voltage values due to the voltage resolution ΔV of the 16-bit NI device. That is, each possible digitized voltage value is separated from its neighboring value by a voltage step ΔV. We see that, relative to the baseline value, the fluctuations deviate (up or down) by at most a few steps. You can explore this feature of digitized data further in Problem 2 at the end of the chapter.

If you are using a "floating" voltage source, an issue peculiar to the differential AI mode may appear here. In the event that your measured voltage undergoes very large variations as time goes on, you may have a problem that results from the use of the floating voltage source. Examples of floating voltage sources are batteries, transformers, and nonreferenced power supplies. To remedy this predicament, reference your floating voltage source to the DAQ device's ground by connecting an extra wire (or a 100 kΩ resistor) from *ai0*'s negative input pin to the DAQ device's AI GND pin.

MCC device users: If the connections from the voltage source to your DAQ device are properly made, the **Scalar** display will report the acquired voltage reading, as shown below for a 2-V DC source:

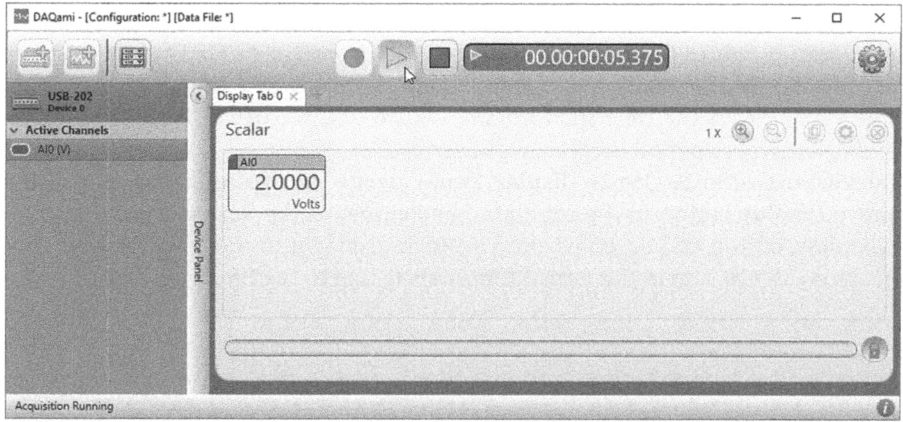

You may find that the **Scalar** reading is very stable, seemingly indicating there is no experimental noise in your system. However, remembering that the voltage resolution of the 12-bit MCC device is 5 mV, this apparent stability simply results from this system's level of sensitivity; i.e., it is not sensitive enough to resolve the experimental noise.

Everyone: Let's now program your DAQ device to digitize an AC voltage input from a function generator. Adjust the settings on a function generator so that it is outputting an analog sine wave with amplitude less than 10 V and a frequency of about 100 Hz. Using the appropriate pins for your particular system, connect the positive and negative (ground) outputs from the function generator to AI differential channel *ai0* on your NI device or AI channel 0 of your MCC device. Note that most function generators are referenced to building ground and so are not "floating" voltage sources.

NI device users: In MAX's **Test Panels** window, make the following changes as shown below: **Mode≫Finite**, **Rate (Hz)≫1000**, and **Samples To Read≫100**. In the **Finite** mode, an analog signal input at the specified AI channel is digitized at N equally spaced times at a rate f_s (i.e., the sampling frequency). The N acquired data points are then plotted. In the **Test Panels** window, **Samples To Read** and **Rate (Hz)** are used to select the values for N and f_s, respectively.

Press the **Start** button. If all goes well, you will see a plot of about 10 cycles of the 100-Hz sine wave.

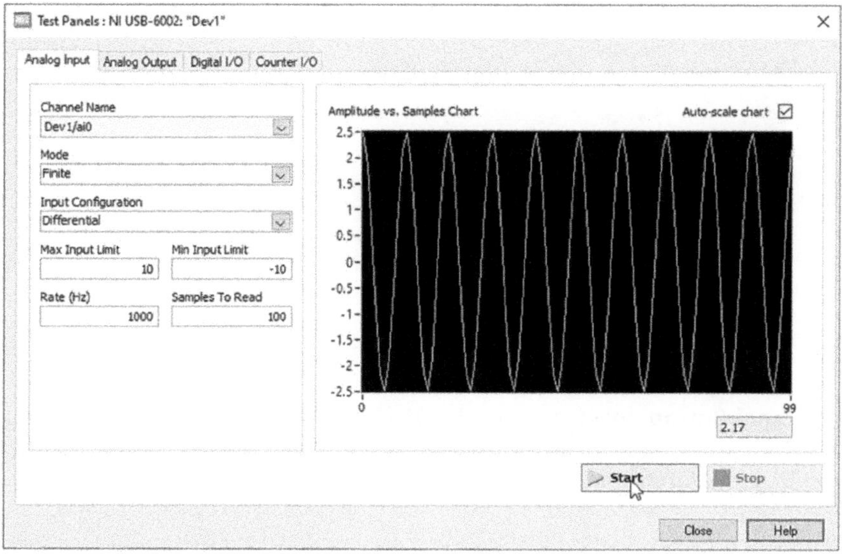

Note that the plot is formed from an array of 100 equally spaced samples and that the x-axis values are the array indices i of the samples, where i runs from 0 to 99. For the chosen sampling frequency of 1000 Hz, neighboring data samples are separated by a time interval of 0.001 s, and so the x-axis covers the time span from $t = 0.000$ to $t = 0.099$ s. Thus, for the 100-Hz sine-wave signal, whose period is 0.01 s, we observe 10 of its cycles.

MCC device users: In DAQami's **Display Panel**, delete the **Scalar** display by clicking on the **Remove Display** button in the top right-hand corner of the **Scalar** display. Then, click on the **Displays** button and drag-and-drop a **Block** display into the **Display Row**. Finally, drag-and-drop the **AI0** icon in the **Active Channels** listing to the **Block** display.

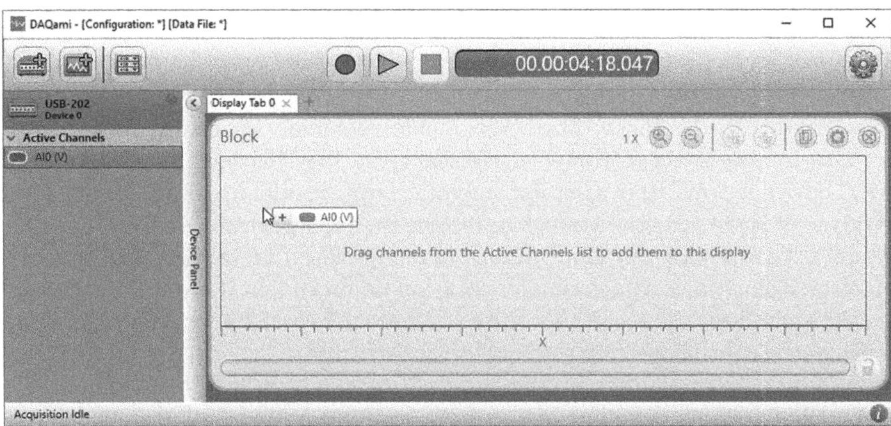

The **Block** display will produce a plot of the acquired waveform from the AI channel 0. Using the **Panel Button**, switch to the **Configuration Panel** and select the **Sample Rates** toolbar-menu item. In the **Sample Rates** window, enter *1000* for the **Analog Samples per second** value. Next, select the **Triggers** toolbar-menu item. In the **Triggers** window, under **Stop Trigger**, select **Source≫Sample Count** and enter *100* for the **Sample Count** value.

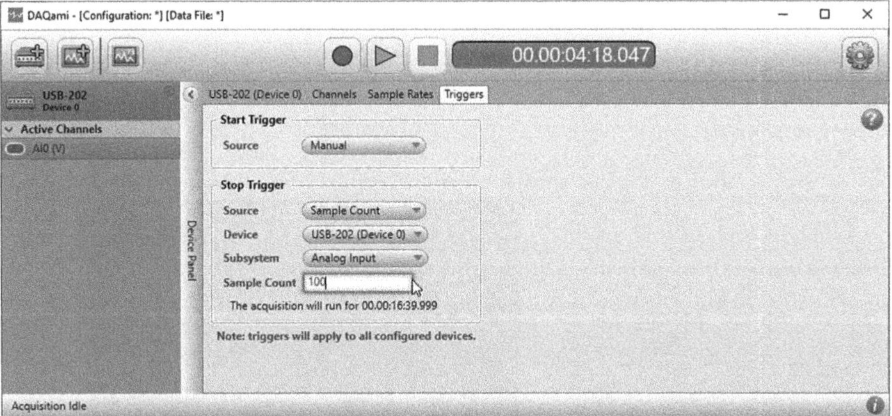

Using the **Panel Button**, toggle back to the **Display Panel**, then press the **Run** button. If all goes well, you will see a plot of about 10 cycles of the 100-Hz sine wave. To rescale the time axis in order to better fill the plot, use the **Zoom In** button.

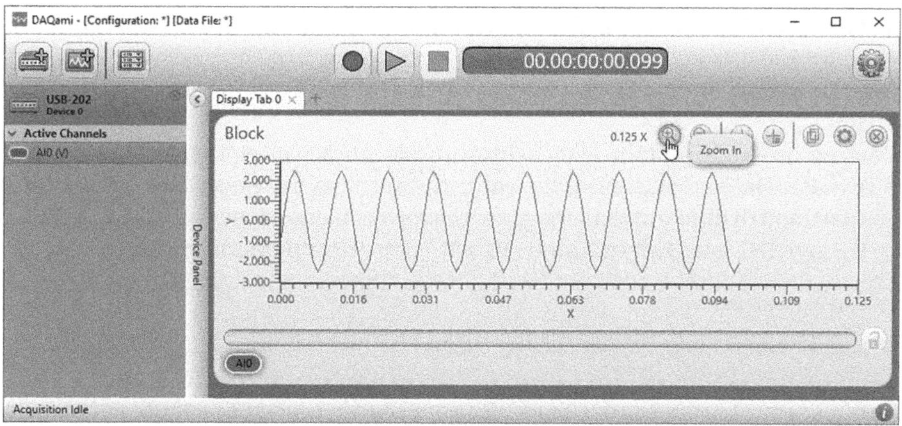

Everyone: On the function generator, set the frequency f of the input sine-wave signal to several values within the range $0 \leq f \leq 500$ Hz and press **Start/Run** to observe the digitized waveform for each frequency setting. You should find that the frequency of each signal in this range is faithfully reproduced by the DAQ device, but that sine-wave shape becomes distorted as the signal frequency approaches 500 Hz (because of the paucity of sampled points per sine-wave cycle at these higher input frequencies). The aliasing effect is explored in Problem 3 at the end of the chapter.

6.7 Analog Output

Multifunction DAQ devices typically have two analog output (AO) channels. Each AO channel can perform n-bit digital-to-analog conversion operations at speeds up to a *maximum update rate* given in samples per second (S/s). The possible analog voltages produced fall within a range from $V_{\min}$ to $V_{\max}$. Within this voltage span $V_{span} \equiv V_{\max} - V_{\min}$, the output analog voltage can be one of 2^n possible values. The voltage resolution ΔV with which a voltage can be produced—that is, the smallest difference between one possible output voltage and the next—is given by $\Delta V = V_{span}/2^n$, i.e., Eq. (6.1). As a typical example, if $n = 16$ and $V_{span} = 20$ V, $\Delta V = 0.3$ mV. The specifications for our NI and MCC DAQ devices are listed in Table 6.2.

On our NI device, the voltage difference associated with AO channel *ao0* is produced between pins AO 0 and AO GND, while for channel *ao1* the output is between pins AO 1 and AO GND. For our MCC device, the voltage difference associated with AO channel 0 is produced between pins AOUT0 and AGND, while for channel 1, the output is between pins AOUT1 and AGND.

Table 6.2 Analog Output (AO) Specifications

Device	Number of AO channels	AO voltage range (V)	Bits	Voltage resolution (mV)	Maximum update rate (S/s)
NI USB-6002	2	± 10	16	0.3	5
MCC USB-202	2	0 to 5	12	1.2	250

6.8 Analog Output Operation Using MAX and DAQami

Everyone: Let's program your DAQ device to output a DC voltage difference at AO channel *ao0* (NI) or 0 (MCC).

NI devices users: To output a DC voltage using MAX, select the **Analog Output** tab in the **Test Panels** window, and then enter the settings as shown below with **Channel Name≫Dev1/ao0** (if different than *Dev1*, use the correct name of your particular DAQ device), **Mode≫Voltage DC**, and **Output Value (V)≫1**. Then press the **Update** button.

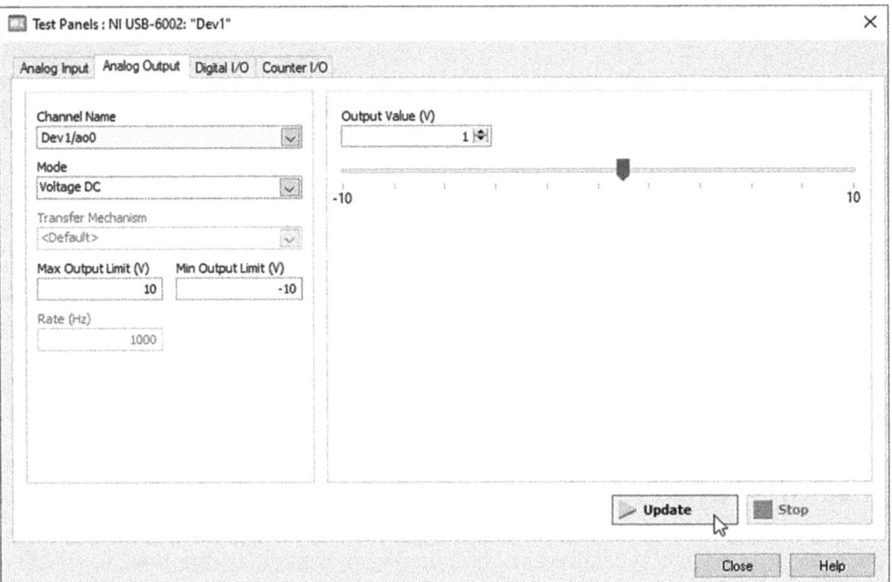

To verify that your DAQ device is outputting a 1-V voltage difference at AO channel *ao0*, connect the leads of a voltmeter to the AO 0 and AO GND pins. Note that this 1-V DC voltage difference will be continuously output at *ao0* until you program the DAQ device to output a different value.

MCC device users: In DAQami's **Display Panel**, delete the **Block** display by clicking on the **Remove Display** button in the top right-hand corner of the **Block** display. Switch to the **Configuration Panel**. Click on the **Channels** menu item, then switch off channel **AI0** by clicking on the **Active** button (to deselect it). Scroll down in the **Channels** menu and switch on AO channel **AO0** by clicking on its **Active** button.

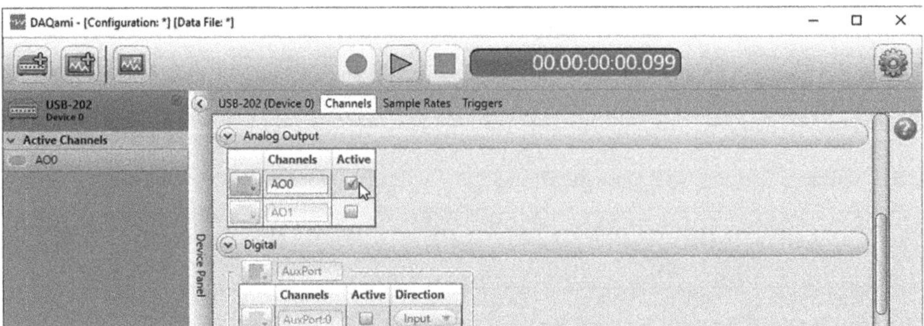

Then, select the **Triggers** menu and set **Start Trigger≫Source≫Manual** and **Stop Trigger≫Source≫Manual**. Click on the **Displays** button and drag-and-drop an **Output** display into the **Display Row**. Finally, drag-and-drop the **AO0** icon in the **Active Channels** listing to the **Output** display. The resulting **Output** display is a slider switch, which allows one to easily select the desired voltage level at AO channel 0.

Press the **Run** button and use the slider switch to select a voltage level within the available range of 0 to 5 V.

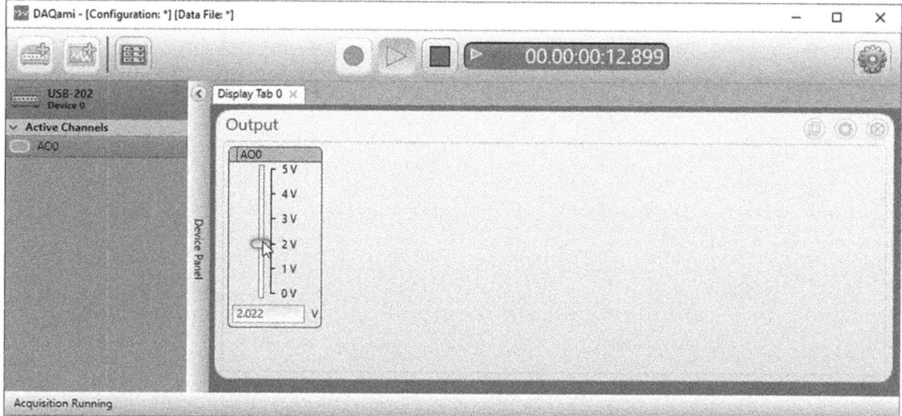

To verify that your DAQ device is outputting the selected voltage difference at AO channel 0, connect the leads of a voltmeter to the AOUT0 and AGND pins. Note that this voltage difference will be continuously output at AO channel 0 until you program the DAQ device to output a different value.

Everyone: For practice in making the correct connection to your DAQ device (or if you don't have a voltmeter available), let's output a voltage on AO channel 0 and read that voltage on AI channel 0.

NI device users: To send the *ao0* output to the *ai0* input, connect a wire from the AO 0 pin to the AI 0 pin. The associated grounding pins, AO GND and AI GND, are connected internally within some DAQ devices so, on those devices, there is no need to connect them externally with a wire. In case your device doesn't make this internal connection, connect a wire between AO GND and AI GND. Select the **Analog Input** tab in **Test Panels**, configure the window as shown below, and press **Start**. If all connections are made properly, you will observe a real-time plot of the voltage difference being output at *ao0*, probably the 1-V DC voltage you programmed in a few moments ago.

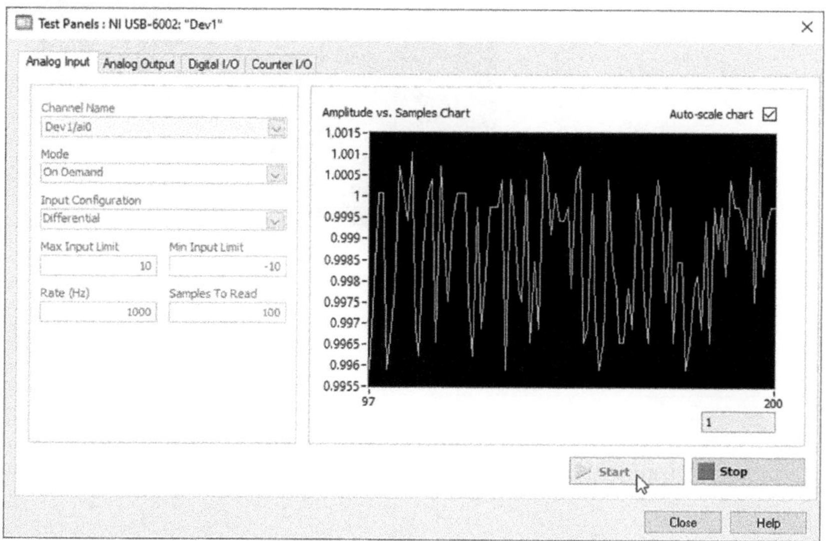

Select the **Analog Output** tab again. This time, we'll program the DAQ device to output an AC voltage at AO channel *ao0*. In the **Test Panels** window, make the following changes shown below: **Mode≫Voltage Sinewave Generation** and **Sinewave Amplitude (V)≫1**. The **Sinewave Frequency (Hz)** is fixed at 1 Hz.

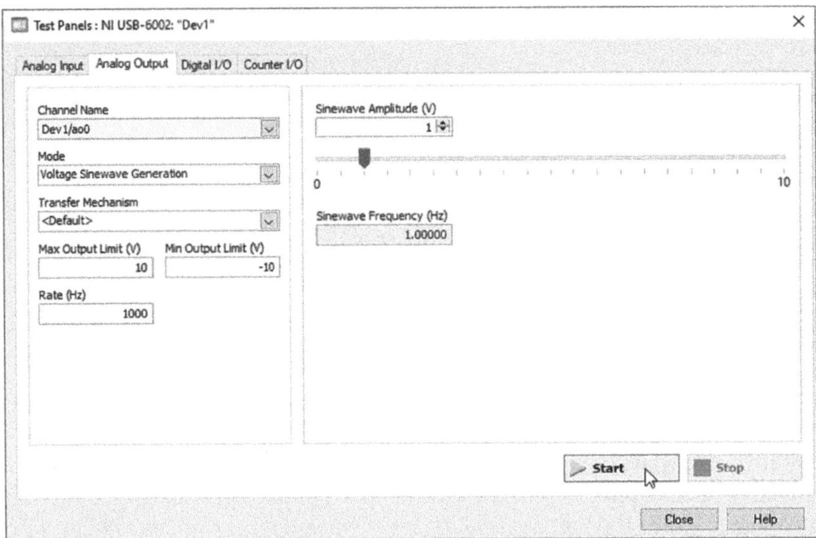

Press the **Start** button. To verify that your DAQ device is outputting a sine wave with an amplitude of 1 V and a frequency of 1 Hz at AO channel *ao0*, select the **Analog Input** tab in **Test Panels**, configure the window as shown below, and press **Start**. With **Rate (Hz)≫1000** and **Samples To Read≫2000**, it will take 2 seconds to acquire the data, which will display two cycles of the 1-Hz sine wave.

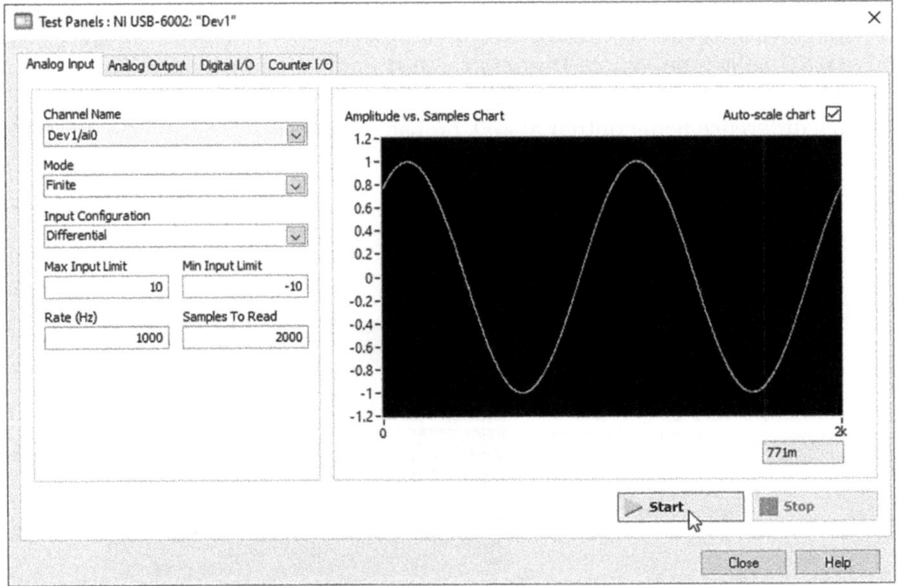

MCC device users: To send the AO channel 0 output to the AI channel 0 input, connect a wire from the analog output AOUT0 pin to the analog input CH0 pin. All AGND grounding pins are connected internally within the DAQ device so there is no need to connect them externally with a wire. In the DAQami window, switch to the **Configuration Panel**. Click on the **Channels** menu item, then switch on channel **AI0** by clicking on the **Active** button and verify that channel **AO0** is still active. With both channels **AI0** and **AO0** active, click on the **Displays** button and drag-and-drop a **Scalar** display into the **Display Row**. The **Display Row** will now be occupied by separate **Output** and **Scalar** displays. Finally, drag-and-drop the **AI0** icon in the **Active Channels** listing to the **Scalar** display. Press the **Run** button and use the slider to select a DC voltage in the range from 0 to 5 V. This voltage will be output at AO channel 0 and read by AI channel 0, whose reading is viewed in the **Scalar** display.

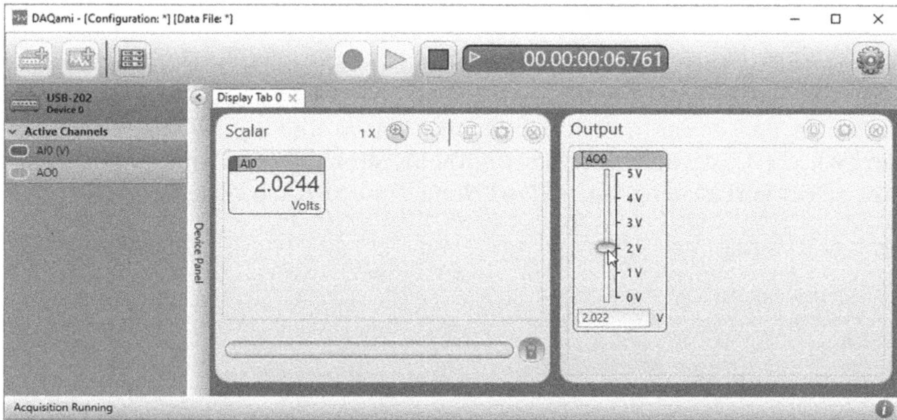

The MCC device has only limited signal generation capability, so DAQami cannot be used to produce a sine wave as in the NI device exercise.

6.9 Digital Input/Output

A digital signal can be in one of two possible states, termed HIGH and LOW. For the digital lines on DAQ devices, these states conform to the *transistor–transistor logic (TTL) standard*, where HIGH and LOW are defined to be within the range 2–5 V and 0–0.8 V, respectively (the exact definitions for TTL output and input states differ slightly, but are both within these ranges). When a digital signal alternates between its two states, the transition from LOW to HIGH is called the *rising edge*, while the HIGH to LOW transition is the *falling edge*.

Multifunction DAQ devices are equipped with a generous collection of *digital input/output (DIO)* pins. Each of these DIO pins is called a *line*, which can be configured to either input or output a digital state, that is, ascertain ("read") whether the pin is in the HIGH or LOW state or set ("write") the pin to the HIGH or LOW state. Our MCC device has eight digital I/O lines, which together form an 8-bit *port*. These lines are labeled DIO0 through DIO7. In addition, a digital TRIG pin is available for triggering data acquisitions (see Use It! project in this chapter). The thirteen digital I/O lines on our NI device, on the other hand, are grouped into three *ports*, where each port is composed of one, four, or eight *lines*. A particular line is specified as P *port.line*, so, for example, P0.3 is line 3 of port 0. Some of the DIO lines have additional capabilities beyond that of regular DIO lines and are denoted as programmable function interface (PFI) lines. PFI lines are used for triggering data acquisition, clock synchronization, counter input, and counter output. Table 6.3 lists the configuration of DIO lines on our representative DAQ devices.

Table 6.3 Digital Input/Output (DIO) Lines' Specifications

Device	Total number of DIO lines	Number of lines			Number of PFI lines
		Port 0	Port 1	Port 2	
NI USB-6002	13	8	4	1	2
MCC USB-202	8	8	—	—	TRIG

6.10 Digital Input/Output Operation Using MAX and DAQami

Everyone: As a first step in controlling the DIO lines on your DAQ device, let's output the HIGH state on a particular pin, then toggle that pin to the LOW state. For this exercise, we'll use the DIO line P0.0 on our NI device or the DIO0 line on our MCC device's.

NI device users: Using MAX, select the **Digital I/O** tab in the **Test Panels** window, and then under the **Select Port** heading, choose **Port Name≫port0** as shown below.

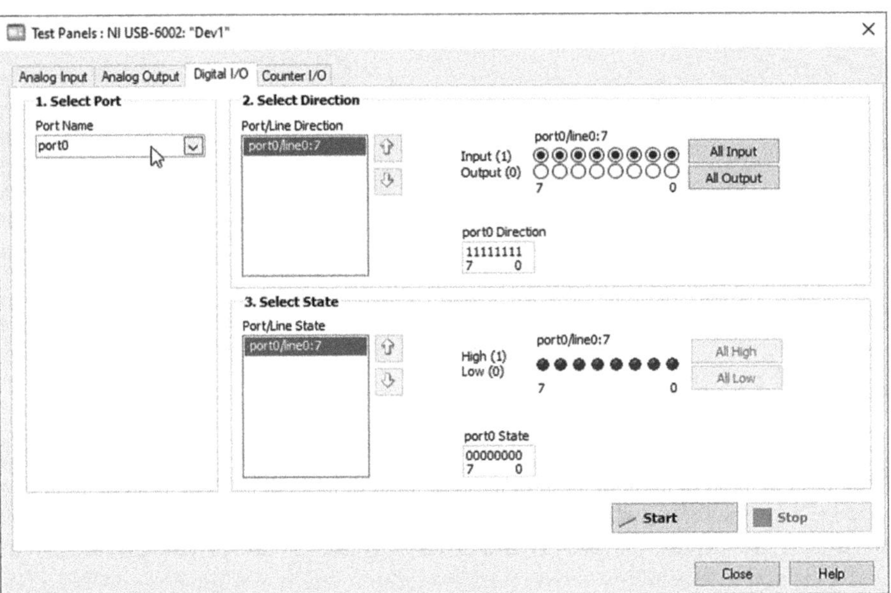

Next, under the **Select Direction** heading, make *P0.0* an output line by clicking on the output button as shown.

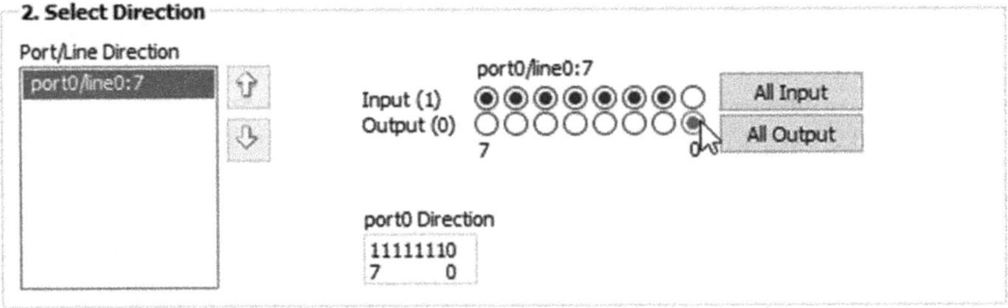

Finally, under the **Select State** heading, set the switch for P0.0 to the HIGH setting.

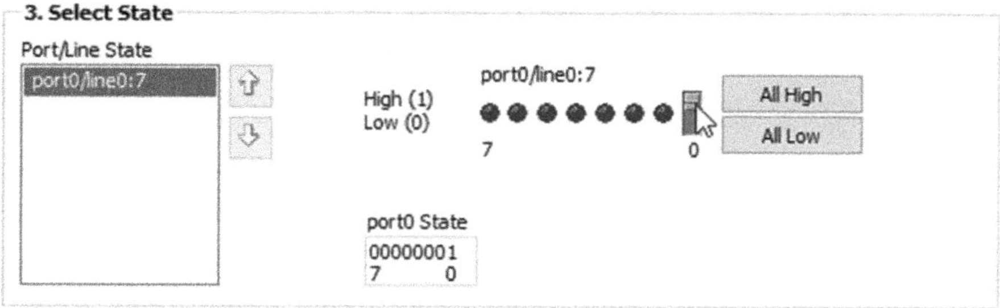

Then press the **Start** button. To verify that your DAQ device is outputting the HIGH state at the designated DIO pin, connect the leads of a voltmeter to the P0.0 and D GND pins. For TTL output signals, the HIGH state corresponds to a voltage difference in the range from 2.7 to 5 V (common values are 3.3 and 5 V). Use MAX to switch P0.0 to the LOW setting. Does the voltmeter now indicate the line is in the LOW state (for TTL output signals, the LOW state is somewhere in the range from 0 to 0.4 V)?

Alternatively, you can read the voltage using your DAQ device (rather than using a voltmeter). For this approach, connect the P0.0 and D GND pins to the positive and negative input pins of AI channel *ai0*. Then, select the **Analog Input** tab in **Test Panels**, configure the window as **Mode≫On Demand**, and press **Start**.

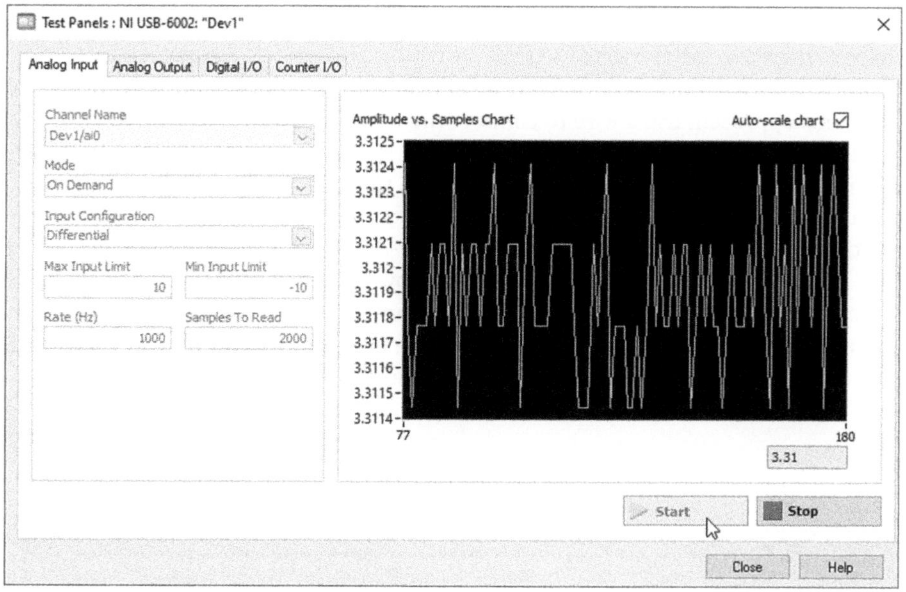

MCC device users: In the **Display Panel** of the DAQami window, remove any existing display and toggle to the **Configuration Panel**. Click on the **Channels** menu item and switch off any active channels. Then, under the **Digital** heading of the **Channels** menu, switch on digital line **DIO0** (labeled as **AuxPort:0**) by clicking on the **Active** button and select **Direction≫ Output**. The DIO0 pin is now active as a digital output. Click on the **Displays** button and drag-and-drop an **Output** display into the **Display Row**. Finally, drag-and-drop the **AuxPort:0** icon in the **Active Channels** listing to the **Output** display. An On/Off switch will appear, which enables you to toggle the **DIO0** pin between its HIGH and LOW states.

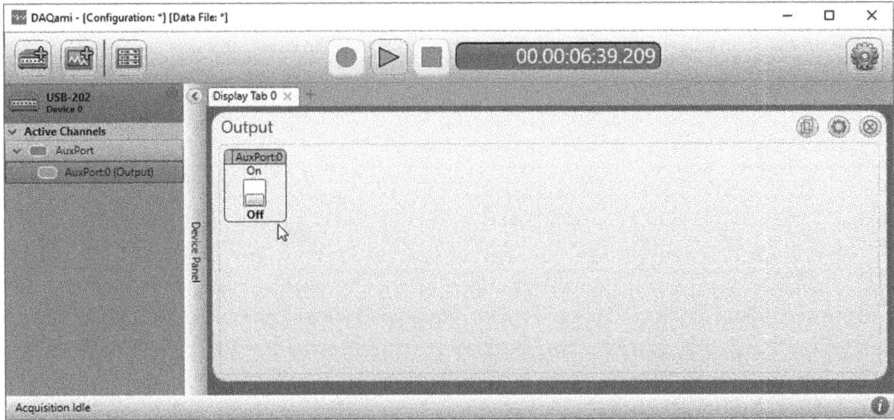

Press the **Run** button and use the switch to select the **On** setting. To verify that your DAQ device is outputting the HIGH state at the designated DIO pin, connect the leads of a voltmeter to the DIO0 and GND (i.e., digital ground) pins. For TTL output signals, the HIGH state corresponds to a voltage difference in the range from 2.7 to 5 V (common values are 3.3 and 5 V). Toggle the DIO0 pin to its LOW state by changing the switch to its **Off** setting. Does the voltmeter now indicate the line is in the LOW state (for TTL output signals, the LOW state is somewhere in the range from 0 to 0.4 V)?

Alternatively, you can read the voltage using your DAQ device (rather than using a voltmeter). For this approach, use a wire to connect the DIO0 pin to the AI channel CH0 pin and another wire to connect the digital ground GND pin to the analog ground AGND pin. In the DAQami window, toggle to the **Configuration Panel**. Click on the **Channels** menu item, then switch on channel **AI0** by clicking on the **Active** button. Both channels **AI0** and **DIO0** should now be active. Click on the **Displays** button and drag-and-drop a **Scalar** display into the **Display Row**. The **Display Row** will now be occupied by separate **Output** and **Scalar** displays. Finally, drag-and-drop the **AI0** icon in the **Active Channels** listing to the **Scalar** display. Press the **Run** button and use the **On/Off** switch to toggle the DIO0 pin output between its HIGH and LOW states, whose associated voltage is viewed in the **Scalar** display.

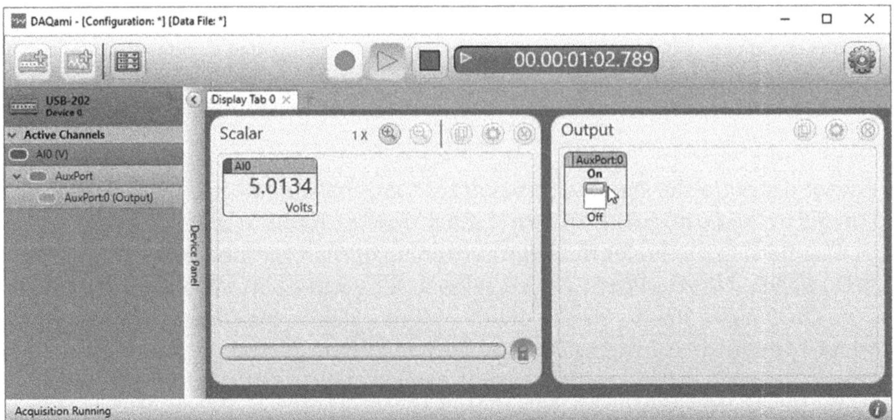

Everyone: Finally, let's write the HIGH state at one digital line—a digital output operation—and read this value at another digital line—a digital input operation.

NI device users: We'll use MAX's **Test Panels** to write the HIGH state at the P0.0 pin and read this value at the P0.1 pin. First, connect these two lines on your DAQ device by attaching a wire to the P0.0 and P0.1 pins; the digital grounds for these two lines will be automatically connected within the DAQ device, so there is no need to wire a ground connection. Second, in **Digital I/O** window, under the **Select Direction** heading, make P0.0 an *output* and P0.1 an *input*; under the **Select State** heading, switch P0.0 to the HIGH position. Now, press the **Start** button. As shown below under the **Select State** heading, the LED indictor for P0.1 will light. Try toggling the P0.0 switch to the LOW state and repeat. Does the LED indicator for P0.1 respond as expected?

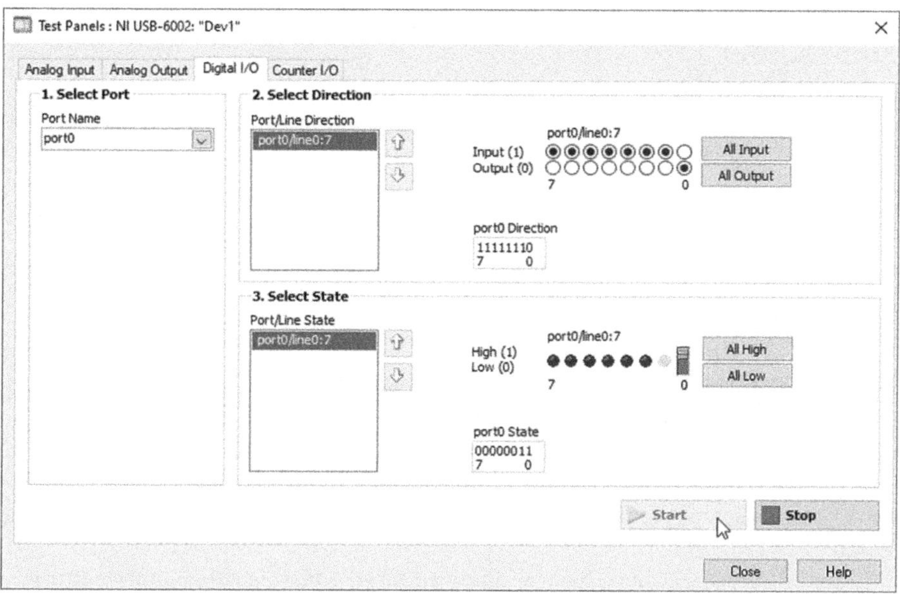

MCC device users: We'll use DAQami to write the HIGH state at the DIO0 pin and read this value at the DIO1 pin. First, connect these two lines on your DAQ device by attaching a wire to the DIO0 and DIO1 pins; the digital grounds for these two lines will be automatically connected within the DAQ device, so there is no need to wire a ground connection. Second, in the **Display Panel** window, there should already be an **Output** display with an **On/Off** switch to control the state of the DIO0 pin as well as a **Scalar** display indicating the voltage value input at channel **AI0**. Switch to the **Configuration Panel**. Click on the **Channels** menu item and deactivate channel **AI0**. Then, under the **Digital** heading, switch on digital line **DIO1** (labeled as **AuxPort:1**) by clicking on the **Active** button and select **Direction≫ Input**. The DIO0 and DIO1 pins are now active as a digital output and input, respectively. Click on the **Displays** button and you should find that the **Scalar** display is empty. Drag-and-drop the **AuxPort:1** icon in the **Active Channels** listing to the **Scalar** display. A LED indicator will appear that, when lit or unlit, indicates whether a HIGH and LOW digital state is being input to the DIO1 pin.

Press the **Run** button and use the **On/Off** switch to toggle the DIO0 pin output between its HIGH and LOW states with the LED being lit and unlit, respectively.

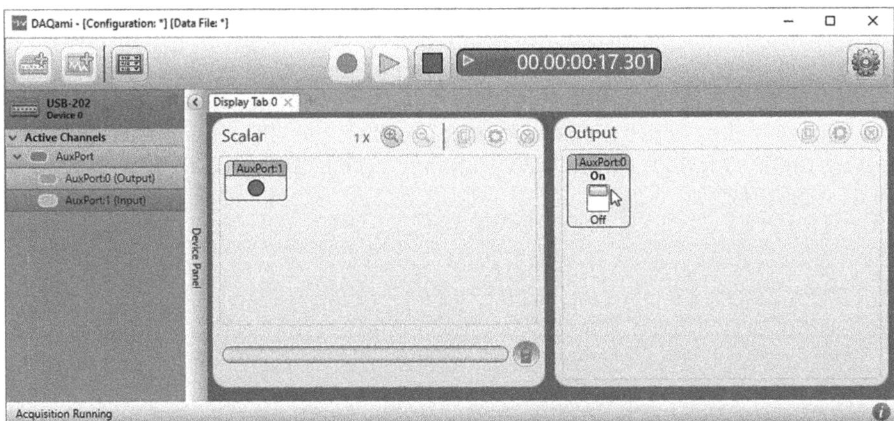

Do It Yourself

Millisecond-Resolution Stopwatch Using MAX and DAQami Multifunction DAQ devices come equipped with one or more counters, which can be used to count how many times an incoming digital signal makes a given transition (e.g., rising edge) between the two digital states. In this exercise, you will use MAX or DAQami to transform the counter in your DAQ device into a millisecond-resolution stopwatch by inputting a 1000-Hz digital signal at its counter pin.

Our NI and MCC devices both have a single counter, which is capable of counting up to 5 million and 1 million digital pulses per second, respectively. First, find the source pin for the counter on your DAQ device (use *counter 0*, if your device has more than one counter). For our NI device, the counter input is pin P2.0/PFI 0, while on our MCC device, it is CTR. Next, configure a function generator to output a TTL digital signal with a frequency of 1000 Hz. The TTL digital signal should always be positive and toggle between (approximately) 0 and, say, 3.5 V. Most function generators create a TTL digital signal at an output labeled *sync* (or sometimes called *TTL* or *trigger out*). If your function generator produces such an output, this is an easy option for obtaining the needed 1000-Hz digital signal. Connect the two lead wires from your function generator's sync output (which carry the 1000-Hz digital signal and its associated ground) to the counter input pin and the digital ground, respectively, of your DAQ device. Digital ground on our NI device is labeled D GND, while it is GND on our MCC device.

NI device users: In MAX, open the **Test Panels** window for your DAQ device, then select the **Counter I/O** tab. Configure *counter 0* on your DAQ device (e.g., *Dev1/ctr0*) in **Mode≫Edge Counting** and select **Edge Source** to be the **PFI0** pin, as shown below. In this mode, *counter 0* will count rising digital edges that are input at the PFI0 pin.

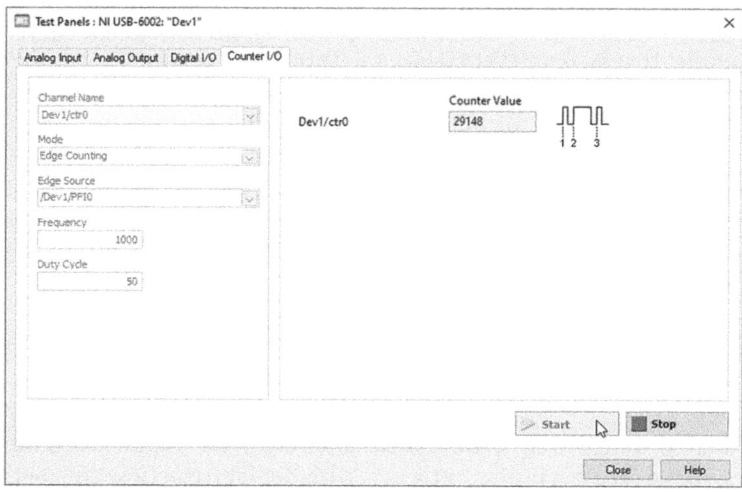

Press the **Start** button. The counter will then count a rising edge once every cycle of the 1000-Hz digital waveform, that is, once every 0.001 s = 1 ms. Thus, at a later time when you press the **Stop** button, the value in the **Counter Value** indicator will equal the number of milliseconds that have elapsed between the two button presses. In the above illustration, that elapsed time was 29,148 ms = 29.148 s.

Try inputting a 1-MHz digital signal at the PFI0 pin, if your function generator is capable of producing that higher frequency. Your system will then be a microsecond-resolution stopwatch.

MCC device users: In the **Display Panel** of the DAQami window, remove any existing display and toggle to the **Configuration Panel**. Click on the **Channels** menu item and switch off any active channels. Then, under the **Counter Input** heading, switch on the counter input **CI0** by clicking on the **Active** button and select **Mode≫Events**, **Count Direction≫Up**, **Units≫Ticks**, and **Initial Count≫0**. The **CI0** pin is now active as a counter input. Switch to the **Sample Rates** menu item and set **Counter Samples per second per channel** to its maximum value (e.g., 50). This setting selects how often the display will be updated in the **Display Panel**, not the maximum input frequency that can be counted by the counter (which is 1 MHz). Finally, select the **Triggers** menu item and set **Start Trigger≫Source≫Manual** and **Stop Trigger≫Source≫Manual**. Click on the **Displays** button and drag-and-drop a **Scalar** display into the **Display Row**. Finally, drag-and-drop the **CI0** icon in the **Active Channels** listing to the **Scalar** display. The box displaying the number of counted ticks (i.e., counted digital pulses) will appear.

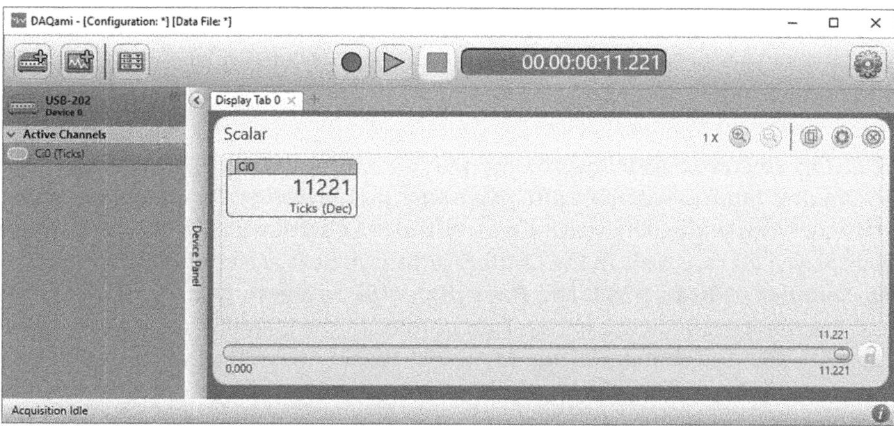

Press the **Run** button. The counter will then register a count once every cycle of the 1000-Hz digital waveform, that is, once every 0.001 s = 1 ms. Thus, at a later time when you press the **Stop** button, the value in the **Scalar** display will equal the number of milliseconds that have elapsed between the two button presses. In the above illustration, that elapsed time was 11,221 ms = 11.221 s.

Try inputting a 1-MHz digital signal at the CTR pin, if your function generator is capable of producing that higher frequency. Your system will then be a microsecond-resolution stopwatch.

Use It!

Configuring Digital Triggering Using MAX and DAQami Digital triggering is a tool commonly used in the data-acquisition process. With digital triggering, the repeated acquisition of a digitized waveform can be synchronized such that each acquisition always begins at the same location on the waveform (e.g., a zero crossing with positive slope), a necessary condition for a digitizing oscilloscope. In this exercise, we will learn how both MAX and DAQami can be used to implement digital triggering of an analog input operation.

In preparation for this exercise, use a function generator to input a 200-Hz sine-wave voltage waveform to analog input channel 0 of your DAQ device.

NI device users: Use the following steps to configure MAX for an analog input operation with a sampling frequency of 10 kHz in which 200 voltage samples are acquired of the sine wave being input at channel *ai0* as follows: First, right-click on the name of your DAQ device and select **Create Task**. . ..

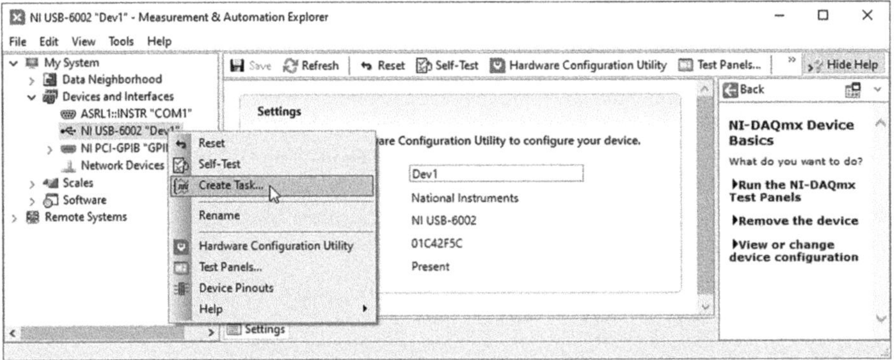

In the dialog windows that follow, choose the following sequence of selections: **Acquire Signals≫Analog Input≫Voltage≫ai0**, name your task *MyVoltageTask*, and then press the **Finish** button. Next, within the central part of the MAX window, click on the **Configuration** tab displayed for your task. In the **Configuration** window, select **Acquisition Mode≫N Samples, Samples to Read≫200**, and **Rate (Hz)≫10k** as shown next. Click on **Save** near the top of the window, then click on the **Run** button. In the graphing region you will see a plot of the 200 digitized samples of the input 200-Hz sine wave that your DAQ device has acquired at a sampling rate of 10 kHz, thus resulting in a plot of four cycles of the 200-Hz sine wave.

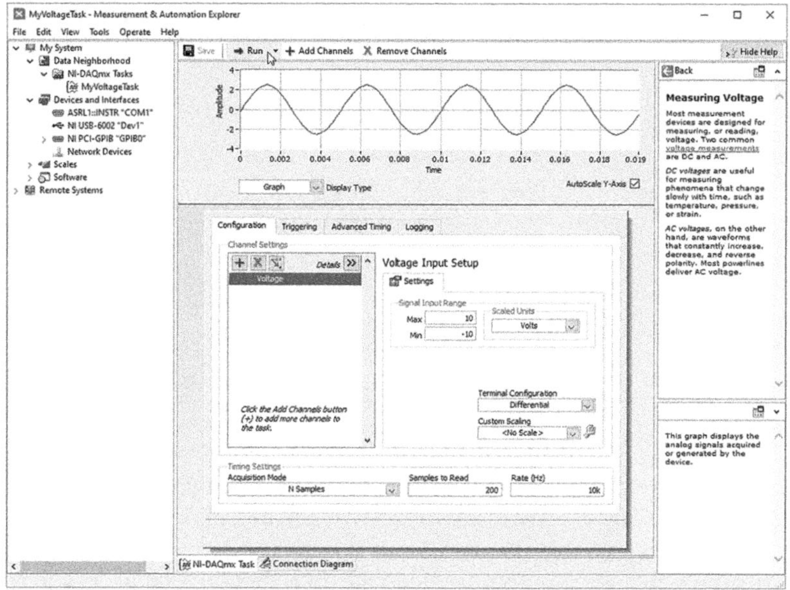

Click on the **Run** button several times. With each click, a 200-sample acquisition will be obtained and plotted. Importantly, note that each acquisition starts on a different location on the sine wave, a result of carrying out the acquisition process without digital triggering.

To activate digital triggering, click on the **Triggering** tab, where you can verify that we have been operating under the option **Trigger Type≫<None>**.

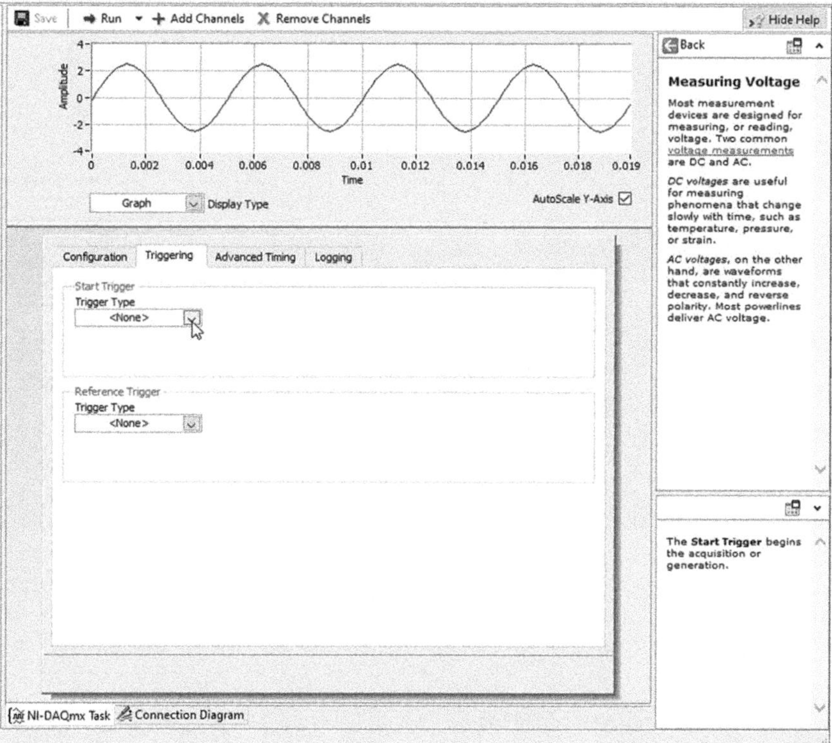

Click on the **Trigger Type** menu item and select **Trigger Type≫Digital Edge**, **Trigger Source≫PFI0**, and **Edge≫Rising**.

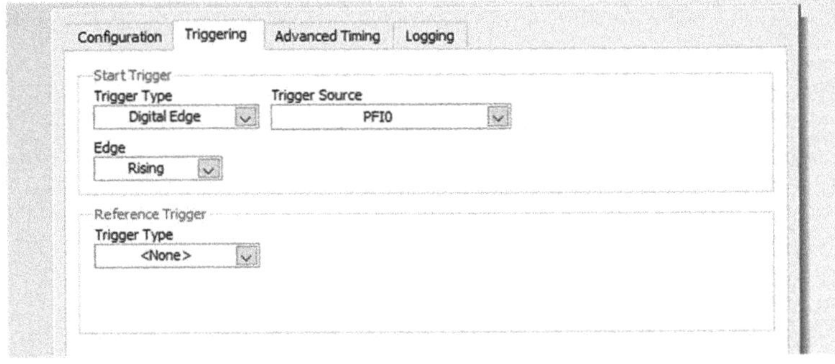

In Chapter 7, we will explain digital triggering in detail. But if you want to see it in action now, keep inputting the 200-Hz sine wave to channel *ai0*, and make the following additional connection: As mentioned in this chapter's Do It Yourself project, most function generators create a TTL digital signal at an output labeled *sync* (or sometimes called *TTL* or *trigger out*), which has the same frequency as the sine wave being output by the function generator. This sync signal is an easy option for obtaining the digital signal needed for triggering. With a cable attached to your function generator's sync output, connect the two lead wires at the cable's other end to the P2.0/PFI 0 pin and D GND, respectively, of your DAQ device. Then, click on the **Run** button repeatedly. You will find that now each acquisition always starts at the same location on the sine wave, e.g., the sine wave's positive-going ("rising") zero-crossing. As mentioned, we are simply demonstrating the digital-triggering function here. An in-depth discussion of digital triggering will be given in Chapter 7.

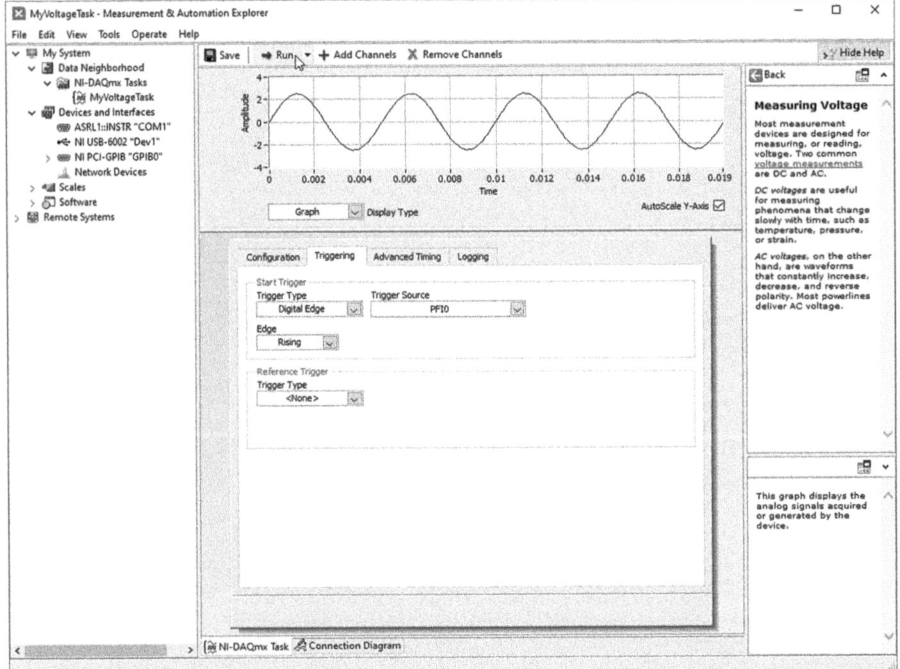

MCC device users: Use the following steps to configure DAQami for an analog input operation with a sampling frequency of 10 kHz in which 200 samples are acquired of the sine wave being input at channel *ai0*: First, in the **Display Panel**, remove any existing display. Toggle to the **Configuration Panel** and click on the **Channels** menu item. There, switch on channel **AI0** by clicking on the **Active** button (and switch off every other channel). Next, click on the **Sample Rates** menu item and enter *10000* for **Analog Samples per second per channel.** Finally, select the **Triggers** menu item. Under **Start Trigger**, choose **Source≫Manual**, while under **Stop Trigger**, choose **Source≫Sample Count** and **Sample Count≫200** as shown below:

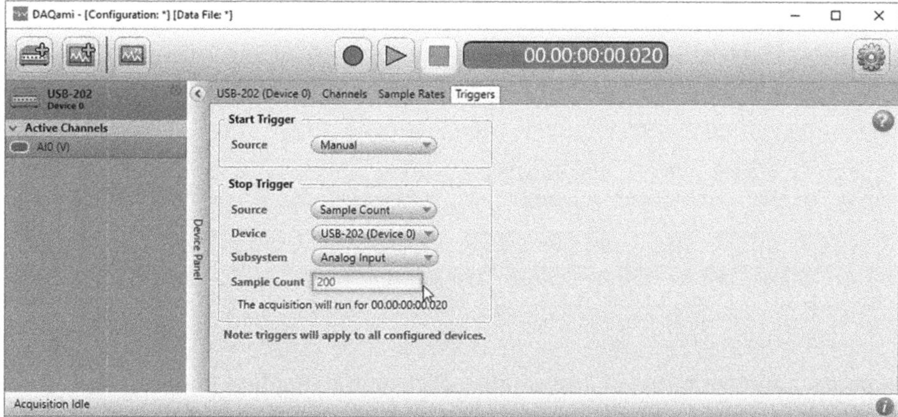

Next, click on the **Displays** button and drag-and-drop a **Block** display into the **Display Row**. Then, drag-and-drop the **AI0** icon in the **Active Channels** listing to the **Block** display. The block display is now configured to produce a waveform plot of the acquired waveform from the analog input channel 0.

Press the **Run** button. In the graphing region, there will be a display of the 200 digitized samples of the input 200-Hz sine wave. Since the samples were acquired at a sampling frequency of 10 kHz, the resulting plot includes four cycles of the 200-Hz sine wave. To rescale the time axis in order to better fill the plot, use the **Zoom In** button.

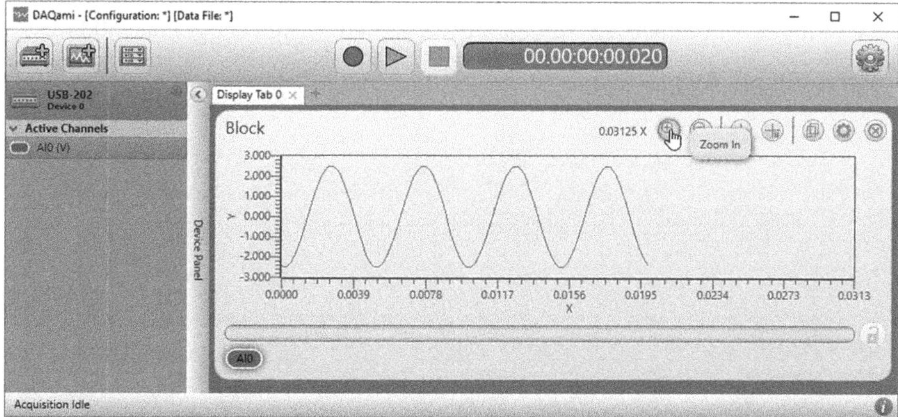

Click on the **Run** button several times. With each click, a 200-sample acquisition will be obtained and plotted. Importantly, note that each acquisition starts on a different location of the sine wave, a result of carrying out the acquisition process without digital triggering.

To activate digital triggering, toggle to the **Configuration Panel** and click on the **Triggers** menu item. Under **Start Trigger**, choose **Source≫External Digital** and **Mode≫Rising Edge**, while under **Stop Trigger**, the selection **Source≫Sample Count** and **Sample Count≫200** remains unchanged as shown below:

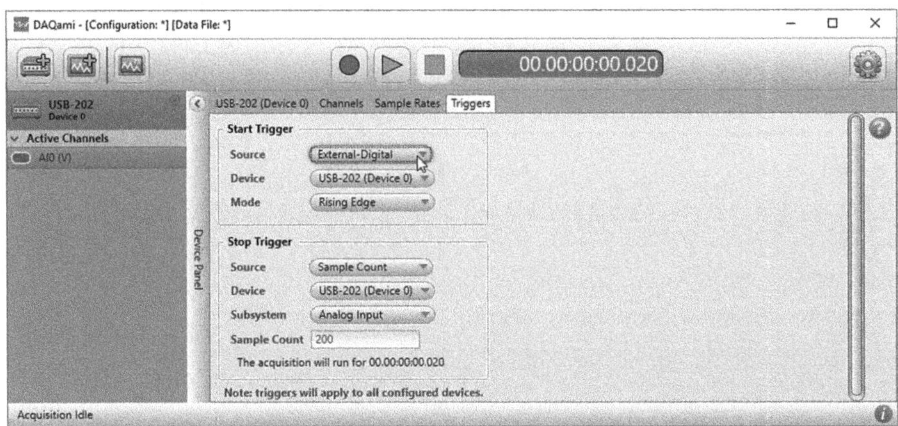

To prepare for another round of acquisitions, click on the **Displays** button.

In Chapter 8, we will explain digital triggering in detail. But if you want to see it in action now, keep inputting the 200-Hz sine wave to AI channel 0, and make the following additional connection: As mentioned in the Do It Yourself project in this chapter, most function generators create a TTL digital signal at an output labeled *sync* (or sometimes called *TTL* or *trigger out*), which has the same frequency as the sine wave being output by the function generator. This sync signal is an easy option for obtaining the digital signal needed for triggering. With a cable attached to your function generator's sync output, connect the two lead wires at the cable's other end to the pins labeled TRIG and GND, respectively, of your DAQ device. Then, click on the **Run** button repeatedly. You will find that now each acquisition always starts at the same location on the sine wave, e.g., the sine wave's positive-going ("rising") zero-crossing. As mentioned, we are simply demonstrating the digital-triggering function here. An in-depth discussion of digital triggering will be given in Chapter 8.

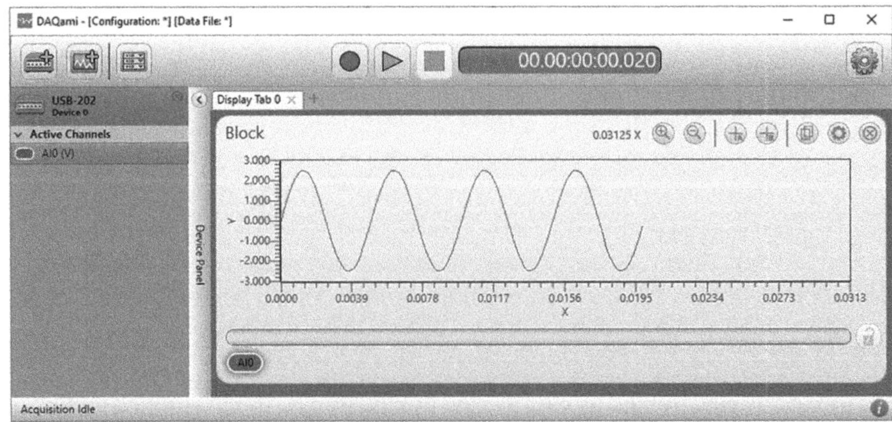

Problems

1. **Derivation of Aliasing Condition** Writing the aliasing condition as $f = \pm f_{alias} + nf_s$, show that $\sin(2\pi f t_i) = \pm\sin(2\pi f_{alias} t_i)$ at all $t_i = i\Delta t$.

2. **Observation of Voltage Resolution** Observe the voltage resolution ΔV of an analog input operation carried out by your DAQ device and verify that your observation is consistent with the prediction of Eq. (6.1).

   NI device users: Attach a DC voltage to AI channel *ai0* and then use MAX's **Test Panels** (under the **Analog Input** tab with **Mode≫On Demand**) to acquire about 100 digitized samples of this input voltage. After stopping the acquisition, inspect the data samples of this "constant" input signal, which in reality will vary slightly due to electronic noise in the experimental setup and, possibly, the signal's slow time-variation. You should find that these samples are distributed in discrete voltage levels. Measure the smallest spacing ΔV between two of these adjacent levels.

   Given the range and resolution of your DAQ device, does your value for ΔV agree with the prediction of Eq. (6.1)? If your DAQ device can operate with a different range, repeat this procedure to verify that ΔV decreases (or increases) as expected.

   MCC device users: Attach a function generator to AI channel 0 and input a low-frequency sine wave with a very small amplitude (e.g., 20 mV), then use DAQami to acquire 100 samples of the AI voltage waveform. Using the **Zoom In/Out** and **Show Display Options≫Axes** controls, scale the axes so the plot is magnified, and the fluctuation of the data sampling is clearly visible. You should find that these samples are distributed in discrete voltage levels. Measure the smallest spacing ΔV between two of these adjacent levels.

   Given the range and resolution of your DAQ device, does your value for ΔV agree with the prediction of Eq. (6.1)? If your DAQ device can operate with a different range, repeat this procedure to verify that ΔV decreases (or increases) as expected.

3. **Observation of Aliasing Effect**
   (a) Using MAX or DAQami, configure your DAQ device for an analog input operation at AI channel 0 in which 100 samples are acquired at a sampling frequency of 1000 Hz. With a function generator, input a $f = 50$ Hz sine wave at AI channel 0 and show that five cycles of the 50-Hz sine wave are acquired as expected. Keeping everything the same, show that by changing the input sine-wave frequency to $f = 1050$ Hz, the digitized sine wave appears to have a frequency of 50 (rather than 1050) Hz. Show that this result is consistent with Eq. (6.2), if $n = 1$.
   (b) Change the sampling frequency to 10 kHz and the number of samples to acquire equal to 150. If a $f = 600$-Hz sine wave is input to AI channel 0, how many cycles of the waveform do you expect to be acquired? Run your acquisition program and show that your expectation is met. Now using Eq. (6.2), show that, with $n = 1$, there are two frequencies that will be aliased as a 600-Hz sine wave. Input these two frequencies at AI channel 0 and verify your prediction. Then, taking $n = 2$, to predict two frequencies that will be aliased as a 600-Hz sine wave and use your DAQ system to verify that prediction.

CHAPTER 7

Data Acquisition Using NI DAQ Device

In this chapter, you will learn how to control a NI data-acquisition (DAQ) device connected to your computer. Here, it is assumed that you have such a DAQ device available, along with a DC voltage source (e.g., battery, power supply, function generator), function generator, voltmeter, oscilloscope, and required cabling. Prior to starting this chapter, you may want to download the latest version of the device driver **NI-DAQmx** from https://www.ni.com/support, if it isn't already present on your computer.

If, while performing the chapter exercises, the communication between the DAQ device and computer breaks down (e.g., caused by an accidental error in your programming), communication can often be restored by either unplugging and then replugging in the USB cabling or restarting your computer.

7.1 Installing Software Drivers for NI DAQ Devices

Equipped with the skills developed in Chapters 1 through 6, you are now ready to write Python programs that automate data-acquisition and generation processes. In this chapter, the programs you create will control the operations of a multifunction data-acquisition device manufactured by the automated test equipment company NI, formerly named National Instruments. Depending on your particular multifunction DAQ device, these operations may include analog-to-digital conversion, digital-to-analog conversion, digital input/output operations, event timing, digital pulse counting, and digital pulse generation.

Control of a NI DAQ device is carried out through the use of low-level driver software called *NI-DAQmx*. NI-DAQmx can be downloaded for free from the NI website, but is only available for the Windows and Linux operating systems. Data-acquisition programs are typically written in higher-level programming languages, including LabVIEW, C, C++, C#, Visual Basic, and Python, in which user-friendly commands call the required low-level NI-DAQmx commands for the desired process. LabVIEW is the high-level graphical programming language that NI has created specifically for control of its NI DAQ devices, so the chain from high-level command to low-level DAQ control follows this path:

The Python-Based Laboratory. John Essick, Oxford University Press. © John Essick (2024). DOI: 10.1093/9780191998478.003.0007

Here, high-level commands within a LabVIEW program are internally converted to proper C-based coding, which is input to the NI-DAQmx driver software. Hidden from the user, LabVIEW itself provides the needed interfacing between the high-level code and NI-DAQmx, facilitating user-friendly control of the NI DAQ device.

With the general-purpose programming language Python, "under-the-hood" interfacing between high-level Python commands and NI-DAQmx is not built-in. Thus, an additional interfacing layer, termed a *Python binding*, is required. This Python binding is called *nidaqmx*, and is available as a free download from PyPI. When using Python to control a NI DAQ device, the chain from high-level command to low-level DAQ control follows the path as shown:

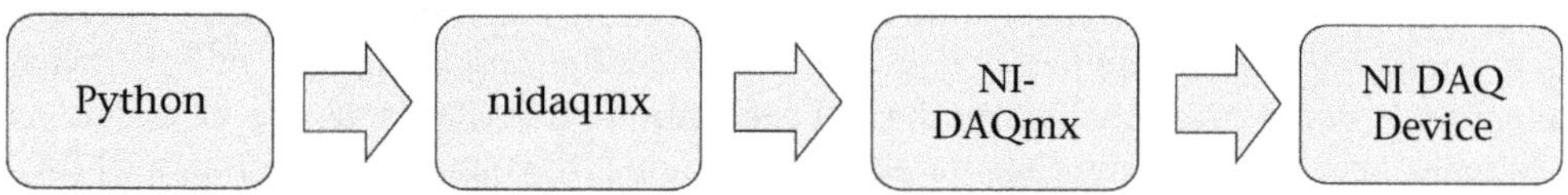

Here, nidaqmx provides a library of high-level Python-based commands, which can be called from a Python program. This library of commands covers the full range of options available on a multifunction NI DAQ device. When one of these commands is called, nidaqmx converts the Python code into proper C-based coding understood by NI-DAQmx, and inputs this converted code into the NI-DAQmx driver software. NI-DAQmx then instructs the NI DAQ device to carry out the desired operation.

Thus, for Python control of a NI DAQ device, two free software packages are required—NI-DAQmx and nidaqmx.

- To download NI-DAQmx, go to https://www.ni.com/downloads and search for "NI-DAQmx."
- To download nidaxmx, use `pip` at the **Command Prompt** as follows:

```
pip install nidaqmx
```

Full information on these two packages and how to download them can be found by searching "Using a NI DAQ Device with Python and NI DAQmx" and "NI-DAQmx Python Documentation."

7.2 DAQmx Basics

In this chapter, you will explore the fundamentals of controlling a NI DAQ device using Python to perform three basic DAQ device operations—analog-to-digital conversion (acquiring a digitized representation of an incoming analog voltage), digital-to-analog conversion (generating an outgoing analog signal from a sequence of digital numbers), and digital line control (setting the line's HIGH/LOW state). In the world of DAQ, these three operations are called *analog input* (*AI*), *analog output* (*AO*), and *digital output* (*DO*), respectively. The chapter's text will give instructions assuming that you are working with our highlighted NI USB-6002 DAQ device. However, most other multifunction DAQ device models from NI are controlled with the same commands and so can be used for the exercises. Please refer to the manual for your specific device in order to adapt to differing technical specifications (e.g., maximum sampling rate, pinout diagram).

A large assortment of data-acquisition (also called *read* or *input*) operations as well as data-generation (aka *write* or *output*) operations are within the capability of any multifunction

DAQ device manufactured by NI. When a programmer chooses to perform one such operation, that choice (e.g., analog input voltage on channel *ai0*), along with its particular timing (e.g., sampling frequency) and triggering (e.g., digital edge trigger) selections, is termed a *task.* To program the execution of a DAQmx task, one follows a template of five sequential steps: *create and configure task, start task, read or write data, stop task, clear task.*

The nidaqmx library contains all of the Python-based commands needed to program the execution of any DAQmx task possible on your NI DAQ device. The commands are implemented in object-oriented programming and so involve classes, along with their associated methods and attributes. When executed, each Python command is translated by nidaqmx into the C-coding understood by the driver software NI-DAQmx. NI-DAQmx then executes the intended step in executing the DAQmx task. Thus, by properly configuring a sequence of the Python commands from the nidaqmx library, the full execution of a DAQmx task can be coded. As an example, here is the Python code for acquiring a single DC voltage sample from a NI DAQ device, which follows the sequential steps: *create and configure task, read data, clear task*:

```
import nidaqmx

task = nidaqmx.Task()
task.ai_channels.add_ai_voltage_chan('Dev1/ai0')

single_sample = task.read()

print(single_sample)

task.close()
```

Here, after importing nidaqmx, the next two lines create and configure the task. Calling the class `Task()` creates a task object called `task`, then the method `ai_channels.add_ai_voltage_chan('Dev1/ai0')` configures the task to be an analog input voltage operation on channel *ai0* of the DAQ device named `Dev1`, which (by NI-DAQmx syntax) is denoted as the string `'Dev1/ai0'`. By default, the input channel is configured in the differential input mode; the argument `terminal_config=TerminalConfiguration.RSE` could be added within the parentheses if, instead, one wished to use the single-ended input mode. Next, the method `read()`, with no argument within its parentheses, acquires ("reads") a single voltage sample from channel *ai0* and assigns the acquired value to `single_sample`. This data value is formatted as a floating-point number. The acquired value `single_sample` is printed and then the task is cleared by the `close()` method. The `close()` command discharges the task, releasing any computing resources such as an allotment of RAM that the task has reserved. Once a task is cleared by `close()`, it cannot be run again; a new similarly defined task must be created and then run.

In some situations, a task needs to be explicitly started and stopped by invoking the `start()` and `stop()` methods, respectively, i.e., using `task.start()` and `task.stop()`. Then, all five steps for executing a task—*create and configure task, start task, read or write data, stop task, clear task*—will appear in the required program. However, if the command `task.read()` is not preceded by `task.start()` and followed by `task.stop()`, `task.read()` will automatically start and, after it has completed its work, stop the task. Hence, the steps of a task can often be written in the abbreviated form: *create and configure task, read data, clear task.*

As we'll discover later, the write operation `task.write()` has a more involved relationship with the start and stop operation. If `task.write()` is used to generate ("write") a *single* sample, this command will automatically start the task; i.e., there is no need to explicitly include the command `task.start()` prior to the `task.write()` command. However, if `task.write()` is used to generate a *multisample* waveform, the commands `task.start()` and `task.stop()` must be explicitly included in the program. As a convenience, the argument `auto_start=True` can be included in `task.write()` to commence the waveform-generation process, taking the place of an explicit `task.start()` command. In addition, if the task is cleared via `task.close()`, as part of its work, this command stops the task if it is running, thus eliminating the need to include `task.stop()` explicitly.

The nidaqmx library contains a wealth of commands, which can be used to program allowed operations on your DAQ device. For example, the commands `ao_channels.add_ao_voltage_channel()`, `di_channels.add_di_chan()`, and `ci_channels.add_ci_count_edges_chan()` are used to configure an analog voltage output, digital input, and counter input operation, respectively, while `timing.cfg_samp_clk_timing()` allows one to program the sampling frequency and number of samples to acquire in an analog input operation.

7.3 Simple Analog Input Operation on DC Voltage

Let's begin by turning your computer into a DC voltmeter. We'll accomplish this feat by creating a Python-based program that carries out the same AI operation that we ran using MAX in Section 6.6. The heart of this DC voltmeter program is the single-sample AI code outlined in Section 7.2. Thus, open IDLE and in the Editor, use **File>>New File** to create a program and save it under the name **simple_ai_operation_ni.py** in **YourNamePythonWork**. Then, code the program as follows, using the shorthand name of your particular DAQ device as found by MAX, which may be different than *Dev1*:

```
import nidaqmx

channel_name = 'Dev1/ai0'

task = nidaqmx.Task()
task.ai_channels.add_ai_voltage_chan(channel_name)

single_sample = task.read()

print(single_sample)

task.close()
```

Save your work. Connect a given DC voltage difference (say, 2 V) to the differential analog input channel *ai0* of your DAQ device. For the NI USB-6002, the high- and low-voltage wires from your DC source should be connected to pins AI 0 and AI 4, respectively. If your voltage source is "floating," use an extra wire (or 100 kΩ resistor) to connect its negative terminal to the DAQ device's AI GND pin (see the discussion in Section 6.6).

Run **simple_ai_operation_ni.py**. The acquired voltage difference will be reported in the Shell. Does your program read the input voltage difference correctly?

Now that we understand how to carry out the data-acquisition portion of our program, giving this program a GUI is an obvious next improvement. Here, we can take advantage of our previous work in building the **gui_creator.py** module, which is stored in **YourNamePythonWork>>Utilities**, and supplied functions that we used in **waveform_simulator.py**. This time around, we have something a bit different to display, though. Rather than displaying a full waveform as created in **waveform_simulator.py**, we have just a single sample value. Thus, our present display need is a simple numeric indicator. Thankfully, in Chapter 5's Use It! project, we added a function named `create_indicator()` to the **gui_creator.py** module that creates such an indicator, so we will make use of that previous work. If you didn't produce `create_indicator()` in Chapter 5, do so now following the instructions in the next paragraph.

In **YourNamePythonWork>>Utilities**, open **gui_creator.py** and, among the function definitions, add the following definition for a function named `create_indicator()`. This function creates a blue-bordered frame within the specified `container` and, within that frame, places a descriptive `label_1` above a second `label_2`. `label_2` will display a string-formatted version of our data value and `label_1` is captioning text. If left unspecified, `label_2` will initialize to its default empty string.

```
def create_indicator(container, label_1,
                     label_2='', char_width=20):

    lbl_1 = tk.Label(container, text=label_1)
    lbl_1.pack()

    frame = tk.Frame(container, bg='blue', bd=2)
    frame.pack()

    lbl_2 = tk.Label(frame, text=label_2, width=char_width)
    lbl_2.pack(fill=tk.X)

    return lbl_2
```

Save your work and then close **gui_creator.py**.

We'll also be using nidaqmx commands in our work ahead, so let's create a module that has handy nidaqmx-related functions that we can call. As with **gui_creator.py**, this nidaqmx-related module will be stored in **YourNamePythonWork>>Utilities**. In the Editor, use **File>>New File** to create a program named **nidaq_operations.py** and save it in **YourNamePythonWork>>Utilities**. After importing `nidaqmx`, first define the function `create_ai_task()`. This function uses the `Task()` class to create the task object named `task` and then configures the task as an AI operation on `channel_name` (i.e., a specified channel on a specified DAQ device). This function returns the `task` object.

```
import nidaqmx

def create_ai_task(channel_name):

    task = nidaqmx.Task()
    task.ai_channels.add_ai_voltage_chan(channel_name)

    return task
```

Save your work and close this module.

Again in the Editor, use **File>>New File** to create a program named **dc_voltmeter_ni.py** and save it in **YourNamePythonWork**. Then enter the following code. Here, functions from the **gui_creator.py** and **nidaq_operations.py** utility modules are used to generate a labeled GUI window, create and configure an analog input voltage task, and produce a labeled indicator to display the acquired DC voltage sample; when using these functions, remember to include the relevent utility module's abbreviated prefix (i.e., `gc` or `no`). A button labeled **Read Sample** will execute a function (to be added in a moment) named `acquire_sample()` when pressed. As with all of our programs, you may wish to optimize the `size` of the operating-system window for your particular computing system.

```
import Utilities.gui_creator as gc
import Utilities.nidaq_operations as no

root = gc.create_window(title='DC Voltmeter',
                        size='250x100')

channel_name = 'Dev1/ai0'
task = no.create_ai_task(channel_name)

lbl_2 = gc.create_indicator(root,
                            label_1='Voltage Reading (V)')

btn = gc.action_button(root, label='Read Sample',
                       action=acquire_sample)

root.mainloop()

task.close()
```

To complete our program, we need to define the function `acquire_sample()`. This function implements the `read()` method with no argument within its parentheses. Written in this manner, `read()` acquires just a single voltage sample, which is formatted as a floating-point number. This voltage sample is given the name `float_data`. The `str()` function is used to convert `float_data` to a string called `string_data` for display as the `label_2` string in the GUI indicator. The definition of `acquire_sample()` must appear before it is mentioned in `action_button()`, or else the following exception will halt execution: `NameError: name 'acquire_sample' is not defined`. Hence, include the definition of `acquire_sample()` as shown next.

```
import Utilities.gui_creator as gc
import Utilities.nidaq_operations as no

root = gc.create_window(title='DC Voltmeter',
                        size='250x100')

channel_name = 'Dev1/ai0'
task = no.create_ai_task(channel_name)
```

```
lbl_2 = gc.create_indicator(root,
                            label_1='Voltage Reading (V)')

def acquire_sample():

    float_data = task.read()
    string_data = str(float_data)

    lbl_2.config(text=string_data)

    return

btn = gc.action_button(root, label='Read Sample',
                       action=acquire_sample)

root.mainloop()

task.close()
```

Save your work, then run **dc_voltmeter_ni.py**. Each time you press the **Read Sample** button, a voltage sample will be acquired by the DAQ device, and the sample value will appear as the `label_2` string in the indicator.

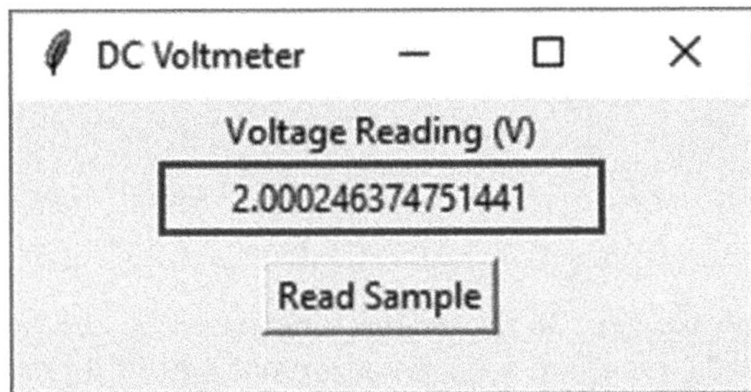

Hopefully you're pleased with your computer-based DC voltmeter, but there is an issue that should raise concern. The voltage value displayed in the indicator is given to many decimal places, well beyond the *digits-of-precision accuracy* of your DAQ device specifications. From Eq. (6.1), we know the voltage resolution ΔV of the n-bit analog-to-digital conversion process performed by your DAQ device is given by $\Delta V = V_{span}/2^n$. For a device such as the NI USB-6002 with $n = 16$ and $V_{span} = 20$ V, $\Delta V = (20\text{V})/2^{16} = 0.3$ mV. Thus, our displayed measurement in units of volts should only have four digits of precision.

Here's how to solve that display problem: In our program, the command `task.read()` reports the sampled voltage as a floating-point number, where the digits of this float are significant up to the fourth decimal place. Digits in the fifth, sixth, and higher places are just numbers left randomly in computer memory. After `task.read()` has acquired the sample, we use `str()` to convert the measured floating-point number to a string, which is then displayed in the GUI indicator. Unfortunately, the `str()` function simply converts all decimal digits, both significant and random, to string characters without being told by us how to properly round the result. There are a few alternate means for carrying out the float-to-string conversion with the correct precision. The newest (Python version 3.6 and higher) and now recommended approach for this process is called *f-strings*. To convert `float_data` to string_data with a precision to the x decimal place, the syntax for this command is as follows:

```
string_data = f'{float_data:.xf}'
```

Confusingly, there are two f's in this command, each with a different meaning. The first f, which is outside the curly brackets and quotation marks, indicates this is an f-string command. The second f is a formatting character, indicating that float_data is a floating-point number.

To correct the display problem in our program, in the acquire_sample() function within **dc_voltmeter_ni.py**, replace the statement string_data = str(float_data) with string_data = f'{float_data:.4f}'. Save your work.

Run **dc_voltmeter_ni.py** and verify that the measured voltage is now displayed with the proper digits of precision.

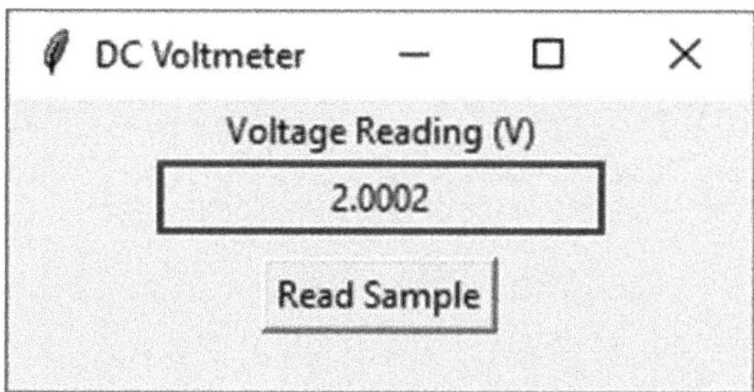

7.4 Live Updating and Threading

While impressed with your computer-based DC voltmeter, perhaps you had hoped that it would be more like a standard voltmeter with live updates. That is, instead of having to press the **Take Sample** button every time you wanted a voltage reading, wouldn't it be better to have a voltmeter that automatically updates measurements after a **Start** button is pressed and continues the updates until told to stop? Let's see if we can do that.

First, we will need a new version of our button, one that toggles between a **Start** and **Stop** state. We outlined how to build a toggling button in Chapter 5's Use It! exercise, but let's now go into greater detail on how to construct this gadget. In the Editor, use **File>>New File** to create a program named **toggle_button.py** and save it in **YourNamePythonWork**. Start coding the program as shown below.

```
import Utilities.gui_creator as gc


root = gc.create_window(title='Toggle Button',
                        size='250x50')

btn = gc.action_button(root, label='Start',
                       action=None)

root.mainloop()
```

Save your work, then run the program. An operating-system window named **Toggle Button** will appear with a button labeled **Start**. When you press the button, nothing will happen because we set `action=None` in the statement that created the button.

Now, going back to the Editor, include the definition of a function named `toggle()` as shown. The push button object `btn` is created by the `Button()` class within the `gc.action_button()` function. Here, we use a method of the `Button()` class called `cget(option)`, which reports the current value of one of `btn`'s options, in this case, its `text` label. The *if... elif* statement toggles the value of the button's `text` label. Remember to make the change `action=toggle` so that the button calls `toggle()` when pressed.

```
import Utilities.gui_creator as gc

def toggle():

    if btn.cget('text') == 'Start':
        btn.config(text='Stop')
    elif btn.cget('text') == 'Stop':
        btn.config(text='Start')

    return


root = gc.create_window(title='Toggle Button',
                        size='250x50')

btn = gc.action_button(root, label='Start',
                       action=toggle)

root.mainloop()
```

Save your work. Run **toggle_button.py**. You will find that the button's label now toggles between **Start** and **Stop** each time it is pressed.

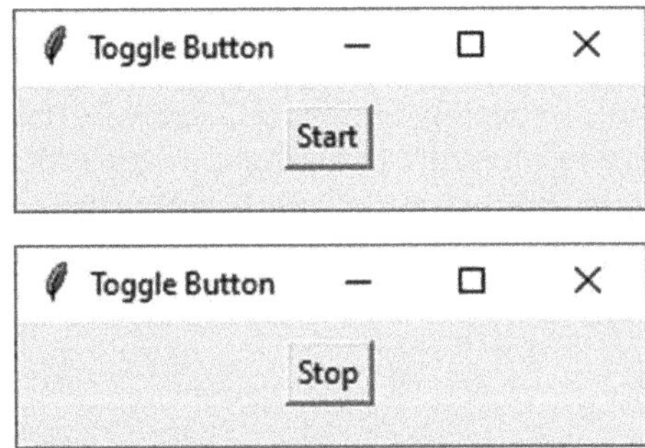

Now that we know how to toggle the button, it seems like it will be an easy task to modify **dc_voltmeter_ni.py** in order to create a live updating DC voltmeter. With **dc_voltmeter_ni.py** open in the Editor, use **File>>Save As...** to create a program named **dc_voltmeter_ni_live.py** and save it in **YourNamePythonWork**. Change the coding as shown. Note that now the argument in `action_button()` has been changed to `label='Start'`. The idea here is that, once the **Start** button is pressed (and the button's `text` label has toggled to `'Stop'`), a `while` loop will begin to execute, taking a single sample with each iteration. The `while`-loop execution will iterate until the button is pressed again, causing its `text` label to toggle to `'Start'`. Many of the commands are included in **toggle_button.py**; consider using copy-and-paste operations to expedite your coding process.

```
import Utilities.gui_creator as gc
import Utilities.nidaq_operations as no

root = gc.create_window(title='DC Voltmeter',
                        size='250x100')

channel_name = 'Dev1/ai0'
task = no.create_ai_task(channel_name)

lbl_2 = gc.create_indicator(root,
                            label_1='Voltage Reading (V)')

def acquire_sample():

    if btn.cget('text') == 'Start':
        btn.config(text='Stop')
    elif btn.cget('text') == 'Stop':
        btn.config(text='Start')

    while btn.cget('text') == 'Stop':
        float_data = task.read()
        string_data = f'{float_data:.4f}'

        lbl_2.config(text=string_data)

    return

btn = gc.action_button(root, label='Start',
                       action=acquire_sample)

root.mainloop()

task.close()
```

Save your work. Run **dc_voltmeter_ni_live.py** and you will find something unexpected happens. The program likely stalls, becoming unresponsive and requiring you to manually abort its execution (e.g., by exiting IDLE/Python). What's going on?

The problem you are observing is common in data-acquisition programs where a computer is being asked to update a GUI at the same time it is supposed to be streaming data from a DAQ device into its memory. In this present case, the GUI-related operation `root.mainloop()` is running an infinite loop, constantly checking for events like button presses, while the DAQ-related operation `read()` is trying to acquire data. Because of how we programmed these two operations, they are trying to run on the same execution path within the computer (termed a *thread*) and end up "blocking" each other's execution.

The solution to this blocking problem is to program the GUI- and DAQ-related operations to execute on separate threads. Python's `threading` module (included in the Python Standard Library) makes this solution easy to program. Modify **dc_voltmeter_ni_live.py** as shown below, where GUI-button- and DAQ-related commands are separated into the two functions `toggle()` and `acquire_sample()`, respectively. Here, GUI-related operations execute on the original thread of the execution, while DAQ-related operations in the `acquire_sample()` function execute on another thread named `t` created by `threading.Thread()`. Remember to add `import threading` to the top of your program as well as to start the new thread with `t.start()` and to change `action=acquire_sample` to `action=toggle`.

```
import Utilities.gui_creator as gc
import Utilities.nidaq_operations as no
import threading

root = gc.create_window(title='DC Voltmeter',
                        size='250x100')

channel_name = 'Dev1/ai0'
task = no.create_ai_task(channel_name)

lbl_2 = gc.create_indicator(root,
                            label_1='Voltage Reading (V)')

def acquire_sample():

    while btn.cget('text') == 'Stop':
        float_data = task.read()
        string_data = f'{float_data:.4f}'

        lbl_2.config(text=string_data)

    return

def toggle():

    if btn.cget('text') == 'Start':
        btn.config(text='Stop')
    elif btn.cget('text') == 'Stop':
        btn.config(text='Start')

    t = threading.Thread(target=acquire_sample)
    t.start()

    return
```

```
btn = gc.action_button(root, label='Start',
                       action=toggle)

root.mainloop()

task.close()
```

Save your work. Run **dc_voltmeter_ni_live.py** and marvel at the live updates to the acquired voltage after the **Start** button is pressed.

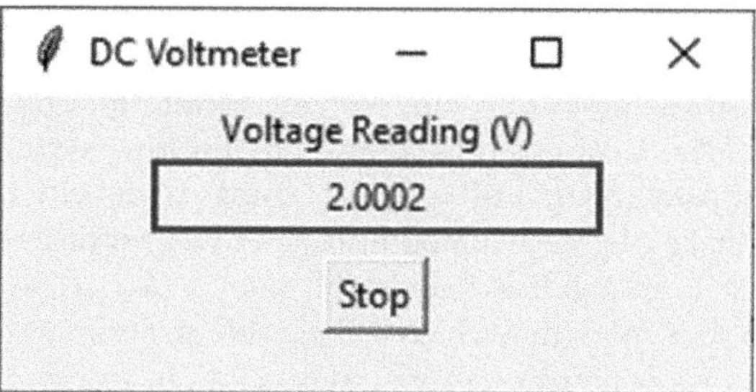

The updates are due to the `while` loop within `acquire_sample()`, which ceases execution when the **Stop** button is pressed. The `return` statement then causes program control to exit `acquire_sample()` and, at that point, the thread upon which `acquire_sample()` was executing is terminated. With the next button-press, a new thread is created upon which `acquire_sample()` resumes execution.

We should pause for a moment to discuss one subtle (but important) feature of this program. As is, the `while` loop is "free running" (i.e., executing as fast as possible), and because of this rapid execution, your computer's processor is having to devote a significant amount of its efforts to running this program, leaving little time for the other operations for which it is responsible, a situation termed *CPU hogging*. To avoid this problem, simply insert a small delay in each `while`-loop iteration. Python's `time` module (included in the Python Standard Library) has the `sleep(delay)` function, which causes the program execution to pause for the time interval `delay` (in units of seconds).

Include a delay of 100 ms in **dc_voltmeter_ni_live.py** between each voltage sample as follows: Place `import time` with the other import statements at the top of the program and add the `time.sleep()` statement in `acquire_sample()` as follows:

```
def acquire_sample():

    while btn.cget('text') == 'Stop':
        float_data = task.read()
        string_data = f'{float_data:.4f}'

        lbl_2.config(text=string_data)

        time.sleep(0.1)

    return
```

Remember to include `import time` at the top of your program. Save your work and then run **dc_voltmeter_ni_live.py**. The voltage updates will still occur plenty fast, ten times a second.

7.5 Hardware-Timed Digital Oscilloscope

In the previous exercise, `task.read()` had no arguments within its parentheses so that, with each execution, it acquired a single voltage reading of the analog signal at a channel input. By putting this command within a repetitive `while` loop and controlling the time per loop iteration with `time.sleep()`, you wrote **dc_voltmeter_live.py**, a program that, with the simple addition of some data storage and plotting capability, one might consider using to acquire and display N data samples of a time-varying waveform. The last sentence outlines the schematic for one of the most useful laboratory monitoring systems—the *digital oscilloscope*. Perhaps you've already thought of some cool things to do with this idea and are ready to go back and start to make the needed modifications to **dc_voltmeter_live.py**..., but unfortunately `time.sleep()` is an Achilles' heel in this plan. `time.sleep()` measures time by accessing a clock within your computer, which will provide an accuracy of somewhere between 1 and 15 millisecond (ms), depending on your system. Thus, by using `time.sleep()` to mark time in the data-taking process, the moment of each voltage digitization will be at best determined with an uncertainty on the order of 0.001 second. For very-low-frequency inputs (say, less than 10 Hz), this level of uncertainty in *time*-axis values might possibly be acceptable for the envisioned oscilloscope's *voltage* versus *time* output, but it will be useless over the higher range of frequencies in which you would want an oscilloscope to operate.

Thankfully, a much better clock called the *sample clock*, with precision timing on the order of 10 nanoseconds (12.5 ns for NI USB-6002), exists on a NI DAQ device. And, even better, commands in the nidaqmx library provide access and control of the sample clock, allowing you to accurately schedule the moment each sample is acquired (or generated) in your experiment. Data-acquisition operations under the control of the sample clock are termed *hardware-timed*, while the scheme outlined above involving the `sleep()` function is called *software-timed*.

Let's explore control of the DAQ device's sample clock as we write a useful digital oscilloscope program. The program we wish to write, called **digital_oscilloscope_ni.py**, will acquire N equally spaced voltage samples of a time-varying analog input and then quickly plot this array of data values. By repeating this process over and over, we'll achieve a real-time display of the waveform input. In writing **digital_oscilloscope_ni.py**, we'll be drawing on components of two previous programs—**waveform_simulator.py** and **dc_voltmeter_ni_live.py**.

In the Editor, use **File>>New File** to create a program named **digital_oscilloscope_ni.py** and save it in **YourNamePythonWork**. Enter the following code, which creates an operating-system window named **Digital Oscilloscope** and places a **Digitizing Parameters** control within a `control_box` frame on the left-hand side of the window and a Matplotlib plot on the right-hand side of the window. A button labeled **Start** is placed below the **Digitizing Parameters** control; for now, make `action=None`. Many of these commands are in **waveform_simulator.py**, so you may wish to copy-and-paste relevant commands from that program.

```
import Utilities.waveform_creator as wc
import Utilities.gui_creator as gc
import Utilities.plot_creator as pc

label_x, label_y = 'Time (ms)', 'Voltage (V)'
```

```
root = gc.create_window(title='Digital Oscilloscope',
                        size='1000x500')
control_box, right_frame = gc.create_frames(root)

var_num, var_samp_freq = gc.digitizing_control(control_box)

ax, canvas = pc.create_plot(right_frame)
pc.update_plot(ax, canvas, [], [], label_x, label_y)

btn = gc.action_button(control_box, label='Start',
                       action=None)

root.mainloop()
```

Save your work and then run the program. If all is coded correctly, the program's GUI should appear (with a blank data plot).

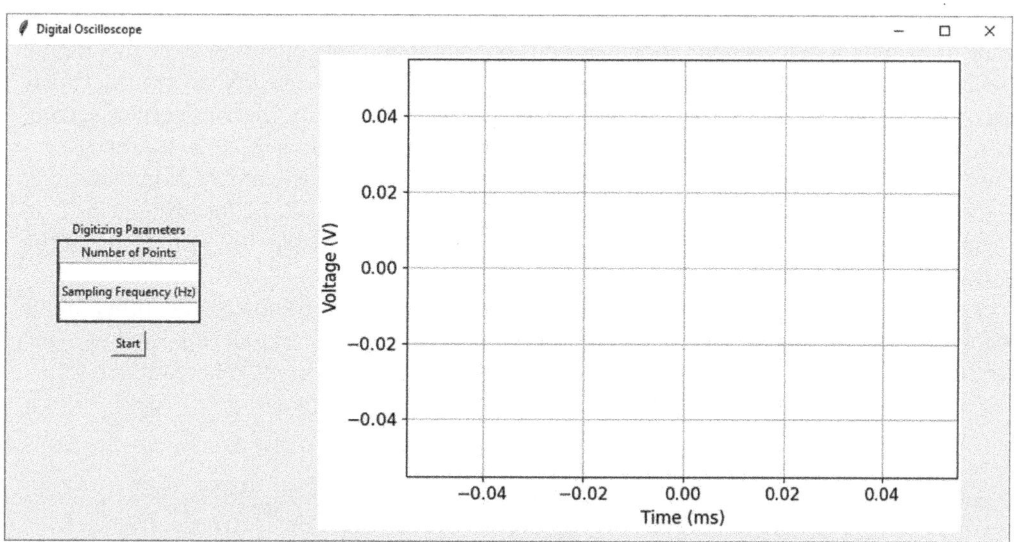

Next, import the `nidaq_operations` module and include the function `task = no.create_ai_task(channel_name)` to create an analog input voltage task and `task.close()` to close that task when done as shown next.

```
import Utilities.waveform_creator as wc
import Utilities.gui_creator as gc
import Utilities.plot_creator as pc
import Utilities.nidaq_operations as no

channel_name = 'Dev1/ai0'
task = no.create_ai_task(channel_name)

label_x, label_y = 'Time (ms)', 'Voltage (V)'
```

```
root = gc.create_window(title='Digital Oscilloscope',
                        size='1000x500')
control_box, right_frame = gc.create_frames(root)

var_num, var_samp_freq = gc.digitizing_control(control_box)

ax, canvas = pc.create_plot(right_frame)
pc.update_plot(ax, canvas, [], [], label_x, label_y)

btn = gc.action_button(control_box, label='Start',
                       action=None)

root.mainloop()

task.close()
```

We now have to decide what we would like to have happen when the **Start** button is pressed. Following what we did in **dc_voltmeter_ni_live.py**, let's have the button label toggle to **Stop**, while launching a new thread that carries out our intended DAQ operation. For this digital oscilloscope program, a function named `acquire_waveform()` will carry out the DAQ operation of acquiring *N* data samples of a time-varying waveform, displaying those acquired data, then repeating that process until the **Stop** button is pressed. Within `acquire_waveform()`, `create_time_array()` from **waveform_creator.py** creates a `num`-element NumPy array `t` of the equal-spaced time values at which data sampling occurs and, by inserting `num` as an argument in the `read()` function, `task.read(num)` acquires those `num` samples from the DAQ device and places them in a NumPy array `y`.

Using what we learned in writing **dc_voltmeter_ni_live.py**, complete the program by changing `action=None` to `action=toggle`, importing the `threading` module, and defining the two functions `toggle()` and `acquire_waveform()`. The function `config_daq_timing()` is a placeholder for now, but will be defined soon.

```
import Utilities.waveform_creator as wc
import Utilities.gui_creator as gc
import Utilities.plot_creator as pc
import Utilities.nidaq_operations as no
import threading

channel_name = 'Dev1/ai0'
task = no.create_ai_task(channel_name)

label_x, label_y = 'Time (ms)', 'Voltage (V)'

root = gc.create_window(title='Digital Oscilloscope',
                        size='1000x500')
control_box, right_frame = gc.create_frames(root)

var_num, var_samp_freq = gc.digitizing_control(control_box)
```

```
ax,canvas = pc.create_plot(right_frame)
pc.update_plot(ax, canvas, [], [], label_x, label_y)

def acquire_waveform():

    while btn.cget('text') == 'Stop':

        num, samp_freq = gc.get_dig_entries(var_num,
                                            var_samp_freq)

        no.config_daq_timing(task, num, 'Finite',
                             samp_freq)

        t = wc.create_time_data(num, samp_freq)
        y = task.read(num)

        pc.update_plot(ax, canvas, 1000*t, y,
                       label_x, label_y, marker='o-')

    return

def toggle():

    if btn.cget('text') == 'Start':
        btn.config(text='Stop')
    elif btn.cget('text') == 'Stop':
        btn.config(text='Start')

    t = threading.Thread(target=acquire_waveform)
    t.start()

    return

btn = gc.action_button(control_box, label='Start',
                       action=toggle)

root.mainloop()

task.close()
```

Remember to place `import threading` module at the top of your program, then save your work.

Finally, in the Editor, open **nidaq_operations.py** in **YourNamePythonWork>>Utilities**, then add the following definition for the function `config_daq_timing()`. Here, `cfg_samp_clk_timing()` is a method from the nidaqmx library's `timing` module, which configures the DAQ device's sample clock to work in one of two modes—finite or continuous—by assigning the argument `sample_mode` to the attribute `AcquisitionType.FINITE` or `AcquisitionType.CONTINUOUS`, respectively. The two possible values for this attribute are found in a module within nidaqmx called `constants`. We will learn about the continuous mode later when creating an AO-based waveform generator, but the present program will operate in the finite mode. In this mode, `cfg_samp_clk_timing ()` sets the sampling frequency at which a finite number of samples is to be acquired in the given task (in this case, an analog input voltage operation):

```
def config_daq_timing(task, num, mode, samp_freq):

    if mode == 'Finite':
        option=nidaqmx.constants.AcquisitionType.FINITE
    elif mode == 'Continuous':
        option=nidaqmx.constants.AcquisitionType.CONTINUOUS

    task.timing.cfg_samp_clk_timing(rate=samp_freq,
                                    sample_mode=option,
                                    samps_per_chan=num)

    return
```

The complete **nidaq_operations.py** module will appear as shown next:

```
import nidaqmx

def create_ai_task(channel_name):

    task = nidaqmx.Task()
    task.ai_channels.add_ai_voltage_chan(channel_name)

    return task


def config_daq_timing(task, num, mode, samp_freq):

    if mode == 'Finite':
        option=nidaqmx.constants.AcquisitionType.FINITE
    elif mode == 'Continuous':
        option=nidaqmx.constants.AcquisitionType.CONTINUOUS

    task.timing.cfg_samp_clk_timing(rate=samp_freq,
                                    sample_mode=option,
                                    samps_per_chan=num)

    return
```

Save your work and close **nidaq_operations.py**.

The N-samples acquisition task we have just programmed is termed a *buffered hardware-timed analog input* operation. In this process, a time-dependent analog signal is digitized at N equally spaced times. The time interval between neighboring samples $\Delta t = 1/f_s$, where f_s is the sampling frequency, is very accurately controlled by a hardware sample clock on the DAQ device. As the data samples are acquired, they are first placed in a memory buffer on the DAQ device and then transferred from there into the computer's memory. This procedure ensures that no data samples are lost when the computer multitasks (i.e., switches among its various chores) during the data-taking process.

Let's test-drive our handywork. Adjust the settings on a function generator so that it is outputting an analog sine wave with amplitude less than 10 V and a frequency of about 50 Hz. Connect the positive and negative (ground) outputs from the function generator to the positive and negative inputs of differential channel *ai0* on your DAQ device. *Note*: Most function generators are referenced to building ground and so are not "floating" voltage sources. Run **digital_oscilloscope_ni.py**. In the **Digitizing Parameters** control, set **Number of Samples** and **Sampling Frequency (Hz)** equal to *100* and *1000*, respectively, then press the **Start** button. If all goes well, you will see a plot of about five cycles of the 50-Hz sine wave.

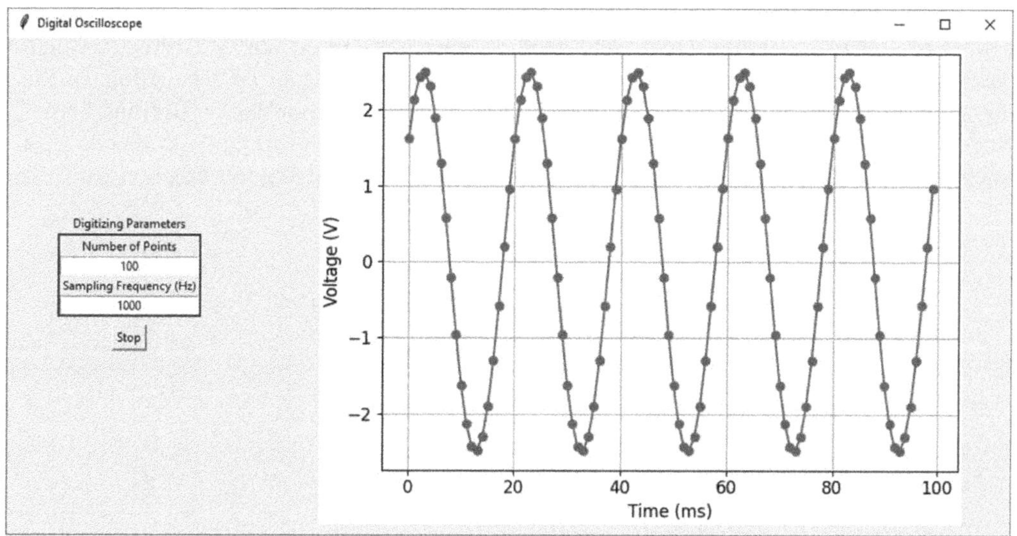

Be impressed that your program successfully acquires and plots a succession of digitized sine waves. However, I'm sure that one flaw in this computer-based instrument is apparent to you. Most likely, the sine-wave plot you are observing does not appear stationary but instead is moving either to the right or to the left, or is just shifting back and forth in a jittery fashion. And, by slightly changing the function generator's sine-wave frequency, the plotted sine wave can be made to move first one direction and then the other. Why is this so?

In your program, `task.read()` is configured so that with each execution, it outputs an array of N equally spaced data samples. Let's call this array of N samples a *trace*. As you know from coding the program, the sine-wave image viewed on the GUI's plot is not just a single trace but is actually a sequence of many traces, where a new trace is generated and plotted with each iteration of the `while` loop. Since nothing in your code forces each trace to begin at an equivalent point on the repetitive incoming waveform, the successive traces are displaced from each other along the *time*-axis, resulting in a motion picture-style moving image or, worse, a jumbled mess. The issue we are broaching here is called *triggering* and, for **digital_oscilloscope_ni.py** to be a useful program, we must build in this capability.

In a commercially available oscilloscope, triggering is accomplished in the following manner. The input signal is monitored by an analog *level-crossing* circuit. The purpose of this circuit is to determine each time the incoming signal passes through a specified threshold voltage level and then immediately trigger the scope's data-acquisition process. Using knobs on the scope's front panel, the oscilloscope user sets the circuit's threshold level and specifies whether acquisition should be initiated when the level is passed through, starting from above (*negative*, or *falling, slope*) or starting from below (*positive*, or *rising, slope*).

Many higher-end NI DAQ devices are capable of performing the analog level-crossing triggering procedure just described; the lower-cost NI USB-6002 does not have this capability. However, we will be able to implement an alternate mode—*digital edge triggering*—in our program because this mode is available on (almost) all NI DAQ devices. At the time of this writing, the exception to this rule is the myDAQ.

To execute a data-acquisition operation under digital edge triggering, one connects a digital signal to the appropriate pins on a DAQ device. Then, by choosing the proper software settings, the desired data-acquisition operation is initiated ("triggered") whenever the digital signal has a rising or, alternatively, falling edge.

Stop **digital_oscilloscope_ni.py**, if it's running. We have a bit of work to do to configure it for digital edge triggering. First, we'll need an option menu on our GUI through which a user can select the desired (rising or falling) digital edge to trigger an acquisition. In the Editor, use **File>>Open**... to open **gui_creator.py** in **YourNamePythonWork>>Utilities**. Somewhere among the other definitions, add the following definition for `trigger_control()`, a function that creates a blue-bordered option menu labeled **Digital Trigger Edge** on our GUI:

```
def trigger_control(container):

    lbl = tk.Label(container, text='Digital Trigger Edge')
    lbl.pack(fill=tk.X)

    frame = tk.Frame(container, bg='blue', bd=2)
    frame.pack()

    var_trig = tk.StringVar()
    options = ['Falling', 'Rising']
    opm = tk.OptionMenu(frame, var_trig, *options)
    opm.config(width=8)
    opm.pack(fill=tk.X)

    return var_trig
```

Also, in **gui_creator.py**, add the following definition for `get_edge_option()`, a function that fetches the selection entered into the **Digital Trigger Edge** option menu:

```
def get_edge_option(var_trig):

    edge_option = var_trig.get()

    return edge_option
```

Save your work and close **gui_creator.py**.

Next, in the Editor, open **nidaq_operations.py** in **YourNamePythonWork>>Utilities**. Add the definition for the function `config_daq_triggering()` to start digital triggering. The required method `start_trigger.cfg_dig_edge_start_trig()` comes from the nidaqmx library's `triggers` module, which sets the desired digital triggering edge on the selected trigger terminal ("pin") of your DAQ device; note the placement of the leading parenthesis for proper line wrapping of this long-named function. The *if... else* statement chooses the nidaqmx-appropriate constant for a rising or falling edge from the `constants` submodule of `nidaqmx`:

```
def config_daq_triggering(task, trig_pin, edge_option):

    if edge_option == 'Rising':
        option = nidaqmx.constants.Edge.RISING
    else:
        option = nidaqmx.constants.Edge.FALLING

    task.triggers.start_trigger.cfg_dig_edge_start_trig(
                        trig_pin, trigger_edge=option)

    return
```

Save your work and close **nidaq_operations.py**.

Finally, a few additions to **digital_oscilloscope_ni.py** are needed to incorporate digital triggering into its operations. First, here's some background needed to understand these changes. A few of the digital input/output lines on a NI DAQ device have additional capabilities beyond that of regular DIO lines and are denoted as programmable function interface (PFI) lines. One of the additional capabilities of a PFI line is the ability to digitally trigger a data-acquisition operation (other capabilities include clock synchronization, counter input, and counter output). The NI USB-6002 device has two PFI lines named PFI 0 and PFI 1, which correspond to the DIO lines P2.0 and P1.1, respectively. To identify the terminal ("pin") on the DAQ device of the PFI line we wish to use, the syntax is `/device_name/terminal_name`. For example, PFI 0 on Device 1 is denoted as `/Dev1/PFI0`. Note the important leading slash, which differs from the syntax for referencing a data-acquisition channel (e.g., `Dev1/ai0`). The lack of a leading slash for a data-acquisition channel reflects the fact that *ai0* is not a single terminal on a DAQ device. For example, on the NI USB-6002, channel *ai0* consist of two terminals—pin AI 0 and AI 4.

In the Editor, open **digital_oscilloscope_ni.py** in **YourNamePythonWork**. Make the following changes: Along with the defining `channel_name` near the top of the program, define the trigger terminal as PFI 0:

```
channel_name = 'Dev1/ai0'
trig_pin = '/Dev1/PFI0'
task = no.create_ai_task(channel_name)
```

Next, along with creating a **Digitizing Parameters** control on the GUI, create a **Digital Trigger Edge** option menu:

```
var_num, var_samp_freq = gc.digitizing_control(control_box)
var_trig = gc.trigger_control(control_box)
```

Finally, in the code block within the `while` loop of the function `acquire_waveform()`, add `get_edge_option()` to fetch the edge option selected in the **Digital Trigger Edge** option menu as well as `config_daq_triggering()` to set the selected edge option on the DAQ device:

```
num, samp_freq = gc.get_dig_entries(var_num,
                                   var_samp_freq)
edge_option = gc.get_edge_option(var_trig)

no.config_daq_timing(task, num, 'Finite',
                     samp_freq)
no.config_daq_triggering(task, trig_pin,
                         edge_option)

t = wc.create_time_data(num, samp_freq)
y = task.read(num)

pc.update_plot(ax, canvas, 1000*t, y,
               label_x, label_y, marker='o-')
```

The complete **digital_oscilloscope_ni.py** program will now appear as follows:

```
import Utilities.waveform_creator as wc
import Utilities.gui_creator as gc
import Utilities.plot_creator as pc
import Utilities.nidaq_operations as no
import threading

channel_name = 'Dev1/ai0'
trig_pin = '/Dev1/PFI0'
task = no.create_ai_task(channel_name)

label_x, label_y = 'Time (ms)', 'Voltage (V)'

root = gc.create_window(title='Digital Oscilloscope',
                        size='1000x500')
control_box, right_frame = gc.create_frames(root)

var_num, var_samp_freq = gc.digitizing_control(control_box)
var_trig = gc.trigger_control(control_box)

ax,canvas = pc.create_plot(right_frame)
pc.update_plot(ax, canvas, [], [], label_x, label_y)

def acquire_waveform():

    while btn.cget('text') == 'Stop':

        num, samp_freq = gc.get_dig_entries(var_num,
                                           var_samp_freq)
        edge_option = gc.get_edge_option(var_trig)
```

```
        no.config_daq_timing(task, num, 'Finite',
                             samp_freq)

        no.config_daq_triggering(task, trig_pin,
                                 edge_option)

        t = wc.create_time_data(num, samp_freq)
        y = task.read(num)

        pc.update_plot(ax, canvas, 1000*t, y,
                       label_x, label_y, marker='o-')

    return

def toggle():

    if btn.cget('text') == 'Start':
        btn.config(text='Stop')
    elif btn.cget('text') == 'Stop':
        btn.config(text='Start')

    t = threading.Thread(target=acquire_waveform)
    t.start()

    return

btn = gc.action_button(control_box, label='Start',
                       action=toggle)

root.mainloop()

task.close()
```

That's it. **digital_oscilloscope_ni.py** is now configured for digital edge triggering. Save your work.

Besides producing a sine wave with frequency *f* at its *main* output, your function generator creates a TTL digital signal of the same frequency *f*, which is available at its *sync* (sometimes called *TTL* or *trigger out*) output. The digital edges of this TTL signal coincide in time with particular points on the sine wave's cycle. For example, on some function generator models, the rising edge of the TTL signal occurs when the sine wave is at its peak; for other models, the rising TTL edge coincides with the positive-going zero-crossing of the sine wave. Thus, the sync output provides a convenient digital signal with which one can control the acquisition of sine-wave data (i.e., always begin at the same equivalent point on the cycle) via digital edge triggering.

Connect the sync output from your function generator to the digital triggering pins on your DAQ device. The sync's positive and negative (ground) terminals should be connected to the PFI 0 and D GND pins, respectively, for the software selection we made in our programming.

Run **digital_oscilloscope_ni.py** with **Number of Samples**, **Frequency (Hz)**, and **Digital Trigger Edge** equal to *100*, *1000*, and *Rising*, respectively, and set the sine-wave frequency on your function generator to approximately 50 Hz. If all goes well, you will see a "stationary" plot of about five cycles of the 50-Hz sine wave. Try switching **Digital Trigger Edge** to *Falling*. Is the resulting change to the waveform plot as you expected?

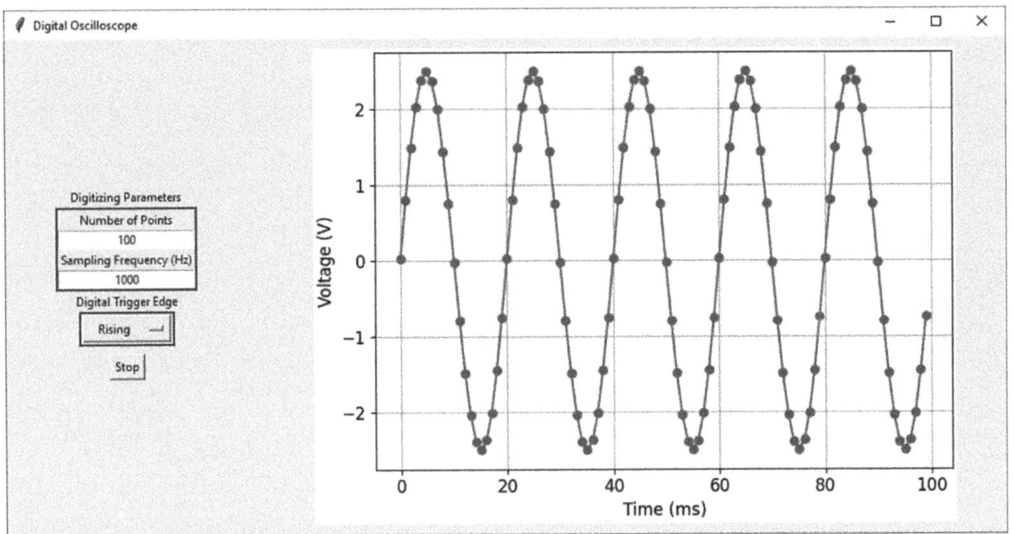

On the function generator, set the frequency f of the input sine-wave signal to several values within the range $0 \leq f \leq 500$ Hz. You should find that frequencies of signals in this frequency range are faithfully reproduced by **digital_oscilloscope_ni.py**, but that sine-wave shape becomes distorted as the signal frequency approaches 500 Hz (because of the paucity of sampled points per sine-wave cycle at these higher input frequencies).

We have set the sampling frequency equal to $f_s = 1000$ Hz; thus, the Nyquist frequency is $f_{nyquist} = f_s/2 = 500$ Hz. For input frequencies greater than the Nyquist frequency, that is, $f > 500$ Hz, we expect the aliasing effect.

With **Number of Samples, Frequency (Hz)**, and **Digital Trigger Edge** equal to *100*, *1000*, and *Rising*, respectively, demonstrate that for $f = 950$ and $f = 1050$ Hz, the digitized sine wave appears to have a frequency of 50 Hz, rather than the actual input frequencies. Also, note that there is a 180° for one of these aliased frequencies (remember digital triggering is on the rising edge of the input frequency). Use Eq. (6.2) with $n = 2$ to predict two other input frequencies f that will appear as a 50-Hz sine wave when digitized because of aliasing, and then demonstrate that your predictions are correct using **digital_oscilloscope_ni.py**.

7.6 DC Voltage Source

Now let's explore the analog output operations available on your DAQ device. To familiarize ourselves with these processes, we'll start by writing a program that simply generates ("writes") a single requested voltage value. This program will implement nidaqmx's `task.write()` command and follow the abbreviated template of sequential steps required to execute a DAQmx task: *create and configure task, write data, clear task*. The abbreviated template is allowed here because `task.write()` automatically executes the `task.start()` and `task.stop()` commands when only a single sample is being written. Here are the Python commands from the nidaqmx library that code the task of simply outputting a single requested voltage value, in this case, 1.0 V:

```
import nidaqmx

channel_name = 'Dev1/ao0'

task = nidaqmx.Task()
task.ao_channels.add_ao_voltage_chan(channel_name)

task.write(1.0)

task.close()
```

Note that this code is very similar to the sample-reading program **simple_ai_operation_ni.py** we wrote previously, except this time every `ai` is replaced by an `ao` and the `read()` method is replaced by the `write()` method.

To streamline coding of our AO program, in the Editor open **simple_ai_operation_ni.py**, then use **File>>Save As**... to create a program named **simple_ao_operation_ni.py** in **YourNamePythonWork**. Next, modify the code to match that shown above and save your work.

Connect the positive and negative (ground) inputs of a voltmeter (or oscilloscope in autotriggering mode) to the AO 0 and AO GND pins of your DAQ device, respectively.

Run **simple_ao_operation_ni.py** and your voltmeter should inform you that the requested voltage difference is being produced by your DAQ device. Try running **simple_ao_operation_ni.py** with it programmed to output a few other values within the allowed range of your DAQ device (−10 to +10 V for the NI USB-6002) to convince yourself that the program is functioning properly.

It would be helpful to have an interactive way to input requested voltage values in this program; in other words, have a GUI. To accomplish that feat, we'll need to define a few new functions in the **gui_creator.py** and **nidaq_operations.py** modules, so let's do that first.

To input a requested voltage value, our GUI will need a solitary control, that is, a single entry widget with an identifying label, along with the means of retrieving its user-entered value. In Chapter 5's Do It Yourself project we added two functions named `create_control()` and `get_control_string()` to the **gui_creator.py** module that provide this needed functionality, so we will make use of that previous work. If you didn't produce these two functions in Chapter 5, do so now following the instructions in the next paragraph.

In the Editor, open **gui_creator.py** in **YourNamePythonWork>>Utilities**. Then, add the following definition for the function `create_control()`. This function first places a descriptive `label` and a blue-bordered `frame` within the specified `container` (e.g., the `root` window), then places an entry widget named `ent` within the `frame`. The value that a user has entered into the entry widget is returned in the string variable `var_ent`.

```
def create_control(container, label, char_width=20):

    lbl = tk.Label(container, text=label)
    lbl.pack()

    frame = tk.Frame(container, bg='blue', bd=2)
    frame.pack()
```

```
    var_ent = tk.StringVar()
    ent = tk.Entry(frame, textvariable=var_ent,
                   justify=tk.CENTER, width=char_width)
    ent.pack(fill=tk.X)

    return var_ent
```

Also, in **gui_creator.py**, define the following function called `get_control_string(var_ent)`, which fetches the string a user has entered in the entry widget:

```
def get_control_string(var_ent):

    control_string = var_ent.get()

    return control_string
```

Save your work and close **gui_creator.py**.

Next, we need a function that creates an analog output task. Since this is a DAQ-related function, in the Editor, open **nidaq_operations.py** in **YourNamePythonWork>>Utilities**. Then, add the following definition for the function `create_ao_task(channel_name)`, which creates a task then configures it as an AO operation on `channel_name` (i.e., a specified channel on a specified DAQ device). This function returns the `task` object. The commands in `create_ao_task()` are the same as in `create_ai_task()`, except `ao` replaces `ai`. Consider a copy-and-paste of `create_ai_task()` as a starting point for coding this AO function.

```
def create_ao_task(channel_name):

    task = nidaqmx.Task()
    task.ao_channels.add_ao_voltage_chan(channel_name)

    return task
```

Save your work and close **nidaq_operations.py**.

We're now ready to create a GUI-equipped DC voltage source. This program will be very similar to the **dc_voltmeter_ni.py** program that you wrote earlier in this chapter, so I recommend the following: In the Editor, open **dc_voltmeter_ni.py** in **YourNamePythonWork**, use **File>>Save As...** to create a program named **dc_voltage_source_ni.py** in **YourNamePythonWork**, then modify the program as shown next. As part of the "write-data" operation, the `float()` command converts the entry-widget string to a floating-point number appropriate for the nidaqmx-related `task.write()` command.

```
import Utilities.gui_creator as gc
import Utilities.nidaq_operations as no

root = gc.create_window('DC Voltage Source',
                        size='300x100')

channel_name = 'Dev1/ao0'
task = no.create_ao_task(channel_name)
```

```
var_ent = gc.create_control(root,
                            label='Input Voltage')

def generate_sample():

    string_data = gc.get_control_string(var_ent)
    float_data = float(string_data)

    task.write(float_data)

    return

btn = gc.action_button(root, label='Write Sample',
                       action=generate_sample)

root.mainloop()

task.close()
```

Save your work. Now run **dc_voltage_source_ni.py**, entering a voltage within the allowed range of your DAQ device's AO capability.

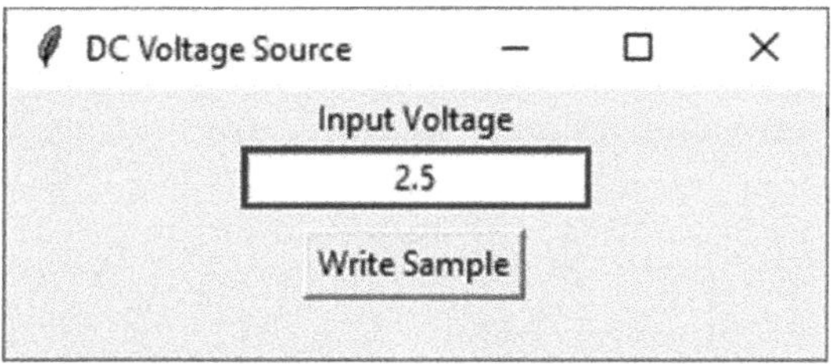

Press the **Write Sample** button and use a voltmeter to verify that the requested voltage difference is being output on channel *ao0* (i.e., between pins AO 0 and AO GND). Try outputting several other positive and negative values using **dc_voltage_source_ni.py**.

When the execution of **dc_voltage_source_ni.py** is stopped (by closing the **DC Voltage Source** window), your NI DAQ device is designed to maintain the last voltage value it generated. Thus, before stopping the program, press **Write Sample** one more time with **Input Voltage** set to *0*, so the channel *ao0* is left in the zero-voltage state.

7.7 Hardware-Timed Waveform Generator

By continually updating an AO channel with appropriately chosen voltage values, one can turn a DAQ device into a waveform generator with a time-dependent voltage output of any desired shape (e.g., sine wave, square wave, triangle wave). To obtain an output voltage whose time dependence accurately represents that desired shape, one must ensure that the AO channel changes ("updates") to the correct voltage values at precisely the right moments. The most accurate method for timing these required updates involves the use of a high-frequency hardware clock and such a clock (i.e., the sample clock) is contained internally on every NI DAQ device. For most NI DAQ devices (including the USB-6002), the sample clock can be used for hardware-timed AO operations and so these DAQ devices can be used to create high-quality analog waveforms (the USB-6009 legacy device does not have the capability for hardware-timed AO operations, and so only the much less accurate method of software-timed waveform generation is possible on that device; see Problem 3).

To understand the command sequence needed for hardware-timed waveform generation, we'll start by writing a GUI-less program that simply produces a 50-Hz sine wave with amplitude of 5 V. To produce this waveform, we will use the maximum update rate available on the NI USB-6002, which is 5 kHz. This program will follow the standard template of sequential steps required to execute a DAQmx task, this time to output ("write") data: *create and configure task, start task, write data, stop task, clear task.*

In the Editor, use **File>>New File** and **File>>Save** to create and save a new program named **simple_ao_waveform_ni.py** in **YourNamePythonWork** and enter the following sequence of commands. Here, we use functions from our **waveform_creator.py** module to create a 500-element NumPy array representing five cycles of a 50-Hz sine wave. In addition, from **nidaq_operations.py**, an AO task is created using `create_ao_task()` and that task is configured as a *hardware-timed continuous analog output* operation (see explanation below) of the 500-element NumPy array at a sampling frequency of 5000 Hz using `config_daq_timing()`. The `write()` method writes this sine-wave array to the DAQ device and then, via the `auto_start=True` option, starts the AO process. In the continuous mode, this sine wave will be generated until `task.stop()` is executed. Thus, `time.sleep(10.)` introduces a 10-second delay between the start and stop commands. Finally, `task.close()` clears the task.

```
import Utilities.waveform_creator as wc
import Utilities.nidaq_operations as no
import time

channel_name = 'Dev1/ao0'
task = no.create_ao_task(channel_name)

num, samp_freq = 500, 5000
a, f, shape = 5., 50., 'Sine'

t = wc.create_time_data(num, samp_freq)
y = wc.create_waveform_data(t, a, f, shape)

no.config_daq_timing(task, num, 'Continuous', samp_freq)

task.write(y, auto_start=True)

time.sleep(10.)

task.stop()

task.close()
```

Save your work. Connect the positive and negative (ground) inputs of an oscilloscope to the AO 0 and AO GND pins of your DAQ device, respectively. Set your scope to the **Normal** or **Auto** triggering mode and the **Trigger Level** to somewhere within the range of the sine-wave variation (i.e., between −5 and +5 V). Run **simple_ao_waveform_ni.py** and, if all is coded correctly, you will view a 50-Hz sine wave continuously generated for 10 seconds.

The waveform you observe on the oscilloscope is the result of programming a hardware-timed continuous analog output process. In this process, a 1D array consisting of a sequence of voltage values of the desired waveform shape is first written from computer memory into a dedicated memory buffer on your DAQ device. Then, when initiated by receipt of the start command, the DAQ device outputs this sequence of voltages at channel *ao0* in a *circular fashion* (i.e., after the last array value is outputted, the next output is the first array value, and the sequence repeats). The circular output of buffer values continues until the `task.stop()` command is received.

Let's now write a program named **waveform_generator_ni.py**, which equips this hardware-timed analog output operation with a GUI. The **waveform_generator_ni.py** program will follow a similar template to **digital_oscilloscope_ni.py**, so you can streamline the necessary coding by using that already-coded program as a starting point.

In the Editor, open **digital_oscilloscope_ni.py**, then use **File>>Save As**... to create a program named **waveform_generator_ni.py** and save it in **YourNamePythonWork**. Modify the task-creation and GUI-creation code so that it appears as shown below. Here, some labeling has been changed and the `trigger_control()` creation function has been replaced by the function that creates a **Waveform Parameters** control within the `control_box` frame.

```
import Utilities.waveform_creator as wc
import Utilities.gui_creator as gc
import Utilities.plot_creator as pc
import Utilities.nidaq_operations as no
import threading

channel_name = 'Dev1/ao0'
task = no.create_ao_task(channel_name)

label_x, label_y = 'Time (ms)', 'Voltage (V)'

root = gc.create_window(title='Waveform Generator',
                        size='1000x500')
control_box, right_frame = gc.create_frames(root)

var_num, var_samp_freq = gc.digitizing_control(control_box)
var_a, var_f, var_shape = gc.waveform_control(control_box)

ax, canvas = pc.create_plot(right_frame)
pc.update_plot(ax, canvas, [], [], label_x, label_y)
```

Next, modify **digital_oscilloscope_ni.py**'s `acquire_waveform()` function to become the following definition of a function called `generate_waveform()`. Here, the code will be familiar from **simple_ao_waveform_ni.py**.

```
def generate_waveform():

    num, samp_freq = gc.get_dig_entries(var_num,
                                        var_samp_freq)
    a, f, shape = gc.get_wav_entries(var_a, var_f,
                                     var_shape)
```

```
        t = wc.create_time_data(num, samp_freq)
        y = wc.create_waveform_data(t, a, f, shape)

        pc.update_plot(ax, canvas, 1000*t, y,
                       label_x, label_y, marker='o-')

        no.config_daq_timing(task, num, 'Continuous',
                             samp_freq)

        task.write(y, auto_start=True)

        return
```

Finally, revise **digital_oscilloscope_ni.py**'s `toggle()` function as shown next. Here, pressing the **Start** button will spawn a thread named `t` upon which the function `generate_waveform()` executes. Through its execution, `generate_waveform()` loads the desired waveform data into the DAQ device's on-board buffer and starts the continuous analog output operation. With its `return` statement, `generate_waveform()` has then completed its execution and so the thread `t` terminates. However, the AO operation continues until the `task.stop()` command is executed through a press of the **Stop** button. The **Stop** button morphs into a **Start** button and the process can be repeated. Throughout, the GUI code runs on the main thread initiated when the program begins its execution.

```
def toggle():

    if btn.cget('text') == 'Start':
        btn.config(text='Stop')
        t = threading.Thread(target=generate_waveform)
        t.start()
    elif btn.cget('text') == 'Stop':
        btn.config(text='Start')
        task.stop()

    return
```

Save your work. The complete **waveform_generator_ni.py** should appear as follows:

```
import Utilities.waveform_creator as wc
import Utilities.gui_creator as gc
import Utilities.plot_creator as pc
import Utilities.nidaq_operations as no
import threading

channel_name = 'Dev1/ao0'
task = no.create_ao_task(channel_name)

label_x, label_y = 'Time (ms)', 'Voltage (V)'
```

```
root = gc.create_window(title='Waveform Generator',
                        size='1000x500')
control_box, right_frame = gc.create_frames(root)

var_num, var_samp_freq = gc.digitizing_control(control_box)
var_a, var_f, var_shape = gc.waveform_control(control_box)

ax,canvas = pc.create_plot(right_frame)
pc.update_plot(ax, canvas, [], [], label_x, label_y)

def generate_waveform():

    num, samp_freq = gc.get_dig_entries(var_num,
                                        var_samp_freq)
    a, f, shape = gc.get_wav_entries(var_a, var_f,
                                     var_shape)

    t = wc.create_time_data(num, samp_freq)
    y = wc.create_waveform_data(t, a, f, shape)

    pc.update_plot(ax, canvas, 1000*t, y,
                   label_x, label_y, marker='o-')

    no.config_daq_timing(task, num, 'Continuous',
                         samp_freq)

    task.write(y, auto_start=True)

    return


def toggle():

    if btn.cget('text') == 'Start':
        btn.config(text='Stop')
        t = threading.Thread(target=generate_waveform)
        t.start()
    elif btn.cget('text') == 'Stop':
        btn.config(text='Start')
        task.stop()

    return


btn = gc.action_button(control_box, label='Start',
                       action=toggle)

root.mainloop()

task.close()
```

Connect the positive and negative (ground) inputs of an oscilloscope to the AO 0 and AO GND pins of your DAQ device, respectively. Set your scope to the **Normal** or **Auto** triggering mode and the **Trigger Level** to somewhere within the range between −5 and +5 V.

Test-drive **waveform_generator_ni.py** by using it to output a 50-Hz sine wave of amplitude 5 V. First, in the **Waveform Parameters** control, set **Amplitude, Frequency (Hz)**, and **Shape** to be *5*, *50*, and *Sine*, respectively. Next, assuming 100 points are required to well define each cycle of the sine wave, the sampling frequency should be 100 times larger than the sine wave's frequency. Thus, for **waveform_generator_ni.py** to create one complete cycle, in the **Digitizing Parameters** control, set **Number of Samples** and **Sampling Frequency (Hz)** to be *100* and *5000*, respectively.

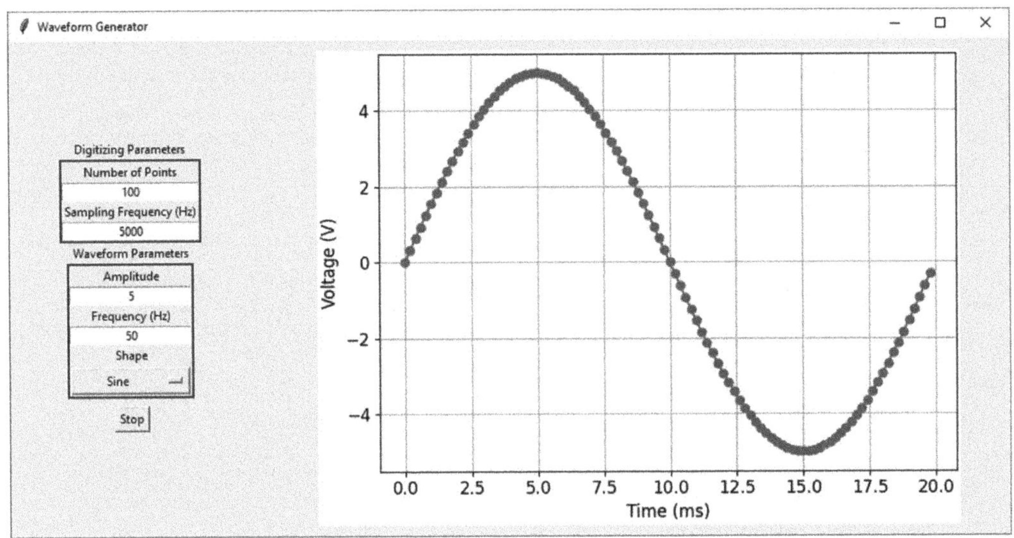

Press the **Start** button. The software waveform produced by `create_waveform_data()` will be displayed on the GUI's plot for you to view. Using your oscilloscope, observe the voltage waveform output at channel *ao0*. Is it a (quasi-) continuous sine wave with a 50-Hz frequency and 5-V amplitude?

Press the **Stop** button. Then, keeping all other parameters the same, change **Frequency (Hz)** (in the **Waveform Parameters** control) to *75*. From the GUI's plot, you will find that `create_waveform_data()` produces 1½ cycles of this waveform. Using your knowledge of the circular fashion in which the DAQ device sequences through the memory buffer values, can you explain the discontinuities that you observe in the voltage waveform being output at channel *ao0*?

With all other settings held the same, try generating sine waves with **Frequency (Hz)** set to *100*, *200*, *250*, and *500*. With each of these choices, `create_waveform_data()` will produce an integer number of complete cycles of the waveform and so produce voltage waveform outputs with no discontinuities. Why does the waveform output at channel *ao0* have noticeable digitizing steps for the higher frequencies? Also, you might enjoy changing **Shape** and **Amplitude** and viewing the resulting waveforms on the oscilloscope.

A final reminder: When program execution is stopped (by closing the **Waveform Generator** window), your NI DAQ device is designed to maintain the last voltage value it generated. Thus, before stopping the program, press **Start** one more time with **Amplitude** set to *0* (and all other parameters set to nonzero values), so the channel *ao0* is left in the zero-voltage state.

Do It Yourself

Lighting LED Using DO Line In this exercise, you will write a program that alternately lights one LED and then another. The signals that toggle these LEDs on and off will be provided by digital output (DO) pins on a DAQ device.

(a) First, light a single LED under computer control as follows: Using the DIO line P0.0 on your DAQ device, construct the hardware circuit shown in Fig. 7.1. In this circuit, when the line goes HIGH, it supplies a DAQ device-dependent voltage difference (common values are 3.3 or 5 V) across the resistor–LED series combination. Assuming this voltage difference is 3 V, the predicted current through the LED will be 3V/ 220 $\Omega \approx$ 14 mA (for a more precise prediction of the current, the voltage drop across the forward-biased LED should be taken into account; for a red LED, this voltage drop is about 1.8 V). If this predicted current is larger than the current-sourcing capability of your DAQ device's digital output line, the current will top out at the maximum allowed value. This maximum value can be found on your DAQ device's datasheet (the USB-6002 is able to supply currents up to 4 mA, which may only dimly light an LED).

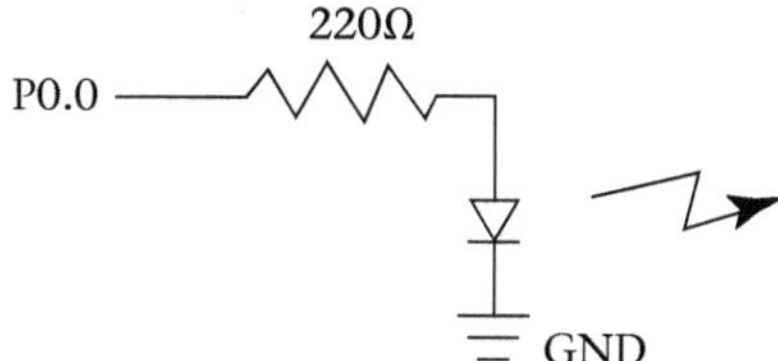

Fig. 7.1 Hardware circuit for Do It Yourself project. The digital I/O line P0.0 of the data-acquisition device is connected to the (current-limiting) resistor in series with LED.

Write a GUI-equipped program called **single_led_ni.py** that can be used to illuminate this single LED via a GUI button. To configure a DO channel, add the following function to **nidaq_operations.py**, which creates a DO task on one or more DIO lines (where each line has its own `channel_name`):

```
def create_do_task(channel_name):

    task = nidaqmx.Task()
    option = nidaqmx.constants.LineGrouping.CHAN_PER_LINE
    task.do_channels.add_do_chan(channel_name,
                                 line_grouping=option)

    return task
```

To write **single_led_ni.py**, you may wish to start with the code in **toggle_button.py**. Then, add coding to create the DO task on a given line identified by the string `channel_name` (e.g., `'Dev1/port0/line0'` for P0.0). Within the `toggle()` function, switch the line between its LOW and HIGH states using the commands `task.write(False)` and `task.write(True)`, respectively.

Run **single_led_ni.py** and show that you can toggle the LED in your hardware circuit on and off using the GUI button.

(b) Next, using the DIO lines P0.0 and P0.1, construct the hardware circuit shown in Fig. 7.2 with two LEDs, which we will call A and B, respectively.

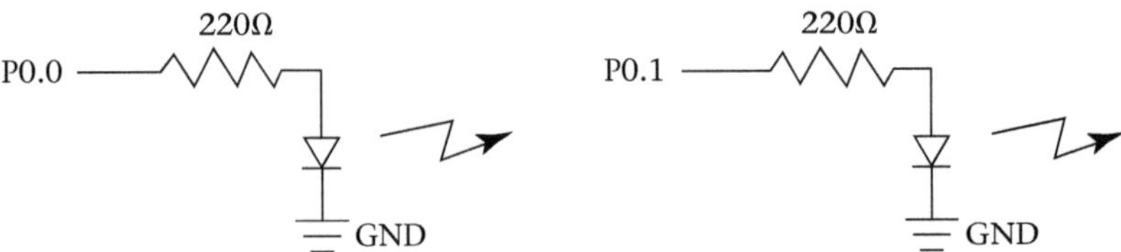

Fig. 7.2 Hardware circuits controlled by alternating LEDs VI. In this figure, digital I/O lines P0.0 and P0.1 are used.

Then, write a GUI-equipped `while`-loop-based program named **toggling_leds_ni.py** such that, when a GUI **Start** button is pressed, a `while` loop commences execution, where each loop iteration lasts 2 seconds. During each loop iteration make *A* lit and *B* dark for 1 second, followed by *A* dark and *B* lit for 1 second.

To write **toggling_leds_ni.py**, you may wish to start with the code in **dc_voltmeter_ni_live.py**. Then, add coding to create the DO task on the two lines P0.0 and P0.1 by specifying `channel_name = 'Dev1/port0/line0,Dev1/port0/line1'`. For a case where two lines are sequential as here, this command can be abbreviated as `channel_name = 'Dev1/port0/line0:1'`. Replace `acquire_sample()` from **dc_voltmeter_ni_live.py** with a new function called `toggle_leds()`. The command to make *A* lit and *B* dark is `task.write(False, True)`, while the command for *A* dark and *B* lit is `task.write(True, False)`.

Use It!

Anti-aliasing Filter By passing an analog signal through a lowpass filter prior to inputting it to a digitizer, the aliasing effect can be suppressed. When used in this manner, the lowpass filter is called an *anti-aliasing filter*. The gain of a filter is defined as its output voltage divided by its input voltage: $G \equiv V_{out}/V_{in}$. An ideal anti-aliasing filter will transmit all signals with frequencies less than the Nyquist frequency to the digitizer at unity gain ($V_{out} = V_{in}$), but will completely block—that is, attenuate to zero gain—all signals with frequencies greater than the Nyquist frequency. The two frequency ranges $f < f_{nyquist}$ and $f > f_{nyquist}$ are called the *passband* and *stopband*, respectively. For real filters, the shift from the passband to the stopband is not abrupt, but takes place over an intermediate range of frequencies called the *transition band*; for frequencies within the transition band, gains are in the range $0 < G < 1$.

Butterworth lowpass filters of various orders n are commonly used as anti-aliasing filters. This class of filter provides "maximally flat" signal transmission (i.e., close to constant unity gain) up to a given cutoff frequency f_c. In the transition region above f_c, an nth-order filter's gain changes by (approximately) a factor of $(1/2)^n$ for each doubling of signal frequency. A doubling of frequency is called an *octave*. For example, if $f_c = 100$ Hz, 200-Hz and 400-Hz signals are one and two octaves above the cutoff frequency, respectively. Thus, for a first-order ($n = 1$) Butterworth filter with $f_c = 100$ Hz (assuming $G \approx 1$ at the cutoff frequency), the gains at 200 and 400 Hz will be $(1/2)^1 = 0.5$ and $(1/2)^1(1/2)^1 = 0.25$, respectively; for an eighth-order ($n = 8$) Butterworth filter with $f_c = 100$ Hz, the gains at 200 and 400 Hz will be $(1/2)^8 = 4 \times 10^{-3}$ and $(1/2)^8(1/2)^8 = 2 \times 10^{-5}$, respectively.

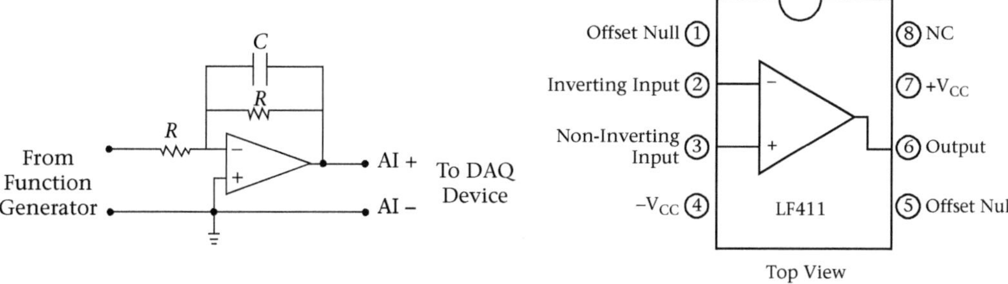

Fig. 7.3 First-order Butterworth lowpass filter. Suggested op-amp is LF411.

To demonstrate the design of an anti-aliasing filter, construct the op-amp circuit shown in Fig. 7.3, which is a first-order Butterworth lowpass filter. The cutoff frequency (also known as the *3 dB frequency*) for this circuit is given by $f_c = 1/2\pi RC$; for your circuit, use $R = 10$ kΩ and $C = 0.1$ μF so that $f_c \approx 150$ Hz. Attach a function generator to the filter's input and connect its output to the analog input channel pins programmed into **digital_oscilloscope_ni.py**. Also, attach the function generator's sync output to the proper DAQ device pins to facilitate digital triggering.

Run **digital_oscilloscope_ni.py** and set **Number of Samples** and **Sampling Frequency (Hz)** equal to *200* and *2000*, respectively, making $f_{nyquist}$ = 1000 Hz. Configure your function generator to input a (smaller than Nyquist frequency) 50-Hz sine wave with amplitude 1 V to the lowpass filter, then press the **Start** button. Since the sine-wave frequency is less than the filter's cutoff frequency, you will see a plot of about five cycles of the unattenuated ($G \approx 1$) 50-Hz sine wave.

Next, input a 1200-Hz sine wave of amplitude 1 V to the filter. This frequency is greater than the Nyquist frequency as well as the filter's cutoff frequency and so, instead of a 1200-Hz waveform, your data-acquisition system will produce an aliased waveform of 800 Hz, but with an amplitude attenuated from the 1-V input value. Since 1200 Hz is three octaves above $f_c \approx 150$ Hz Hz, we expect the gain due to this first-order filter to be approximately $(1/2)^1(1/2)^1(1/2)^1 = 0.125$. Thus, this simple filter suppresses aliasing, but not to the level that we might hope. Since 1200 Hz exceeds $f_{nyquist}$, this input sine wave is in the stopband and so the gain would ideally be much closer to zero.

To be optimally effective, an anti-aliasing filter will receive a large-amplitude signal within the stopband and attenuate it to below the DAQ system's smallest resolvable voltage level. Thus, for a 16-bit digitizer with a range of ±10 V and voltage resolution $\Delta V = 0.3$ mV, the optimal gain for this filter within its stopband will be on the order of $G \equiv V_{out}/V_{in} = 0.0003\text{V}/10\text{V} = 3 \times 10^{-5}$. From our above calculations, we see that an eighth-order Butterworth filter can achieve this order of attenuation for frequencies greater than two octaves above f_c. Such high-order Butterworth filters are available as low-cost integrated circuits. For example, the Maxim MAX291 chip is an eighth-order Butterworth lowpass filter, whose cutoff frequency can be chosen anywhere within the range from 0.1 Hz to 25 kHz. When placed in the circuit shown in Fig. 7.4, the cutoff frequency (in Hz) is given by $f_c = 333/C$, where C is in nF. Build this circuit with $C = 1$ nF so that $f_c \approx 300$ Hz and demonstrate its ability to completely suppress the aliasing of a 1200-Hz input sine wave (1200 Hz is two octaves above 300 Hz).

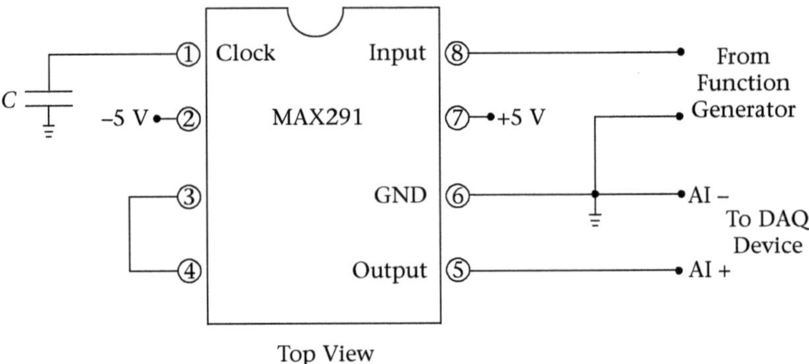

Fig. 7.4 MAX291 eighth-order Butterworth lowpass filter. Capacitor *C* determines cutoff frequency.

Problems

1. **dual_channel_digital_oscilloscope_ni.py** Build a two-channel digital oscilloscope, which acquires and plots digitalized waveforms on both AI channels *ai0* and *ai1*. With **digital_oscilloscope_ni.py** open, use **File>>Save As**... to create the new program and save it in **YourNamePythonWork**. The only changes required are the following:
   - To specify that sampling is to be done on the two channels *ai0* and *ai1*, define `channel_name` as the string `'Dev1/ai0:1'`. Because the two channels are sequential, we can use this abbreviated form of `channel_name`. The general "long form" of this specification is the string `'Dev1/ai0, Dev1/ai1'`.
   - In `acquire_waveform()`, the command `y = task.read(num)` remains as is. Now, however, instead of `y` being a list with `num` elements, consisting of the `num` samples acquired on a single channel, `y` is a two-element list, where each element is a `num`-element list. That is, `y` is a list of lists, where the first element of `y`, i.e., `y[0]`, is itself a `num`-element list consisting of the `num` samples acquired on channel *ai0*, while the second element of `y`, i.e., `y[1]`, is a `num`-element list consisting of the `num` samples acquired on channel *ai1*.
   - For the display, a plot of two curves is needed. Because this plot is rapidly updated within the `while` loop, it's best to write a dedicated plotting routine. In **plot_creator.py**, add the following function and then use it in your program:

```
def update_two_curve_plot(ax, canvas, x, y, z,
                          label_x, label_y,
                          marker='-'):

    size = 12

    ax.clear()
    ax.tick_params(axis='both', labelsize=size)
    ax.set_xlabel(label_x, fontsize=size)
    ax.set_ylabel(label_y, fontsize=size)
    ax.grid()
```

```
    ax.plot(x, y, marker)
    ax.plot(x, z, marker)
    canvas.draw()

    return
```

On your DAQ device, attach a voltage waveform from a function generator to AI channel *ai0* and the function generator's sync output's digital signal to the AI channel *ai1*. Additionally, attach the sync output's digital signal to the PFI 0 channel of your DAQ device. Run your program and verify that it simultaneously displays digitally triggered traces of the two inputs.

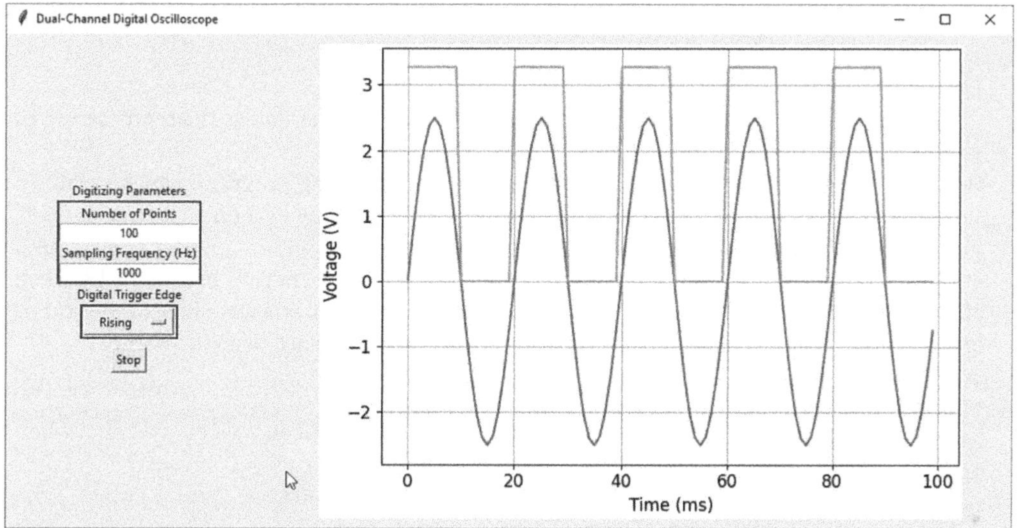

2. **tmp36_thermometer_ni.py** The Analog Devices TMP36 temperature sensor (available from https://www.adafruit.com and https://www.sparkfun.com) is a low-cost three-terminal semiconductor device, which can be used to measure temperatures in the range from −40 to 125°C with accuracy of +1C. As shown in Fig. 7.5, to operate the TMP36, simply connect a power supply voltage to pin 1 (this voltage can be anything in the range from +2.7 to 5.5 V) and ground to pin 3. Then measure the voltage difference V between pin 2 and pin 3. The temperature T in degrees Celsius is then given by $T = 100 \times V - 50$.

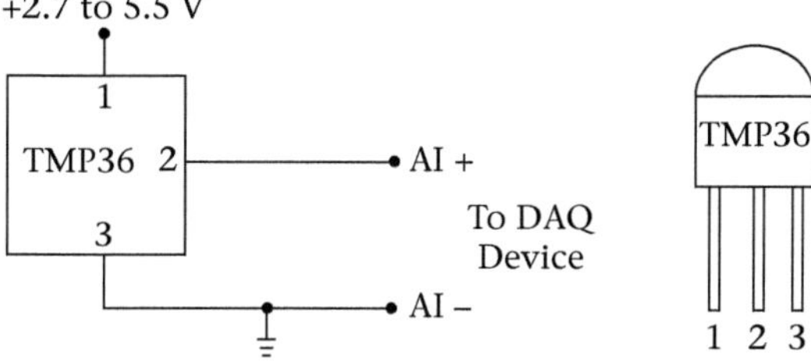

Fig. 7.5 TMP36 temperature sensor.

Connect a powered TMP36 device to the AI differential channel *ai0* pins of your DAQ device. Using in **dc_voltmeter_ni_live.py** as your guide, write a GUI-equipped program named **tmp36_thermometer_ni.py** that, every 250 ms until a **Stop Button** is pressed, reads the voltage *V*, converts this value to the corresponding temperature in Celsius, and then displays this temperature in a labeled GUI indicator.

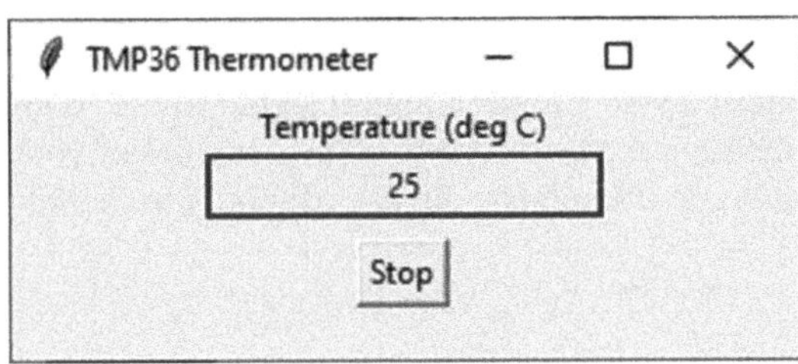

Run your program and use it to measure room temperature as well as the temperature of your skin.

3. **sinewave_generator_ni_software_timed.py** Build the following program, which creates a sine-wave voltage output using the software timing provided by Python's `time.sleep()` function. As the `while` loop iterates the code, the function $y(x) = 2 + \sin x$ is evaluated at N equally spaced arguments x in each sine-wave cycle. The spacing in radians between adjacent arguments is $2\pi/N$, where N is the number of samples per cycle. Since each `while`-loop iteration takes 0.01 second, its 1000 iterations will take place over 10 seconds.

```
import Utilities.nidaq_operations as no
import math
import time

channel_name = 'Dev1/ao0'
task = no.create_ao_task(channel_name)
task.start()
n=10

for i in range(1000):
    x = 2*math.pi*i/n
    y = 2 + math.sin(x)
    task.write(y)
    i +=1
    time.sleep(0.01)

task.close()
```

Connect the AO 0 and AO GND pins of your DAQ device to the positive and negative (ground) inputs of an oscilloscope, respectively. Then run your program with samples per cycle set to `n=10` so that each sine cycle is represented by ten voltage levels. Given that each `while`-loop iteration takes 10 ms, the period of the observed sine wave should be (10 ms/update) $\times$ (10 updates) $=$ 100 ms, making its frequency 10 Hz. You will find that, although the underlying sine pattern is apparent, the ten voltage levels used to form this waveform are too well resolved to give the appearance of a continuous function.

Try increasing samples per cycle to `n=20`, then `n=30`, and so on. Since the `while`-loop iteration time remains constant, as you increase the samples per cycle, the frequency of the output waveform will decrease. You will need to adjust your oscilloscope's *Time/Div* control appropriately to view a complete cycle of the waveform. At what value for samples per cycle does the output waveform start to take on a continuous appearance?

This problem can be done with all NI DAQ devices, including the USB-6009 legacy device. However, you will find the generated sine wave has a frequency a bit smaller than expected due to the inherent accuracy of your computer's clock, which is the time base for Python's `time.sleep()` function as well as the time taken to execute the `task.write(y)` command in each `while`-loop iteration.

4. **frequency_meter_ni.py** Build a frequency meter program that determines the frequency of a digital signal input to your DAQ device as follows.
   (a) Add the function given below to **nidaq_operation.py**, which configures a counter input task on your DAQ device to count (by default) rising digital edges at the counter's input pin. The particular counter is specified by `channel_name` and the counter's input pin by `edge_source`; example choices for these two quantities are `Dev1/ctr0` and `/Dev1/PFI0`, respectively.

```
def create_ci_task(counter_name, edge_source):

    task = nidaqmx.Task()
    task.ci_channels.add_ci_count_edges_chan(counter_name)
    task.ci_channels.ci_count_edges_term = edge_source

    return task
```

   (b) Using **dc_voltmeter_ni_live.py** as your model, write a GUI-equipped program called **frequency_meter_ni.py** that displays the frequency of a digital waveform input at the counter's input pin. Measure the frequency as follows: Within each iteration of a `while` loop, start the counting task, use `time.sleep()` to define the counting interval to be 1.0 second, read the number of counts N accumulated during the 1.0-second interval, stop the task, then display N in a GUI indicator. This process is coded as follows:

```
task.start()
time.sleep(1.0)
count = task.read()
task.stop()
string_data = str(count)

lbl_2.config(text=string_data)
```

   (c) Configure the program's GUI so that a **Start** button initiates the `while` loop as it morphs into a **Stop** button. Pressing the **Stop** button ceases the `while` loop's operation.

After you have completed your program, program a function generator to output a certain frequency, say, 1000 Hz. Then, connect the sync output of the function generator to the input pin of counter 0 (and GROUND) on your DAQ device, and run **frequency_meter_ni.py** to measure the frequency of the sync output's digital signal. You may find a slight error in the measured frequency due to the inherent accuracy of your computer's clock, which is the time base for Python's `time.sleep()` function.

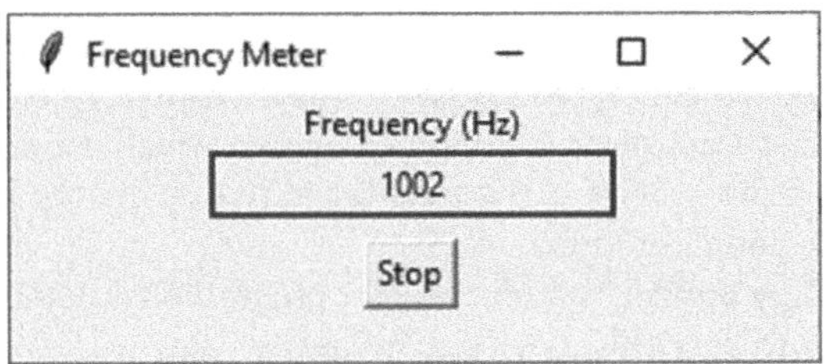

5. **music_box_ni.py** Write a program that plays the scale of A major, with each note sounding for 500 ms. The eight notes in this scale have the following frequencies in hertz: 440.0, 493.9, 554.4, 587.3, 659.3, 740.0, 830.6, and 880.0. Store these frequencies in a Python list called `freqs`. In your program, use a `for` loop to input each frequency `freq` one at a time into a function called `generate_waveform(freq)`, calculate a sine wave with frequency `freq`, and then output that waveform at AO channel *ao0* for 0.5 s. Keep in mind the maximum AO update rate and the size of the on-board buffer for your DAQ device, which is 5 kHz and 2047 samples for the USB-6002 unit, respectively.

   To play the scale, attach an amplifier and speaker to the AO channel such as shown in Fig. 7.6, and then run your program.

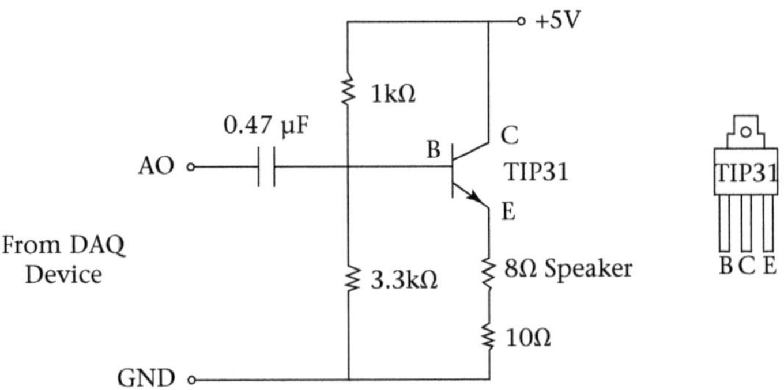

Fig. 7.6 Amplifier circuit for Problem 5.

6. **data_logger_ni.py** Starting with Chapter 4's **sinewave_animation_blit_gui.py**, use **File>>Save As...** to create **ni_data_logger.py**, then modify the function `animate()` so that the quantity `y_i` is a single voltage sample obtained from your NI DAQ device, rather than the calculated value `y_i = a*math.sin(2.0*math.pi* f*t_i)`. Configure the program to acquire one voltage sample every 0.5 seconds. Input a low-frequency (e.g., 0.1 Hz) waveform (e.g., a triangle wave as shown below) from a function generator to channel *ai0* of your DAQ device to demonstrate a real-time plot of this input signal over a time span of 50 seconds.

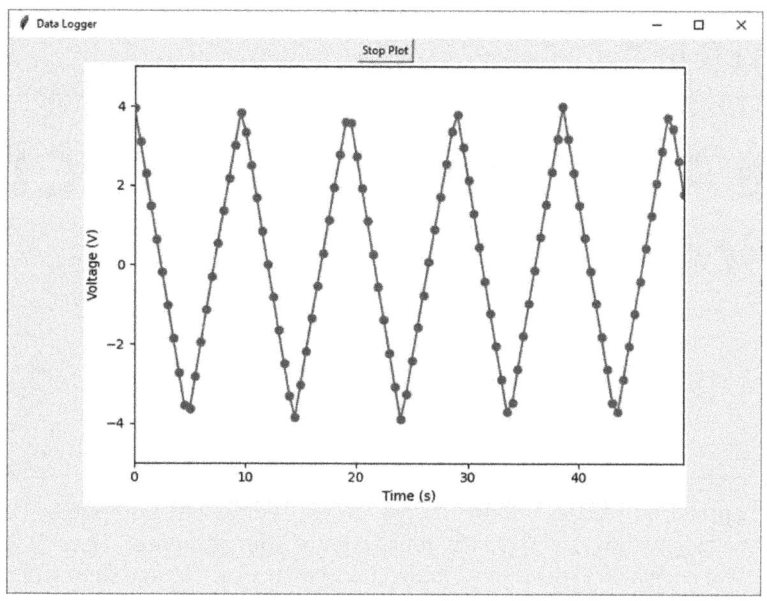
Data Logger
Stop Plot
4
2
0
−2
−4
Voltage (V)
0
10
20
30
40
Time (s)

CHAPTER 8

Data Acquisition Using MCC DAQ Device

In this chapter, you will learn how to control a Measurement Computing Corporation (MCC) data-acquisition (DAQ) device connected to your computer. Here, it is assumed that you have such a DAQ device available, along with a DC voltage source (e.g., battery, power supply, function generator), function generator, voltmeter, (optional) oscilloscope, and required cabling. MCC DAQ devices are distributed by Digilent, a company based in Pullman, WA and owned by NI.

If, while performing the chapter exercises, the communication between the DAQ device and computer breaks down (e.g., caused by an accidental error in your programming), communication can often be restored either by unplugging and then replugging the USB cabling, or restarting your computer.

8.1 Installing Software Driver for MCC DAQ Devices

Equipped with the skills developed in Chapters 1 through 6, you are now ready to write Python programs that automate data-acquisition and generation processes. In this chapter, the programs you create will control the operations of a multifunction data-acquisition (DAQ) device manufactured by Measurement Computing Corporation (MCC), a leading manufacturer of value-priced data-acquisition hardware and software. Depending on your particular DAQ device, these operations may include analog-to-digital conversion, digital-to-analog conversion, digital input/output operations, event timing, digital pulse counting, and digital pulse generation.

Control of a MCC DAQ device is carried out through the use of driver software called the *Universal Library* (UL), which is written in the C programming language. The Universal Library provides a collection of C-formatted functions that can be directly incorporated as high-level ("user-friendly") commands into a custom-written C application program. In this scheme, when a particular high-level command is called by the C application program, the Universal Library translates that command into the required low-level code that then instructs the MCC DAQ device to carry out the desired process. Thus, UL is the link between a command in a C application program and the low-level control of a MCC DAQ device as shown here:

The Python-Based Laboratory. John Essick, Oxford University Press. © John Essick (2024). DOI: 10.1093/9780191998478.003.0008

To control a MCC DAQ device through a Python application program, a software package called *mcculw* is used. This driver software provides Python-based renderings of the C-based functions in the UL and is termed a *Python binding* (often called a *wrapper*) in that it provides an interfacing layer at its input that converts Python commands into the C coding understood by UL. The mcculw package includes two main modules—*ul* and *enums*. The ul module consists of the Python-based functions mentioned above, where the Python conversion code ("wrapping") and UL functions are stitched together "under-the hood"; i.e., from the programmer's point of view, it appears like Python functions are used to communicate directly with the MCC DAQ device. In Python, an *enum* (short for enumerated data type) is a text variable that maps to a particular integer. The enums module comprises all of the enums used to configure a MCC DAQ device. For example, the enum `ScanOptions.SCALEDATA` configures a MCC DAQ device for reporting digitized samples in the units of volts and is associated with the decimal integer 16 (or, equivalently, the binary number 10000_2). In summary, mcculw provides the link between a command in a Python application program and the low-level control of a MCC DAQ device as shown here:

In addition to mcculw, another software program called *InstaCal* is required to control a MCC DAQ device. As we'll see, InstaCal is used to initially discover and configure the DAQ device.

To obtain the two required free software packages:

- Download InstaCal by going to https://files.digilent.com/#downloads/.
- Download mcculw by using `pip` at the Command Prompt as follows:

```
pip install mcculw
```

Full information on these two packages and how to download them from either PyPI or GitHub can be found by searching "MCC Software Download." The software mcculw is only supported on the Windows operating system. A driver named *uldaq* supports the Linux operating system.

8.2 mcculw Basics

In this chapter, you will explore the fundamentals of controlling a MCC DAQ device using Python to perform three basic DAQ device operations—analog-to-digital conversion (acquiring a digitized representation of an incoming analog voltage), digital-to-analog conversion (generating an outgoing analog signal from a sequence of digital numbers), and digital line control (setting the line's HIGH/LOW state). In the world of DAQ, these three operations are called *analog input* (*AI*), *analog output* (*AO*), and *digital output* (*DO*), respectively. The chapter's text will give instructions assuming that you are working with our highlighted MCC USB-202

DAQ device. However, most other multifunction DAQ device models from MCC are controlled with the same commands and so can be used for the exercises. Please refer to the manual for your specific device in order to adapt to differing technical specifications (e.g., maximum sampling rate, pinout diagram).

A large assortment of data-acquisition (also called *read* or *input*) operations as well as data-generation (aka *write* or *output*) operations are within the capability of any multifunction DAQ device manufactured by MCC. When a programmer chooses to perform one such operation, that choice includes the type of task (e.g., analog input voltage on AI channel 0), along with its particular timing (e.g., sampling frequency) and triggering (e.g., digital edge trigger) selections. The ul and enums modules contain all of the Python-based commands needed to program the execution of any DAQ operation possible on your MCC DAQ device. The commands are in the form of Python functions that possess a set of arguments. The arguments allow the user to select various options related to the DAQ operation such as identifying the intended DAQ device on which the operation will take place or setting the sampling frequency. As an example, the Python code for acquiring a single DC voltage sample from a MCC DAQ device using the `v_in()` function from the ul module is as follows:

```
import mcculw.ul
import mcculw.enums

board_num = 0
channel = 0
ai_range = mcculw.enums.ULRange.BIP10VOLTS

single_sample = mcculw.ul.v_in(board_num, channel,
                               ai_range)
```

Here, after importing the `ul` and `enums` modules from the `mcculw` package, the next three lines specify the particular DAQ device, analog input channel, and range for the DAQ operation through the `board_num`, `channel`, `ai_range` arguments of the `v_in()` function, respectively. In this example, the DAQ device and AI channel are both identified by the integers `0`; the enum `ULRange.BIP10VOLTS` selects a bipolar 10-V range for the operation, i.e., ± 10 V. These arguments are then input to the function `v_in(board_num,channel,ai_range)`, which is found in the ul module. When the `v_in()` function executes, it acquires ("reads") a single voltage sample from AI channel 0 on the specified DAQ device and assigns the acquired value to `single_sample`. This data value is formatted as a floating-point number and is in the units of volts.

The ul and enums modules contain a wealth of commands, which can be used to program allowed operations on your DAQ device. For example, the commands `a_out()`, `d_in()` and `c_in()` are used to configure a single-sample analog voltage output, digital input, and counter input operation, respectively, while `ai_in_scan()` allows one to program the sampling frequency and number of samples to acquire a waveform in an AI operation.

8.3 DAQ Device Configuration Using InstaCal

Before we control a MCC DAQ device using Python programming, we must first run the configuration program called *InstaCal*. When run, InstaCal detects your DAQ device, assigns computing resources to it, and records its work in a configuration file named **CB.CFG**. Then, when carrying out DAQ operations using functions from the ul module, the Universal Library functions use the **CB.CFG** file to determine what devices are active and how they are configured.

With your DAQ device connected to your computer (e.g., DAQ device's USB cable plugged into one of your computer's USB ports), open InstaCal. In the **InstaCal** window that appears, either select **Install>>Refresh** in the drop-down menu or click on the **Refresh Boards** icon in the toolbar. A dialog window called **Plug and Play Board Detection** will pop up. In that dialog window, select the name of your DAQ device, then press **OK**. The dialog window will close and return you to the **InstaCal** widow where, in the **PC Board List**, the name of your DAQ device and its identification number is now given. Note the identification number (it's an integer) as we'll use it in our Python programs. To verify your DAQ device is properly communicating with your computer, right-click on its name in the **PC Board List** and select **Configure**.... In the **Board Configuration** dialog window that pops up, press the **Flash LED** button. If your connection is good, an LED on your DAQ device will flash several times. To close the dialog window, press **OK**.

Close InstaCal. The configuration file **CB.CFG** has been created as a result of running InstaCal. If you'd like to view this configuration file, it's stored in the **C:\ProgramData\Measurement Computing\DAQ** folder. If the folders are hidden, you will need to instruct Windows to show hidden items.

8.4 Simple Analog Input Operation on DC Voltage

Let's begin by turning your computer into a DC voltmeter. We'll accomplish this feat by creating a Python-based program that carries out the same analog input operation that we ran using DAQami in Section 6.6. The heart of this DC voltmeter program is the single-sample AI code outlined in Section 8.3. Thus, open IDLE and in the Editor, use **File>>New File** to create a program and save it under the name **simple_ai_operation_mcc.py** in **YourNamePythonWork**. Then, code the program as follows, using the board number of your particular DAQ device as found through InstaCal, which may be different than 0:

```
import mcculw.ul as mu
import mcculw.enums as me

board_num, channel = 0, 0
ai_range = me.ULRange.BIP10VOLTS

single_sample = mu.v_in(board_num, channel, ai_range)

print(single_sample)
```

Save your work. Connect a given DC voltage difference (say, 2 V) to the AI channel 0 of your DAQ device. For the MCC USB-6002, the high- and low-voltage wires from your source should be connected to CH0 and AGND, respectively. The AI channels of the MCC USB-202 are all single-ended, so there is no problem if your DC voltage is "floating" (see the discussion in Section 6.6).

Run **simple_ai_operation_mcc.py**. The acquired voltage difference will be reported in the Shell. Does your program read the input voltage difference correctly?

Now that we understand how to carry out the data-acquisition portion of our program, giving this program a GUI is an obvious next improvement. Here, we can take advantage of our previous work in building the **gui_creator.py** module, which is stored in **YourNamePythonWork>>Utilities**, and supplied functions that we used in **waveform_simulator.py**. This time around, we have something a bit different to display, though.

Rather than displaying a full waveform as created in **waveform_simulator.py**, we have just a single sample value. Thus, our present display need is a simple numeric indicator. Thankfully, in Chapter 5's Use It! project, we added a function named `create_indicator()` to the **gui_creator.py** module that creates such an indicator, so we will make use of that previous work. If you didn't produce `create_indicator()` in Chapter 5, do so now following the instructions in the next paragraph.

In **YourNamePythonWork>>Utilities**, open **gui_creator.py** and, among the function definitions, add the following definition for a function named `create_indicator()`. This function creates a blue-bordered frame within the specified `container` and, within that frame, places a descriptive `label_1` above a second `label_2`. `label_2` will display a string-formatted version of our data value and `label_1` is captioning text. If left unspecified, `label_2` will initialize to its default empty string.

```
def create_indicator(container, label_1,
                     label_2='', char_width=20):
    lbl_1 = tk.Label(container, text=label_1)
    lbl_1.pack()

    frame = tk.Frame(container, bg='blue', bd=2)
    frame.pack()

    lbl_2 = tk.Label(frame, text=label_2, width=char_width)
    lbl_2.pack(fill=tk.X)

    return lbl_2
```

Save your work and then close **gui_creator.py**.

In the Editor, use **File>>New File** to create a program named **dc_voltmeter_mcc.py** in **YourNamePythonWork**. Then enter the following code. Here, functions from the **gui_creator.py** module are used to create a labeled GUI window that contains a labeled indicator to display the acquired DC voltage sample. A button labeled **Read Sample** will execute a function (to be added in a moment) named `acquire_sample()` when pressed. As with all of our programs, you may wish to optimize the `size` of the operating-system window for your particular computing system.

```
import Utilities.gui_creator as gc
import mcculw.ul as mu
import mcculw.enums as me

board_num, channel = 0, 0
ai_range = me.ULRange.BIP10VOLTS

root = gc.create_window(title='DC Voltmeter',
                        size='200x100')

lbl_2 = gc.create_indicator(root,
                            label_1='Voltage Reading (V)')

btn = gc.action_button(root, label='Read Sample',
                       action=acquire_sample)

root.mainloop()
```

To complete our program, we need to define the function `acquire_sample()`. This function implements the command `v_in()` to acquire a single voltage sample, which is formatted as a floating-point number given the name `float_data`. The `str()` function is used to convert `float_data` to a string called `string_data` for display as the `label_2` string in the GUI indicator. The definition of `acquire_sample()` must appear before it is mentioned in `action_button()`, or else the following exception will halt execution: `NameError: name 'acquire_sample' is not defined`. Hence, include the definition of `acquire_sample()` as shown next.

```
import Utilities.gui_creator as gc
import mcculw.ul as mu
import mcculw.enums as me

board_num, channel = 0, 0
ai_range = me.ULRange.BIP10VOLTS

root = gc.create_window(title='DC Voltmeter',
                        size='200x100')

lbl_2 = gc.create_indicator(root,
                            label_1='Voltage Reading (V)')


def acquire_sample():

    float_data = mu.v_in(board_num, channel, ai_range)
    string_data = str(float_data)

    lbl_2.config(text=string_data)

    return

btn = gc.action_button(root, label='Read Sample',
                       action=acquire_sample)

root.mainloop()
```

Save your work, then run **dc_voltmeter_mcc.py**. Each time you press the **Read Sample** button, a voltage sample will be acquired by the DAQ device, and the sample value will appear as the `label_2` string in the indicator.

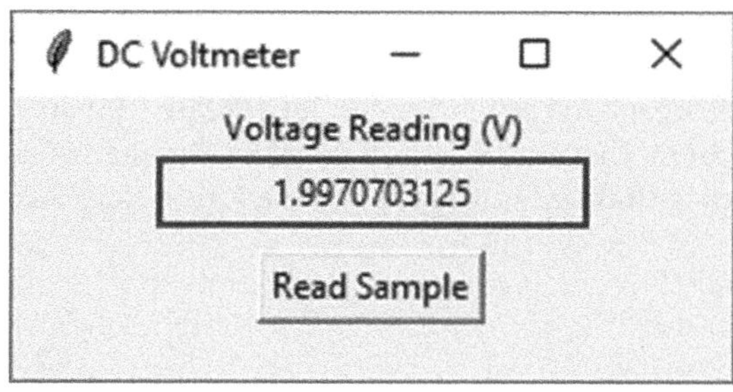

Hopefully you're pleased with your computer-based DC voltmeter, but there is an issue that should raise concern. The voltage value displayed in the indicator is given to many decimal places, well beyond the *digits-of-precision accuracy* of your DAQ device specifications. From Eq. (6.1), we know the voltage resolution ΔV of the n-bit analog-to-digital conversion process performed by your DAQ device is given by $\Delta V = V_{span}/2^n$. For a device such as the MCC USB-202 with $n = 12$, $\Delta V = (20\,\text{V})/2^{12} = 5$ mV. Thus, our displayed measurement in units of volts should only have three digits of precision.

Here's how to solve that display problem: In our program, the command `v_in()` reports the sampled voltage as a floating-point number, where the digits of this float are significant up to the third decimal place. Digits in fourth, fifth, sixth, and higher places are just numbers left randomly in computer memory. After `v_in()` has acquired the sample, we use `str()` to convert the measured floating-point number to a string, which is then displayed in the GUI indicator. Unfortunately, the `str()` function simply converts all decimal digits, both significant and random, to string characters without being told by us to properly round the result. There are a few alternate means for carrying out the float-to-string conversion with the correct precision. The newest (Python version 3.6 and higher) and now recommended approach for this process is called *f-strings*. To convert `float_data` to `string_data` with a precision to the `x` decimal place, the syntax for this command is as follows:

```
string_data = f'{float_data:.xf}'
```

Confusingly, there are two `f`'s in this command, each with a different meaning. The first `f`, which is outside the curly brackets and quotation marks, indicates this is an f-string command. The second `f` is a formatting character, indicating that `float_data` is a floating-point number.

To correct the display problem in our program, in the `acquire_sample()` function within **dc_voltmeter_mcc.py**, replace the statement `string_data = str(float_data)` with `string_data = f'{float_data:.3f}'`. Save your work.

Run **dc_voltmeter_mcc.py** and verify that the measured voltage is now displayed with the proper digits of precision.

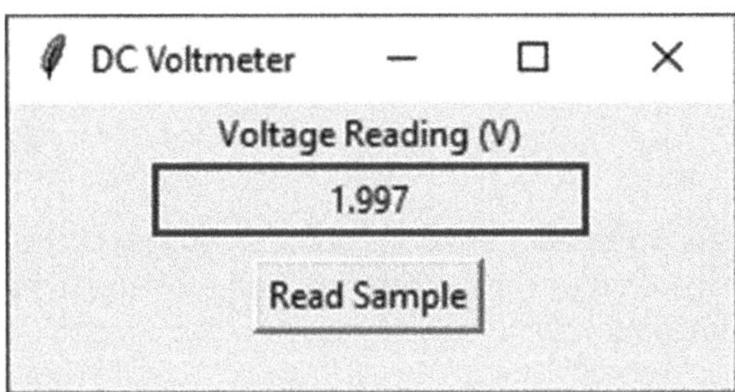

In our work ahead, we'll be using commands from the ul and enums modules of the mcculw package. As a method for organizing our programs, let's create a module that has handy data-acquisition-related functions that involve these mcculw commands and store this module in **YourNamePythonWork>>Utilities**. In the Editor, use **File>>New File** to create a program named **mccdaq_operations.py** and save it in **YourNamePythonWork>>Utilities**. After importing the `ul` and `enums` modules, first define the function `chan_range()`, which configures the MCC USB-202 for one of the two ranges we will encounter with this DAQ device. This function will assign the correct enum associated with a given `range_option`, where there are two possible values for `range_option`—`'bipolar_10'` or `'unipolar_5'`—that are matched with the enums `ULRange.BIP10VOLTS` and `ULRange.UNI5VOLTS` in the `enums` module, respectively.

```
import mcculw.ul as mu
import mcculw.enums as me

def chan_range(range_option):

    if range_option == 'bipolar_10':
        range_enum = me.ULRange.BIP10VOLTS
    elif range_option == 'unipolar_5':
        range_enum = me.ULRange.UNI5VOLTS

    return range_enum
```

Next, add the function `read_sample()`, which acquires a single AI sample from a MCC DAQ device:

```
import mcculw.ul as mu
import mcculw.enums as me

def chan_range(range_option):

    if option == 'bipolar_10':
        range_enum = me.ULRange.BIP10VOLTS
    elif option == 'unipolar_5':
        range_enum = me.ULRange.UNI5VOLTS

    return range_enum


def read_sample(board_num, channel, ai_range):

    float_data = mu.v_in(board_num, channel, ai_range)

    return float_data
```

Save your work and close **mccdaq_operations.py**.

Open **dc_voltmeter_mcc.py** and modify it as shown below. Note that the only changes are in the import statements and in the commands used to assign `ai_range` and `float_data`. When using functions from **mccdaq_operations.py**, remember to include this utility module's abbreviated prefix `mo`.

```
import Utilities.gui_creator as gc
import Utilities.mccdaq_operations as mo

board_num, channel = 0, 0
ai_range = mo.chan_range('bipolar_10')

root = gc.create_window(title='DC Voltmeter',
                        size='250x100')
lbl_2 = gc.create_indicator(root,
                            label_1='Voltage Reading (V)')


def acquire_sample():

    float_data = mo.read_sample(board_num, channel,
                                ai_range)
    string_data = f'{float_data:.3f}'

    lbl_2.config(text=string_data)

    return

btn = gc.action_button(root, label='Read Sample',
                       action=acquire_sample)

root.mainloop()
```

Save your work. Run this finalized version of **dc_voltmeter_mcc.py** and verify that the measured voltage is properly displayed.

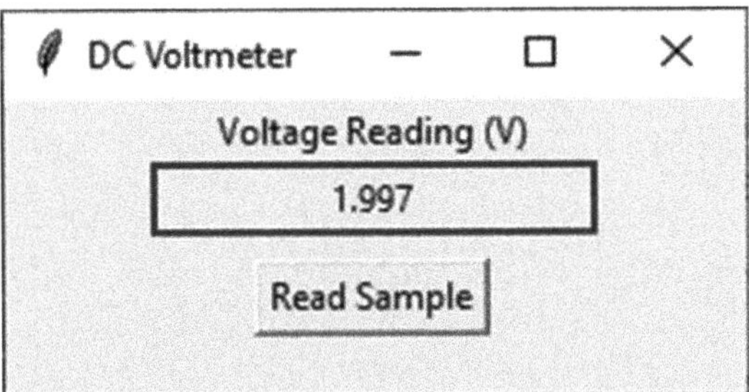

8.5 Live Updating and Threading

While impressed with your computer-based DC voltmeter, perhaps you had hoped that it would be more like a standard voltmeter with live updates. That is, instead of having to press the **Take Sample** button every time you wanted a voltage reading, wouldn't it be better to have a voltmeter that automatically updates measurements after a **Start** button is pressed and continues the updates until told to stop? Let's see if we can do that.

First, we will need a new version of our button, one that toggles between a **Start** and **Stop** state. We outlined how to build a toggling button in Chapter 5's Use It! exercise, but let's now go into greater detail on how to construct this gadget. In the Editor, use **File≫New File** to create a program named **toggle_button.py** and save it in **YourNamePythonWork**. Start coding the program as shown below.

```
import Utilities.gui_creator as gc


root = gc.create_window(title='Toggle Button',
                        size='250x50')

btn = gc.action_button(root, label='Start',
                       action=None)

root.mainloop()
```

Save your work, then run the program. An operating-system window named **Toggle Button** will appear with a button labeled **Start**. When you press the button, nothing will happen because we set `action=None` in the statement that created the button.

Now, going back to the Editor, include the definition of a function named `toggle()` as shown. The push button object `btn` is created by the `Button()` class within the `gc.action_button()` function. Here, we use a method of the `Button()` class called `cget(option)`, which reports the current value of one of `btn`'s options, in this case, its `text` label. The *if*. . . *elif* statement toggles the value of the button's `text` label. Remember to make the change `action=toggle` so that the button calls `toggle()` when pressed.

```
import Utilities.gui_creator as gc

def toggle():

    if btn.cget('text') == 'Start':
        btn.config(text='Stop')
    elif btn.cget('text') == 'Stop':
        btn.config(text='Start')

    return


root = gc.create_window(title='Toggle Button',
                        size='250x50')

btn = gc.action_button(root, label='Start',
                       action=toggle)

root.mainloop()
```

Save your work. Run **toggle_button.py**. You will find that the button's label now toggles between **Start** and **Stop** each time it is pressed.

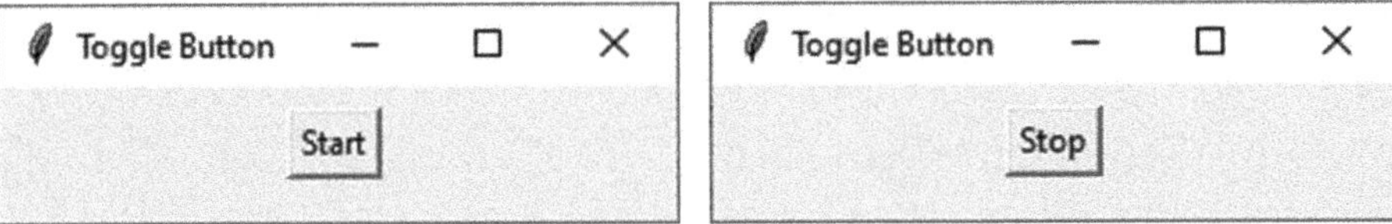

Now that we know how to toggle the button, it seems like it will be an easy task to modify **dc_voltmeter_mcc.py** in order to create a live updating DC voltmeter. With **dc_voltmeter_mcc.py** open in the Editor, use **File>>Save As**. . . to create a program named **dc_voltmeter_mcc_live.py** and save it in **YourNamePythonWork**. Change the coding as shown. Note that now the argument in `action_button()` has been changed to `label='Start'`. The idea here is that, once the **Start** button is pressed (and the button's `text` label has toggled to `'Stop'`), a `while` loop will begin to execute, taking a single sample with each iteration. The `while`-loop execution will iterate until the button is pressed again, causing its `text` label to toggle to `'Start'`. Many of the commands are included in **toggle_button.py**; consider using copy-and-paste operations to expedite your coding process.

```
import Utilities.gui_creator as gc
import Utilities.mccdaq_operations as mo

board_num, channel = 0, 0
ai_range = mo.chan_range('bipolar_10')

root = gc.create_window(title='DC Voltmeter',
                        size='250x100')

lbl_2 = gc.create_indicator(root,
                            label_1='Voltage Reading (V)')

def acquire_sample():

    if btn.cget('text') == 'Start':
        btn.config(text='Stop')
    elif btn.cget('text') == 'Stop':
        btn.config(text='Start')

    while btn.cget('text') == 'Stop':
        float_data = mo.read_sample(board_num, channel,
                                    ai_range)
        string_data = f'{float_data:.3f}'

        lbl_2.config(text=string_data)
```

```
    return


btn = gc.action_button(root, label='Start',
                       action=acquire_sample)

root.mainloop()
```

Save your work. Run **dc_voltmeter_mcc_live.py** and you will find something unexpected happens. The program likely stalls, becoming unresponsive and requiring you to manually abort its execution (e.g., by exiting IDLE/Python). What's going on?

The problem you are observing is common in data-acquisition programs where a computer is being asked to update a graphical user interface at the same time it is supposed to be streaming data from a DAQ device into its memory. In this present case, the GUI-related operation `root.mainloop()` is running an infinite loop, constantly checking for events like button presses, while the DAQ-related operation `read()` is trying to acquire data. Because of how we programmed these two operations, they are trying to run on the same execution path within the computer (termed a *thread*) and end up "blocking" each other's execution.

The solution to this blocking problem is to program the GUI- and DAQ-related operations to execute on separate threads. Python's `threading` module (included in the Python Standard Library) makes this solution easy to program. Modify **dc_voltmeter_mcc_live.py** as shown below, where GUI-button-related and DAQ-related commands are separated into the two functions `toggle()` and `acquire_sample()`, respectively. Here, GUI-related operations execute on the original thread of the execution, while DAQ-related operations in the `acquire_sample()` function execute on another thread named `t` created by `threading.Thread()`. Remember to add `import threading` to the top of your program as well as to start the new thread with `t.start()` and to change `action=acquire_sample` to `action=toggle`.

```
import Utilities.gui_creator as gc
import Utilities.mccdaq_operations as mo
import threading

board_num, channel = 0, 0
ai_range = mo.chan_range('bipolar_10')


root = gc.create_window(title='DC Voltmeter',
                        size='250x100')

lbl_2 = gc.create_indicator(root,
                            label_1='Voltage Reading (V)')
```

```
def acquire_sample():

    while btn.cget('text') == 'Stop':
        float_data = mo.read_sample(board_num, channel,
                                    ai_range)
        string_data = f'{float_data:.3f}'

        lbl_2.config(text=string_data)

    return

def toggle():

    if btn.cget('text') == 'Start':
        btn.config(text='Stop')
    elif btn.cget('text') == 'Stop':
        btn.config(text='Start')

    t = threading.Thread(target=acquire_sample)
    t.start()

    return

btn = gc.action_button(root, label='Start', action=toggle)

root.mainloop()
```

Save your work. Run **dc_voltmeter_mcc_live.py** and marvel at the live updates to the acquired voltage after the **Start** button is pressed.

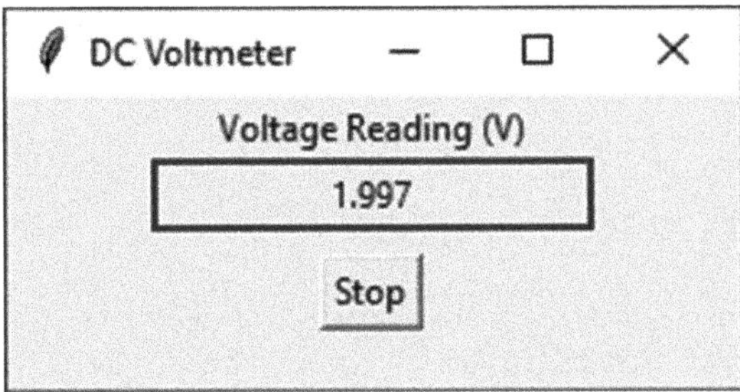

The updates are due to the `while` loop within `acquire_sample()`, which ceases execution when the **Stop** button is pressed. The `return` statement then causes program control to exit `acquire_sample()` and, at that point, the thread upon which that function was executing is terminated. With the next button-press, a new thread is created upon which `acquire_sample()` resumes execution.

We should pause for a moment to discuss one subtle (but important) feature of this program. As is, the `while` loop is "free running" (i.e., executing as fast as possible), and because of this rapid execution, your computer's processor is having to devote a significant amount of its efforts to running this program, leaving little time for the other operations for which it is responsible, a situation termed *CPU hogging*. To avoid this problem, simply insert a small delay in each `while`-loop iteration. Python's `time` module (included in the Python Standard Library) has the `sleep(delay)` function, which causes the program execution to pause for the time interval `delay` (in units of seconds).

Include a delay of 100 ms in **dc_voltmeter_mcc_live.py** between each voltage sample as follows: Place `import time` with the other import statements at the top of the program and add the `time.sleep()` statement in `acquire_sample()` as follows:

```
def acquire_sample():

    while btn.cget('text') == 'Stop':
        float_data = mo.read_sample(board_num, channel,
                                    ai_range)
        string_data = f'{float_data:.3f}'

        lbl_2.config(text=string_data)

        time.sleep(0.1)

    return
```

Remember to include `import time` at the top of your program. Save your work and then run **dc_voltmeter_mcc_live.py**. The voltage updates will still occur plenty fast, ten times a second.

8.6 Hardware-Timed Digital Oscilloscope

In the previous exercise, `v_in()` acquired a single voltage reading of the analog signal at a channel input. By putting this command within a repetitive `while` loop and controlling the time per loop iteration with `time.sleep()`, you wrote **dc_voltmeter_mcc_live.py**, a program that, with the simple addition of some data storage and plotting capability, one might consider using to sample and display N data samples of a time-varying waveform. The last sentence outlines the schematic for one of the most useful laboratory monitoring systems—the *digital oscilloscope*. Perhaps you've already thought of some cool things to do with this idea and are ready to go back and start to make the needed modifications to **dc_voltmeter_mcc_live.py**..., but unfortunately `time.sleep()` is an Achilles' heel in this plan. `time.sleep()` measures time by accessing a clock within your computer, which will provide an accuracy of somewhere between 1 and 15 millisecond (ms), depending on your system. Thus, by using `time.sleep()` to mark time in the data-taking process, the moment of each voltage digitization will be at best determined with an uncertainty on the order of 0.001 second. For very-low-frequency inputs (say, less than 10 Hz), this level of uncertainty in *time*-axis values might possibly be acceptable for the envisioned oscilloscope's *voltage* versus *time* output, but it will be useless over the higher range of frequencies in which you would want an oscilloscope to operate.

Thankfully, a much better clock called the *sample clock*, with precision timing on the order of 10 nanoseconds (14 ns for MCC USB-202), exists on a Measurement Computing DAQ device. And, even better, commands in the ul and enums modules provide access and control of the sample clock, allowing you to accurately schedule the moment each sample is acquired (or generated) in your experiment. DAQ operations under the control of the sample clock are termed *hardware-timed*, while the scheme outlined above involving the `sleep()` function is called *software-timed*.

Let's explore control of the DAQ device's sample clock as we write a useful digital oscilloscope program. The program we wish to write, called **digital_oscilloscope_mcc.py**, will acquire N equally spaced voltage samples of a time-varying analog input and then quickly plot this array of data values. By repeating this process over and over, we'll achieve a real-time display of the waveform input.

The heart of **digital_oscilloscope_mcc.py** will be the function `a_in_scan()` from the ul module, whose form is as follows:

```
mcculw.ul.a_in_scan(board_num, low_chan, high_chan,
                    num_points, rate, ai_range, memhandle,
                    scan_options)
```

As you can see, with eight parameters, this function has many moving parts, so let's take a few moments to understand its operation and the meaning of its various parameters. Remembering that a multifunction DAQ device typically has many AI channels (e.g., the MCC USB-202 has eight), the overall purpose of `a_in_scan()` is to perform a *scan* of a specified number of AI channels on the DAQ device identified by `board_num`. The AI channels are scanned in sequence, starting and ending at the AI channels denoted by the integers `low_chan` and `high_chan`, respectively. At a particular moment, the scan quickly sequences through the channels, acquiring a voltage sample at each. `a_in_scan()` repeats this process at the specified sampling frequency `rate` until it acquires ("reads") N samples at each channel. The parameter `num_points` is the total number of samples acquired during a scan and so equals the number of samples acquired per channel times the number of scanned channels, as given here:

```
num_points = N x (high_chan-low_chan +1)
```

The argument `ai_range` has the meaning we've seen before; it sets the range of the AI channels.

Finally, we have the two arguments `memhandle` and `scan_options`, each of which requires some detailed explanation. First, `scan_options` is a 32-bit number, best thought of as a binary number. Each of the 32 bits of this binary number performs as a *flag*, which can turn off (if *0*) or on (if *1*) an associated option of the DAQ device. For example, when a MCC DAQ device performs 12-bit digitization of an input voltage, a raw integer value results, which is in the range from 0 to $2^{12} - 1 = 4,095$. The fifth bit of `scan_options` configures how the DAQ device reports its digitized samples, whether as raw integer values (fifth bit set to *0*) or scaled in the units of volts (fifth bit set to *1*). As a convenience, the enums module contains text-based enums that automatically translate to the correct binary number. That is, the enum

`ScanOptions.SCALEDATA` is associated with the 32-bit binary number $0......10000_2$ (i.e., all zeros, except in the fifth bit) and so can be used to turn on voltage-scaling of acquired AI samples.

Second, we have the argument `memhandle`. Here, the manner in which `a_in_scan()` stores its acquired samples departs from our previous experience. For example, when we acquired a single voltage sample using `v_in()`, we wrote the following:

```
single_sample = mcculw.ul.v_in(board_num, channel,
                               ai_range)
```

In this statement, when the function `v_in()` executes, it returns its acquired sample, which is assigned to the object `single_sample`. The function `a_in_scan()` works differently. Instead of returning its acquired samples in the standard way in which functions return their results when called, `a_in_scan()` places its acquired samples in a predefined array in computer memory. Thus, before calling `a_in_scan()`, one must create the array of memory (called a *buffer*). Basic Python itself does not allow the level of control over memory management required for this task, so the buffer is created using a module from the Python Standard Library called *ctypes*. This module allows one to invoke C programming commands from within a Python program, which will produce the needed buffer. Here's a brief background: In C, an array of values is stored in a block of contiguous memory. The array has a known number of bytes, so to specify its location, one needs only to know the memory location of the first (index-zero) element. Then, by knowing the format of each value (e.g., two-byte integer, eight-byte float) and the fact that the memory locations are contiguous, one can work out the location of any other element in the array. The memory location of the array's first element and the formatting of the array values are contained in an object called the *pointer*. With that background, here's the three-step process for creating the buffer used by the `a_in_scan()` function:

```
import mcculw.ul as mu
import ctypes


memhandle = mu.scaled_win_buf_alloc(num_points)

ptr = ctypes.POINTER(ctypes.c_double)

ctypes_array = ctypes.cast(memhandle, ptr)
```

In this sequence of three steps, the function `scaled_win_buf_alloc(num_points)` from the ul module allocates a buffer in memory for `num_points` data elements, where each value is eight bytes in size. At this point, the data elements are not formatted, they are each just 64 bits in memory (remember, one byte equals eight bits), and the buffer is assigned a name `memhandle`, where a *handle* is an abstract quantity that identifies the block of memory; i.e., it is not a pointer. Next, a command from the `ctypes` module called `POINTER()` creates a pointer object assigned the name `ptr`. Its argument is an enum from the `ctype` module, which formats the pointer as a double-precision (eight-byte) floating-point number. At this point, `ptr` is a *void pointer*; i.e., it is created and configured as a particular data type, but it is not pointing to a particular array yet. Finally, the `cast(memhandle,ptr)` from the ctypes module assigns the pointer `ptr` to the array with the handle `memhandle` and that array is assigned the name `ctypes_array`.

Okay, that was a lot. Let's test-drive `a_in_scan()` with a program that will outline the data-acquisition portion of our digital oscilloscope program. We will assume that a single analog signal is being acquired at AI channel CH0, so `low_chan = high_chan = 0`. In the Editor, use **File>>New File** to create a program named **ai_scan_operation_mcc.py** in **YourNamePythonWork**, then enter the following code, which is explained below:

```
import mcculw.ul as mu
import mcculw.enums as me
import Utilities.waveform_creator as wc
import matplotlib.pyplot as plt
import ctypes as ct

board_num, low_chan, high_chan = 0, 0, 0
ai_range = me.ULRange.BIP10VOLTS

rate=1000
num_points=100

memhandle = mu.scaled_win_buf_alloc(num_points)

ptr = ct.POINTER(ct.c_double)
ctypes_array = ct.cast(memhandle, ptr)

scan_options = me.ScanOptions.SCALEDATA

mu.a_in_scan(board_num, low_chan, high_chan,
             num_points, rate, ai_range,
             memhandle, scan_options)

num, samp_freq = num_points, rate

t = wc.create_time_data(num, samp_freq)
y = [ctypes_array[i] for i in range(num_points)]

plt.plot(t, y)
plt.show()

mu.win_buf_free(memhandle)
```

In this code, first the `num_points`-element C-type array named `ctypes_array` is created as a memory buffer. Then, `a_in_scan()` acquires `num_points` samples with a sampling frequency `rate` and stores these samples in `ctypes_array`. Next, the following list comprehension is used to fill a `num_points`-element Python list named `y` with the `num_points` samples in `ctypes_array`:

```
y =[ctypes_array[i] for i in range(num_points)]
```

The acquired samples are then plotted in a pop-up window by the function `plot()`, found in `matplotlib.pyplot`. At the conclusion of this program, the function `win_buf_free(memhandle)` from the ul module releases the allotment of RAM that the buffer had reserved.

Save your work. To test out this program, adjust the settings on a function generator so that it is outputting an analog sine wave with amplitude less than 10 V and a frequency of about 50 Hz. Connect the positive and negative (ground) outputs from the function generator to the positive and negative inputs of AI channel CH0 and AGND on your DAQ device, respectively. *Note:* Most function generators are referenced to building ground and so are not "floating" voltage sources. Now, run **ai_scan_operation_mcc.py**. If all goes well, you will see a Matplotlib plot of about five cycles of the 50-Hz sine wave.

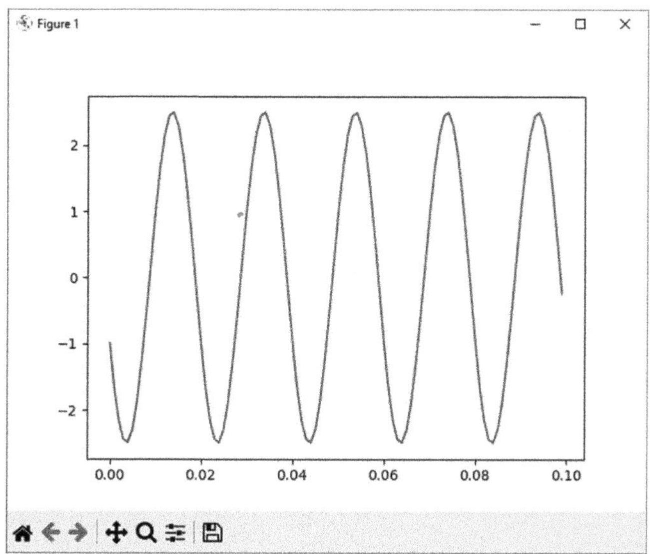

Before going further, let's add three data-acquisition functions to our **mccdaq_operations.py** module, one that creates the memory buffer, another that acquires a waveform on a given analog input channel, and a third that releases the memory buffer. In the Editor, open **mccdaq_operations.py** in **YourNamePythonWork>>Utilities**. Add the import statement `import ctypes as ct` at the top of the module, then the following definition for the three functions `create_buffer()`, `read_waveform()`, and `free_buffer()`:

```
def create_buffer(num_points):

    memhandle = mu.scaled_win_buf_alloc(num_points)

    ptr = ct.POINTER(ct.c_double)
    ctypes_array = ct.cast(memhandle, ptr)

    return memhandle, ctypes_array
```

```
def read_waveform(board_num, low_chan, high_chan,
                  num_points, rate, ai_range,
                  memhandle, ctypes_array):



    scan_options = me.ScanOptions.SCALEDATA

    mu.a_in_scan(board_num, low_chan, high_chan,
                 num_points, rate, ai_range,
                 memhandle, scan_options)

    y = [ctypes_array[i] for i in range(num_points)]

    return y


def free_buffer(memhandle):

    mu.win_buf_free(memhandle)

    return
```

The complete **mccdaq_operations.py** module will appear as shown next:

```
import mcculw.ul as mu
import mcculw.enums as me
import ctypes as ct

def chan_range(range_option):

    if range_option == 'bipolar_10':
        range_enum = me.ULRange.BIP10VOLTS
    elif range_option == 'unipolar_5':
        range_enum = me.ULRange.UNI5VOLTS

    return range_enum


def read_sample(board_num, channel, ai_range):

    float_data = mu.v_in(board_num, channel, ai_range)

    return float_data
```

```
def create_buffer(num_points):

    memhandle = mu.scaled_win_buf_alloc(num_points)

    ptr = ct.POINTER(ct.c_double)
    ctypes_array = ct.cast(memhandle, ptr)

    return memhandle, ctypes_array


def read_waveform(board_num, low_chan, high_chan,
                  num_points, rate, ai_range,
                  memhandle, ctypes_array):

    scan_options = me.ScanOptions.SCALEDATA

    mu.a_in_scan(board_num, low_chan, high_chan,
                 num_points, rate, ai_range,
                 memhandle, scan_options)

    y = [ctypes_array[i] for i in range(num_points)]

    return y


def free_buffer(memhandle):

    mu.win_buf_free(memhandle)

    return
```

Save your work and close **mccdaq_operations.py**.

We're now ready to write **digital_oscilloscope_mcc.py**, a program that will supply a GUI to the data-acquisition coding we just developed. For this program, along the code from **ai_scan_operation_mcc.py**, we'll also be drawing on components of two other previous programs—**waveform_simulator.py**, and **dc_voltmeter_mcc_live.py**.

In the Editor, use **File>>New File** to create a program named **digital_oscilloscope_mcc.py** and save it in **YourNamePythonWork**. Enter the following code, which creates an operating window named **Digital Oscilloscope** and places a **Digitizing Parameters** control within a `control_box` frame on the left-hand side of the window and a Matplotlib plot on the right-hand side of the window. A button labeled **Start** is placed below the **Digitizing Parameters** control; for now, make `action=None`. Many of these commands are in **waveform_simulator.py**, so you may wish to copy-and-paste relevant commands from that program.

```
import Utilities.waveform_creator as wc
import Utilities.gui_creator as gc
import Utilities.plot_creator as pc

label_x, label_y = 'Time (ms)', 'Voltage (V)'

root = gc.create_window(title='Digital Oscilloscope',
                        size='1000x500')
control_box, right_frame = gc.create_frames(root)

var_num, var_samp_freq = gc.digitizing_control(control_box)

ax, canvas = pc.create_plot(right_frame)
pc.update_plot(ax, canvas, [], [], label_x, label_y)

btn = gc.action_button(control_box, label='Start',
                       action=None)

root.mainloop()
```

Save your work and then run the program. If all is coded correctly, the program's GUI should appear (with a blank data plot).

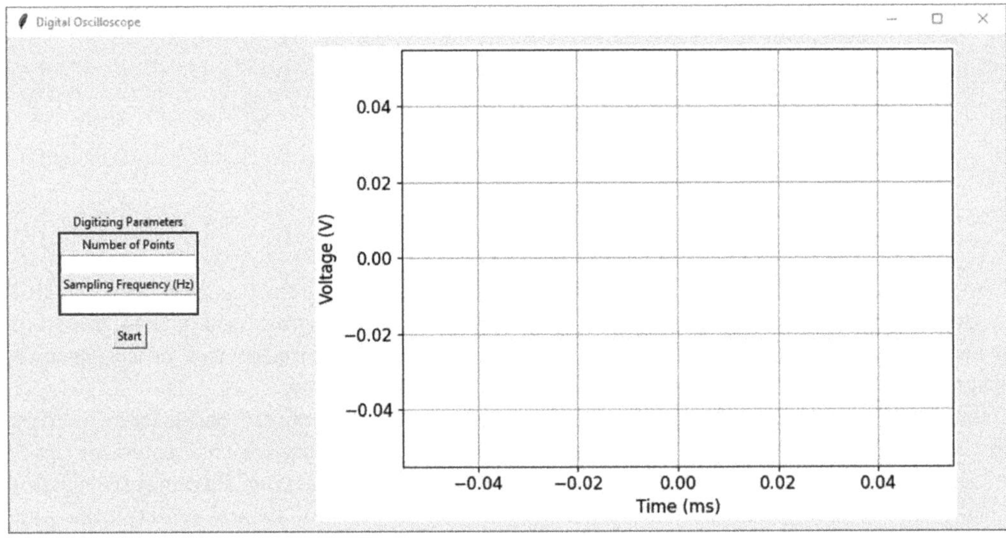

Next, at the top of the program, add the `import Utilities.mccdaq_operations as mo` and `import threading` statement as well as the DAQ device parameter assignments as shown below:

```
import Utilities.waveform_creator as wc
import Utilities.gui_creator as gc
import Utilities.plot_creator as pc
import Utilities.mccdaq_operations as mo
import threading

board_num, low_chan, high_chan = 0, 0, 0
ai_range = mo.chan_range('bipolar_10')
```

We now have to decide what we would like to have happen when the **Start** button is pressed. Following what we did in **dc_voltmeter_mcc_live.py**, let's have the button label toggle to **Stop**, while launching a new thread that carries out our intended DAQ operation. For this digital oscilloscope program, a function named `acquire_waveform()` will carry out the DAQ operation of acquiring N data samples of a time-varying waveform, displaying that acquired data, then repeating that process until the **Stop** button is pressed.

Using what we learned in writing **dc_voltmeter_mcc_live.py**, modify the program by changing `action=None` to `action=toggle` as well as by defining the two functions `toggle()` and `acquire_waveform()`. The function `acquire_waveform()` is a placeholder for now, but will be defined soon.

```
import Utilities.waveform_creator as wc
import Utilities.gui_creator as gc
import Utilities.plot_creator as pc
import Utilities.mccdaq_operations as mo
import threading

board_num, low_chan, high_chan = 0, 0, 0
ai_range = mo.chan_range('bipolar_10')

label_x, label_y = 'Time (ms)', 'Voltage (V)'

root = gc.create_window(title='Digital Oscilloscope',
                        size='1000x500')
control_box, right_frame = gc.create_frames(root)

var_num, var_samp_freq = gc.digitizing_control(control_box)

ax, canvas = pc.create_plot(right_frame)
pc.update_plot(ax, canvas, [], [], label_x, label_y)

def acquire_waveform():

    return
```

```
def toggle():

    if btn.cget('text') == 'Start':
        btn.config(text='Stop')
    elif btn.cget('text') == 'Stop':
        btn.config(text='Start')

    t = threading.Thread(target=acquire_waveform)
    t.start()

    return

btn = gc.action_button(control_box, label='Start',
                              action=toggle)

root.mainloop()
```

Save your work.

Finally, the `acquire_waveform()` function, which executes when the **Start** button is pressed, will carry out the data-acquisition process we developed in **ai_scan_operation_mcc.py.** This function first fetches the values that have been entered into the **Digitizing Parameters** control, which are then used to create the proper-sized memory buffer in which the acquired samples will be stored. Also, `create_time_array()` from **waveform_creator.py** creates a `num`-element NumPy array `t` of the equal-spaced time values at which data sampling occurs. Then, the `while` loop initiates and continues to execute. With each `while`-loop iteration, `read_waveform()` acquires the requested number of samples from the DAQ device, which are returned in a Python list `y`. The function `update_plot()` then plots *y* versus *t* on the GUI plot. When the **Stop** button is pressed, the `while` loop terminates execution and `free_buffer()` releases the RAM that was allocated to the buffer.

```
def acquire_waveform():

    num, samp_freq = gc.get_dig_entries(var_num,
                                        var_samp_freq)

    num_points, rate = num, int(samp_freq)

    memhandle, ctypes_array = mo.create_buffer(num_points)

    t = wc.create_time_data(num, samp_freq)

    while btn.cget('text') == 'Stop':

        y = mo.read_waveform(board_num, low_chan,high_chan,
```

```
                             num_points, rate, ai_range,
                             memhandle, ctypes_array)

            pc.update_plot(ax, canvas, 1000*t, y,
                           label_x, label_y, marker='o-')

    mo.free_buffer(memhandle)

    return
```

Complete **digital_oscilloscope_mcc.py** by adding the definition of the `acquire_waveform()` function. The finalized program is shown next:

```
import Utilities.waveform_creator as wc
import Utilities.gui_creator as gc
import Utilities.plot_creator as pc
import Utilities.mccdaq_operations as mo
import threading

board_num, low_chan, high_chan = 0, 0, 0
ai_range = mo.chan_range('bipolar_10')

label_x, label_y = 'Time (ms)', 'Voltage (V)'

root = gc.create_window(title='Digital Oscilloscope',
                        size='1000x500')
control_box, right_frame = gc.create_frames(root)

var_num, var_samp_freq = gc.digitizing_control(control_box)

ax, canvas = pc.create_plot(right_frame)
pc.update_plot(ax, canvas, [], [], label_x, label_y)

def acquire_waveform():

    num, samp_freq = gc.get_dig_entries(var_num,
                                        var_samp_freq)

    num_points, rate = num, int(samp_freq)

    memhandle, ctypes_array = mo.create_buffer(num_points)

    t = wc.create_time_data(num, samp_freq)

    while btn.cget('text') == 'Stop':
```

```
        y = mo.read_waveform(board_num, low_chan,high_chan,
                             num_points, rate, ai_range,
                             memhandle, ctypes_array)

        pc.update_plot(ax, canvas, 1000*t, y,
                       label_x, label_y, marker='o-')

    mo.free_buffer(memhandle)

    return


def toggle():

    if btn.cget('text') == 'Start':
        btn.config(text='Stop')
    elif btn.cget('text') == 'Stop':
        btn.config(text='Start')

    t = threading.Thread(target=acquire_waveform)
    t.start()

    return

btn = gc.action_button(control_box, label='Start',
                       action=toggle)

root.mainloop()
```

Save your work.

The N-samples acquisition task we have just programmed is termed a *buffered hardware-timed analog input* operation. In this process, a time-dependent analog signal is digitized at N equally spaced times. The time interval between neighboring samples $\Delta t = 1/f_s$, where f_s is the sampling frequency, is very accurately controlled by a hardware sample clock on the DAQ device. As the data samples are acquired, they are first placed in a memory buffer on the DAQ device and then transferred from there into the computer's memory buffer. This procedure assures that no data samples are lost when the computer multitasks (i.e., switches among its various chores) during the data-taking process.

Let's test-drive our handywork. Adjust the settings on a function generator so that it is outputting an analog sine wave with amplitude less than 10 V and a frequency of about 50 Hz. Connect the positive and negative (ground) outputs from the function generator to the positive and negative inputs of AI channel CH0 and AGND on your DAQ device, respectively. *Note:* Most function generators are referenced to building ground and so are not "floating" voltage sources. Run **digital_oscilloscope_mcc.py**. In the **Digitizing Parameters** control, set **Number of Samples** and **Sampling Frequency (Hz)** equal to *100* and *1000*, respectively, then press the **Start** button. If all goes well, you will see a plot of about five cycles of the 50-Hz sine wave.

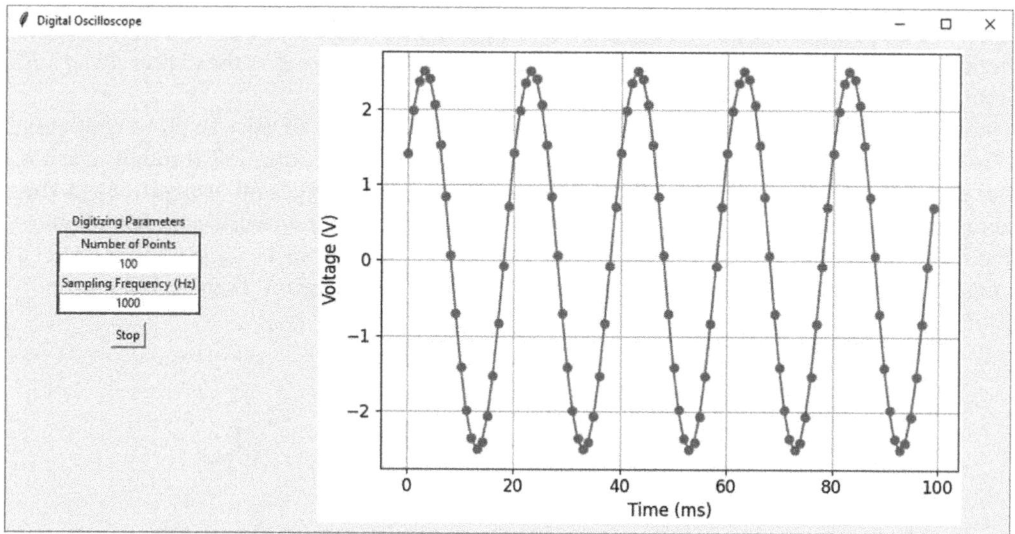

Be impressed that your program successfully acquires and plots a succession of digitized sine waves. However, I'm sure that one flaw in this computer-based instrument is apparent to you. Most likely, the sine-wave plot you are observing does not appear stationary but instead is moving either to the right or to the left, or is just shifting back and forth in a jittery fashion. And, by slightly changing the function generator's sine-wave frequency, the plotted sine wave can be made to move first one direction and then the other. Why is this so?

In your program, `read_waveform()` is configured so that with each execution, it acquires an array of N equally spaced data samples. Let's call this array of N samples a *trace*. As you know from coding the program, the sine-wave image viewed on the GUI's plot is not just a single trace but is actually a sequence of many traces, where a new trace is generated and plotted with each iteration of the `while` loop. Since nothing in your code forces each trace to begin at an equivalent point on the repetitive incoming waveform, the successive traces are displaced from each other along the *time*-axis, resulting in a motion picture-style moving image or, worse, a jumbled mess. The issue we are broaching here is called *triggering* and, for **digital_oscilloscope_mcc.py** to be a useful program, we must build in this capability.

In a commercially available oscilloscope, triggering is accomplished in the following manner. The input signal is monitored by an analog *level-crossing* circuit. The purpose of this circuit is to determine each time the incoming signal passes through a specified threshold voltage level and then immediately trigger the scope's data-acquisition process. Using knobs on the scope's front panel, the oscilloscope user sets the circuit's threshold level and specifies whether acquisition should be initiated when the level is passed through, starting from above (*negative*, or *falling*, *slope*) or starting from below (*positive*, or *rising*, *slope*).

Many higher-end MCC DAQ devices are capable of performing the analog level-crossing triggering procedure just described; the lower-cost MCC USB-202 does not have this capability. However, we will be able to implement an alternate mode—*digital edge triggering*—in our program because this mode is available on many MCC DAQ devices, including the MCC USB-202.

To execute a data-acquisition operation under digital edge triggering, one connects a digital signal to the appropriate pins on a DAQ device. Then, by choosing the proper software settings, the desired data-acquisition operation is initiated ("triggered") whenever the digital signal has a rising or, alternatively, falling edge.

Stop **digital_oscilloscope_mcc.py**, if it's running. We have a bit of work to do to configure it for digital edge triggering. First, we'll need an option menu on our GUI through which a user can select the desired (rising or falling) digital edge to trigger an acquisition. In the Editor, use **File>>Open**... to open **gui_creator.py** in **YourNamePythonWork>>Utilities**. Somewhere among the other definitions, add the following definition for `trigger_control()`, a function that creates a blue-bordered option menu labeled **Digital Trigger Edge** on our GUI:

```
def trigger_control(container):

    lbl = tk.Label(container,text='Digital Trigger Edge')
    lbl.pack(fill=tk.X)

    frame = tk.Frame(container, bg='blue', bd=2)
    frame.pack()

    var_trig = tk.StringVar()
    options = ['Falling', 'Rising']
    opm = tk.OptionMenu(frame, var_trig, *options)
    opm.config(width=8)
    opm.pack(fill=tk.X)

    return var_trig
```

Also, in **gui_creator.py**, add the following definition for `get_edge_option`, a function that fetches the selection entered into the **Digital Trigger Edge** option menu:

```
def get_edge_option(var_trig):

    edge_option = var_trig.get()

    return edge_option
```

Save your work and close **gui_creator.py**.

Next, in the Editor, open **mccdaq_operations.py** in **YourNamePythonWork>>Utilities**. Add the following definition for the function `config_digital_triggering()` to configure digital triggering. The required command `set_trigger(board_num,trig_type,0,0)` comes from the mcculw's `ul` module, which sets the desired digital triggering edge (via the `trig_type` argument) on the selected DAQ device. The third and fourth arguments are relevant to devices capable of analog level-crossing triggering and so we simply set them both to *0*. The *if . . . else* statement chooses the appropriate constant from the `enums` module for `trig_type` in order to select a rising or falling edge:

```
def config_digital_triggering(board_num, edge_option):

    if edge_option == 'Rising':
        trig_type = me.TrigType.TRIG_POS_EDGE
    else:
        trig_type = me.TrigType.TRIG_NEG_EDGE

    mu.set_trigger(board_num, trig_type, 0, 0)

    return
```

In addition, we need to configure the DAQ device for digital triggering by setting the digital-triggering flag in the 32-bit `scan_options` constant. Recall that we have already set the fifth bit of `scan_options` to *1*, configuring the DAQ device to report its digitized samples scaled in the units of volts. As a convenience, the enums module contains text-based enums that automatically translate to the correct binary number and so our selection was made by simply writing `scan_options = ScanOptions.SCALEDATA`. Now, in addition to setting the flag for scaled data, we wish to set the flag for digital triggering, which is the fifteenth bit, i.e., $0......100000000000000_2$. The enum that encodes this binary number is `ScanOptions.EXTTRIGGER`. As a trick for setting both of these flags, we apply the *bitwise OR* operation to the two enums. The bitwise OR is a binary operator that produces a ("resultant") binary number by evaluating an OR operation at each corresponding bit of two binary numbers (called the *operands*). At the *n*th bit, if at least one of the operands' bits is *1*, the *n*th bit of the resultant is *1*; otherwise, it is *0*. Thus, the resultant of the bitwise OR operation between the two binary numbers represented by `ScanOptions.SCALEDATA` and `ScanOptions.EXTTRIGGER` is the 32-bit binary number $0......100000000010000_2$, exactly the binary number we need to set the two desired flags. In Python, the bitwise OR operator is denoted by the vertical pipe symbol (|). In **mccdaq_operations.py**, modify the `read_waveform()` function as shown below, in order to activate digital triggering:

```
def read_waveform(board_num, low_chan, high_chan,
                  num_points, rate, ai_range,
                  memhandle, ctypes_array):

    enum_1 = me.ScanOptions.SCALEDATA
    enum_2 = me.ScanOptions.EXTTRIGGER

    scan_options = enum_1 | enum_2


    mu.a_in_scan(board_num, low_chan, high_chan,
                 num_points, rate, ai_range,
                 memhandle, scan_options)

    y = [ctypes_array[i] for i in range(num_points)]

    return y
```

Save your work and close **mccdaq_operations.py**.

Finally, a few additions to **digital_oscilloscope_mcc.py** are needed to incorporate digital triggering into its operations. In the Editor, open **digital_oscilloscope_mcc.py** in **YourNamePythonWork.** Make the following changes: Along with creating a **Digitizing Parameters** control panel on the GUI, create a **Digital Trigger Edge** option menu:

```
var_num, var_samp_freq = gc.digitizing_control(control_box)
var_trig = gc.trigger_control(control_box)
```

Finally, within the code block above the `while` loop in the function `acquire_waveform()`, add `get_edge_option()` to fetch the edge option selected in the **Digital Trigger Edge** option menu as well as `config_digital_triggering()` to set the selected edge option on the DAQ device. The block of code above the `while` loop will be as shown next:

```
    num, samp_freq = gc.get_dig_entries(var_num,
                                        var_samp_freq)
    edge_option = gc.get_edge_option(var_trig)

    num_points, rate = num, int(samp_freq)

    mo.config_digital_triggering(board_num, edge_option)

    memhandle, ctypes_array = mo.create_buffer(num_points)

    t = wc.create_time_data(num, samp_freq)
```

The complete **digital_oscilloscope_mcc.py** program is as follows:

```
import Utilities.waveform_creator as wc
import Utilities.gui_creator as gc
import Utilities.plot_creator as pc
import Utilities.mccdaq_operations as mo
import threading

board_num, low_chan, high_chan = 0, 0, 0
ai_range = mo.chan_range('bipolar_10')

label_x, label_y = 'Time (ms)', 'Voltage (V)'

root = gc.create_window(title='Digital Oscilloscope',
                        size='1000x500')
control_box, right_frame = gc.create_frames(root)

var_num, var_samp_freq = gc.digitizing_control(control_box)
var_trig = gc.trigger_control(control_box)

ax, canvas = pc.create_plot(right_frame)
pc.update_plot(ax, canvas, [], [], label_x, label_y)
```

```
def acquire_waveform():

    num, samp_freq = gc.get_dig_entries(var_num,
                                        var_samp_freq)
    edge_option = gc.get_edge_option(var_trig)

    num_points, rate = num, int(samp_freq)

    mo.config_digital_triggering(board_num, edge_option)

    memhandle, ctypes_array = mo.create_buffer(num_points)

    t = wc.create_time_data(num, samp_freq)


    while btn.cget('text') == 'Stop':

        y = mo.read_waveform(board_num, low_chan, high_chan,
                             num_points, rate, ai_range,
                             memhandle, ctypes_array)

        pc.update_plot(ax, canvas, 1000*t, y,
                       label_x, label_y, marker='o-')

    mo.free_buffer(memhandle)

    return


def toggle():

    if btn.cget('text') == 'Start':
        btn.config(text='Stop')
    elif btn.cget('text') == 'Stop':
        btn.config(text='Start')

    t = threading.Thread(target=acquire_waveform)
    t.start()

    return

btn = gc.action_button(control_box, label='Start',
                       action=toggle)

root.mainloop()
```

That's it. **digital_oscilloscope_mcc.py** is now configured for digital edge triggering. Save your work.

Besides producing a sine wave with frequency f at its *main* output, your function generator creates a TTL digital signal of the same frequency f, which is available at its *sync* (sometimes called *TTL* or *trigger out*) output. The digital edges of this TTL signal coincide in time with particular points on the sine wave's cycle. For example, on some function generator models, the rising edge of the TTL signal occurs when the sine wave is at its peak; for other models, the rising TTL edge coincides with the positive-going zero-crossing of the sine wave. Thus, the sync output provides a convenient digital signal with which one can control the acquisition of sine-wave data (i.e., always begin at the same equivalent point on the cycle) via digital edge triggering.

Connect the sync output from your function generator to the digital triggering pins on your DAQ device. The sync's positive and negative (ground) terminals should be connected to the TRIG and GND pins, respectively, for the software selection we made in our programming.

Run **digital_oscilloscope_mcc.py** with **Number of Samples, Frequency (Hz),** and **Digital Trigger Edge** equal to *100*, *1000*, and *Rising*, respectively, and set the sine-wave frequency on your function generator to approximately 50 Hz. If all goes well, you will see a "stationary" plot of about five cycles of the 50-Hz sine wave. Press the **Stop** button, then try switching **Digital Trigger Edge** to *Falling*. After you press the **Start** button, is the resulting change to the waveform plot as you expected?

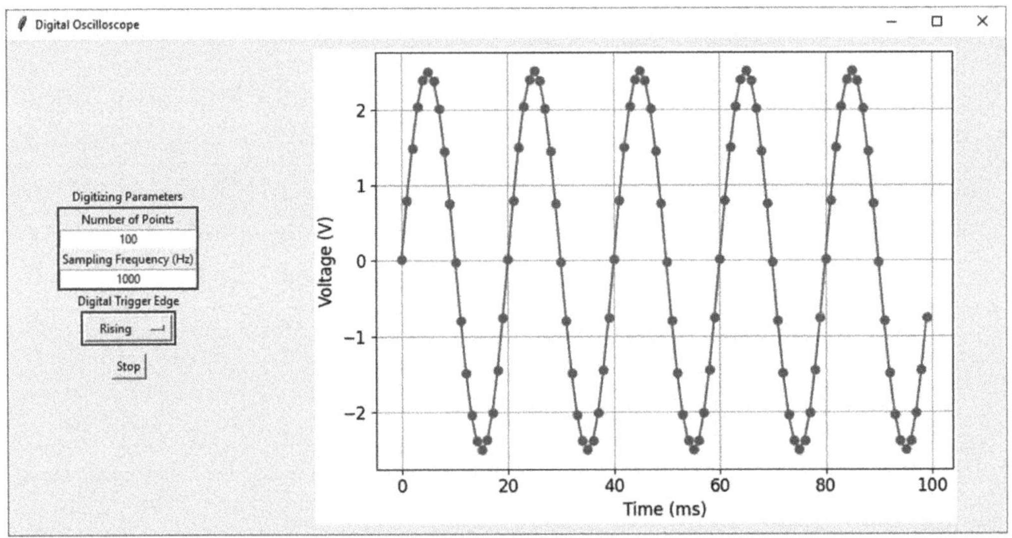

On the function generator, set the frequency f of the input sine-wave signal to several values within the range $0 \leq f \leq 500$ Hz. You should find that frequencies of signals in this frequency range are faithfully reproduced by **digital_oscilloscope_mcc.py**, but that sine-wave shape becomes distorted as the signal frequency approaches 500 Hz (because of the paucity of sampled points per sine-wave cycle at these higher input frequencies).

We have set the sampling frequency equal to $f_s = 1000$ Hz; thus, the Nyquist frequency is $f_{nyquist} = f_s/2 = 500\,\text{Hz}$. For input frequencies greater than the Nyquist frequency, that is, $f > 500$ Hz, we expect the aliasing effect.

With **Number of Samples, Frequency (Hz),** and **Digital Trigger Edge** equal to *100*, *1000*, and *Rising*, respectively, demonstrate that for $f = 950$ Hz and $f = 1050$, the digitized sine wave appears to have a frequency of 50 Hz, rather than the actual input frequencies. Also, note that there is a 180° for one of these aliased frequencies (remember digital triggering is on the rising edge of the input frequency). Use Eq. (6.2) with $n = 2$ to predict two other input frequencies f that will appear as a 50-Hz sine wave when digitized because of aliasing, and then demonstrate that your predictions are correct using **digital_oscilloscope_mcc.py**.

As a final note, we have configured the buffer for a certain number of samples, the sampling frequency, and the trigger edge prior to entering the data-acquiring `while` loop. If, say, you decide you'd like to change the number of acquired samples, press the **Stop** button, then enter the newly desired value in **Number of Samples** entry box. When you press the **Start** button a properly size buffer for your new number of samples will be created and data acquisition under that new choice of parameter will commence.

8.7 DC Voltage Source

Now let's explore the analog output (AO) operations available on your DAQ device. To familiarize ourselves with these processes, we'll start by writing a program that simply generates ("writes") a single desired voltage value. This program will implement the `v_out()` command in the mcculw's `ul` module. Here are the Python commands that code the operation of simply outputting a single requested voltage value, in this case, 1.0 V:

```
import mcculw.ul
import mcculw.enums

board_num, channel = 0, 0
ao_range = mcculw.enums.ULRange.UNI5VOLTS

data_value = 1.0

mcculw.ul.v_out(board_num, channel, ao_range, data_value)
```

Note that this code is very similar to the sample-reading program **simple_ai_operation_mcc.py** we wrote previously, except this time `v_in` and `ai_range` are replaced by `v_out` and `ao_range`, respectively. The MCC USB-202 DAQ device is capable of only generating output voltages in the unipolar range of 0 to +5 V; hence, `ao_range` is assigned the enum `ULRange.UNI5VOLTS`.

To streamline coding of our AO program, in the Editor open **simple_ai_operation_mcc.py**, then use **File>>Save As...** to create a new program named **simple_ao_operation_mcc.py** in **YourNamePythonWork**. Next, modify the code to match that shown above and save your work.

Connect the positive and negative (ground) inputs of a voltmeter (or oscilloscope in autotriggering mode) to the AOUT0 and AGND pins of your DAQ device, respectively.

Run **simple_ao_operation_mcc.py** and your voltmeter should inform you that the requested voltage difference is being produced by your DAQ device. Try running **simple_ao_operation_mcc.py** with it programmed to output a few other values within the allowed range of your DAQ device (0 to +5 V for the MCC USB-202) to convince yourself that the program is functioning properly.

It would be helpful to have an interactive way to input requested voltage values in this program; in other words, have a GUI. To accomplish that feat, we'll need to define a few new functions in the **gui_creator.py** and **mccdaq_operations.py** modules, so let's do that first.

To input a requested voltage value, our GUI will need a solitary control, that is, a single entry widget with an identifying label, along with the means of retrieving its user-entered value. In Chapter 5's Do Iy Yourself project, we added two functions named `create_control()` and `get_control_string()` to the **gui_creator.py** module that provide this needed functionality, so we will make use of that previous work. If you didn't produce these two functions in Chapter 5, do so now following the instructions in the next paragraph.

In the Editor, open **gui_creator.py** in **YourNamePythonWork>>Utilities**. Then, add the following definition for the function `create_control()`. This function first places a descriptive `label` and a blue-bordered `frame` within the specified `container` (e.g., the `root` window), then places an entry widget named `ent` within the `frame`. The value that a user has entered into the entry widget is returned in the string variable `var_ent`.

```
def create_control(container, label, char_width=20):

    lbl = tk.Label(container, text=label)
    lbl.pack()

    frame = tk.Frame(container, bg='blue', bd=2)
    frame.pack()

    var_ent = tk.StringVar()
    ent = tk.Entry(frame, textvariable=var_ent,
                   justify=tk.CENTER, width=char_width)
    ent.pack(fill=tk.X)

    return var_ent
```

Also, in **gui_creator.py**, define the following function called `get_control_string(var_ent)`, which fetches the string a user has entered in the entry widget.

```
def get_control_string(var_ent):

    control_string = var_ent.get()

    return control_string
```

Save your work as you close **gui_creator.py**.

Next, we need a function that creates an analog output task. Since this is a DAQ-related function, in the Editor, open **mccdaq_operations.py** in **YourNamePythonWork>>Utilities**. Then, add the following definition for the function `write_sample()`, which carries out an AO operation on a specified channel on a specified DAQ device:

```
def write_sample(board_num, channel, ao_range, data_value):

    mu.v_out(board_num, channel, ao_range, data_value)

    return
```

Save your work and close **mccdaq_operations.py**.

We're now ready to create a GUI-equipped DC voltage source. This program will be very similar to the **dc_voltmeter_mcc.py** program that you wrote earlier in this chapter, so I recommend the following: In the Editor, open **dc_voltmeter_mcc.py**, then use **File>>Save As...** to create a program named **dc_voltage_source_mcc.py** in **YourNamePythonWork**. Next, modify the program as shown here.

```
import Utilities.gui_creator as gc
import Utilities.mccdaq_operations as mo


board_num, channel = 0, 0
ao_range = mo.chan_range('unipolar_5')

root = gc.create_window(title='DC Voltage Source',
                        size='300x100')

var_ent = gc.create_control(root,
                            label='Input Voltage')

def generate_sample():

    string_data = gc.get_control_string(var_ent)
    data_value = float(string_data)

    mo.write_sample(board_num, channel, ao_range,
                    data_value)

    return


btn = gc.action_button(root, label='Write Sample',
                       action=generate_sample)

root.mainloop()
```

Save your work. Now run **dc_voltage_source_mcc.py**, entering a voltage within the allowed range of your DAQ device's AO capability.

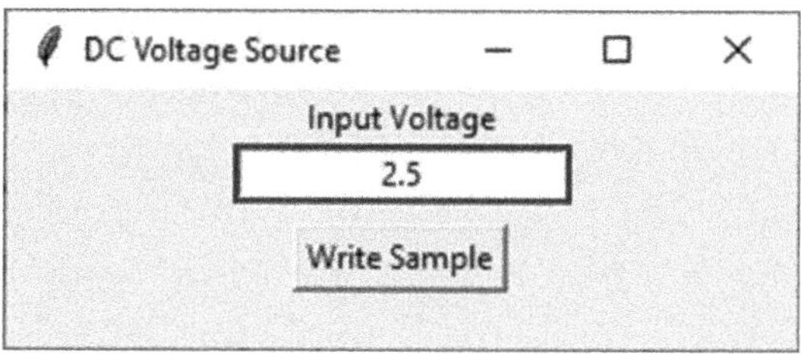

Press the **Write Sample** button and use a voltmeter to verify that the requested voltage difference is being output on AO channel 0 (i.e., between pins AOUT0 and AGND). Try outputting several other values using **dc_voltage_source.py**.

When the execution of **dc_voltage_source.py** is stopped (by closing the **DC Voltage Source** window), your MCC DAQ device is designed to maintain the last voltage value it generated. Thus, before stopping the program, press **Write Sample** one more time with **Input Voltage** set to *0*, so the channel AOUT0 is left in the zero-voltage state.

8.8 Hardware-Timed Waveform Generator

By continually updating an AO channel with appropriately chosen voltage values, one can turn a DAQ device into a waveform generator with a time-dependent voltage output of any desired shape (e.g., sine wave, square wave, triangle wave). To obtain an output voltage whose time dependence accurately represents that desired shape, one must ensure that the AO channel changes ("updates") to particular values at precisely the right moments. The most accurate method for timing these required updates involves the use of a high-frequency hardware clock and such a clock (i.e., the sample clock) is contained internally on many MCC DAQ devices. Unfortunately, AO operations on our highlighted USB-202 device cannot be controlled by a hardware clock, so only the much less accurate method of software-timed waveform generation is possible on that device; see Problem 3. However, the higher-end, but still affordable USB-231 DAQ device from MCC does have this capability. Thus, for this section, we will assume you have access to that device (or another MCC product with hardware-timed AO capability) in order to show how to create high-quality analog waveforms.

To understand the command sequence needed for hardware-timed waveform generation, we'll start by writing a GUI-less program that simply produces a 50-Hz sine wave with amplitude of 5 V. This program will require two additions to **mccdaq_operations.py** in **YourNamePythonWork>>Utilities**, so open that file in the Editor and add the following two functions:

```
def write_waveform(board_num, low_chan, high_chan,
                   num_points, rate, ao_range, memhandle):

    enum_1 = me.ScanOptions.BACKGROUND
    enum_2 = me.ScanOptions.CONTINUOUS
    enum_3 = me.ScanOptions.SCALEDATA

    scan_options = enum_1 | enum_2 | enum_3

    mu.a_out_scan(board_num, low_chan, high_chan,
                  num_points, rate, ao_range, memhandle,
                  scan_options)

    return


def stop_waveform(board_num):
```

```
    function_type = me.FunctionType.AOFUNCTION
    mu.stop_background(board_num, function_type)

    return
```

The function `write_waveform()` uses the ul module's `a_out_scan()` command to initiate the production of a waveform using a *hardware-timed continuous analog output* operation. The shape of the waveform is defined by a C-type array with `num_points` elements, whose location in memory is given by the handle `memhandle`. The parameter `rate` gives the update frequency of the AO operation producing the waveform element by element. Three options are chosen, one of which we've encountered before—`SCALEDATA`. As before, the `SCALEDATA` option indicates that the values in the C-type array are in the unit of volts. The `CONTINUOUS` option instructs the DAQ device to continually output the waveform (in a few moments, more explanation will be given on how this is accomplished) and the `BACKGROUND` option allows the AO operation, once initiated, to run independently, while the Python program execution immediately goes to the next line of the program. Finally, when called at a later time, `stop_waveform()` will stop the production of the waveform. Save **mccdaq_operations.py** as you close it.

In the Editor, use **File>>New File** to create a new program named **simple_ao_waveform_mcc.py** and save it in **YourNamePythonWork**. Then, enter the following sequence of commands. Here, we use functions from our **waveform_creator.py** module to create a 500-element NumPy array representing five cycles of a 50-Hz sine wave. In addition, from **mccdaq_operations.py**, `create_buffer()` is used to create the C-type array named `ctypes_array` in your computer's memory and the `for` loop loads the sine-wave data from a NumPy array into `ctypes_array`. Next, `write_waveform()` transfers the sine-wave data contained in `ctypes_array` to the DAQ device and then starts the AO process with an update rate of 5 kHz (the maximum update rate for the USB-231). In the continuous mode, the sine wave will be generated until `stop_waveform()` is called. Thus, `time.sleep(10.)` introduces a 10-second delay between the start and stop of the AO process.

```
import Utilities.waveform_creator as wc
import Utilities.mccdaq_operations as mo
import time

board_num, low_chan, high_chan = 1, 0, 0
ao_range = mo.chan_range('bipolar_10')

num, samp_freq = 500, 5000
a, f, shape = 5., 50., 'Sine'

t = wc.create_time_data(num, samp_freq)
y = wc.create_waveform_data(t, a, f, shape)

num_points, rate = num, samp_freq

memhandle, ctypes_array = mo.create_buffer(num_points)
```

```
for i in range(num_points):
    ctypes_array[i] = y[i]

mo.write_waveform(board_num, low_chan, high_chan,
                  num_points, rate, ao_range, memhandle)

mo.free_buffer(memhandle)

time.sleep(10.)

mo.stop_waveform(board_num)
```

Save your work. Connect the positive and negative (ground) inputs of an oscilloscope to the AOUT0 and AGND pins of your DAQ device, respectively. Set your scope to the **Normal** or **Auto** triggering mode and the **Trigger Level** to somewhere within the range of sine-wave variation (i.e., between −5 and +5 V). Run **simple_ao_waveform_mcc.py** and, if all is coded correctly, you will view a 50-Hz sine wave continuously generated for 10 seconds.

The waveform you observe on the oscilloscope is the result of `a_out_scan()` programming a *hardware-timed continuous analog output* process. In this process, a 1D array consisting of a sequence of voltage values of the desired waveform shape is first written from computer memory into a dedicated memory buffer on your DAQ device. Then, when initiated by receipt of the start command, the DAQ device outputs this sequence of voltages at analog output channel 0 in a *circular fashion* (i.e., after the last array value is outputted, the next output is the first array value, and the sequence repeats). The circular output of the DAQ device's on-board buffer values continues until the `stop_background()` command within `mo.stop_waveform()` is received.

Let's now write a program named **waveform_generator_mcc.py**, which equips this hardware-timed AO operation with a GUI. The **waveform_generator_mcc.py** program will follow a similar template to **digital_oscilloscope_mcc.py**, so you can streamline the necessary coding by using that already-coded program as a starting point.

In the Editor, open **digital_oscilloscope_mcc.py**, then use **File>>Save As**... to create a program named **waveform_generator_mcc.py** and save it in **YourNamePythonWork.** Modify the task- and GUI-creation code so that it appears as shown below. Here, the `import time` command has been added, some labeling has been changed, and the `trigger_control()` creation function has been replaced by the function that creates a **Waveform Parameters** control within the `control_box` frame.

```
import Utilities.waveform_creator as wc
import Utilities.gui_creator as gc
import Utilities.plot_creator as pc
import Utilities.mccdaq_operations as mo
import threading

board_num, low_chan, high_chan = 0, 0, 0
ao_range = mo.chan_range('bipolar_10')

label_x, label_y = 'Time (ms)', 'Voltage (V)'
```

```
root = gc.create_window(title='Waveform Generator',
                        size='1000x500')
control_box, right_frame = gc.create_frames(root)

var_num, var_samp_freq = gc.digitizing_control(control_box)
var_a, var_f, var_shape = gc.waveform_control(control_box)

ax, canvas = pc.create_plot(right_frame)
pc.update_plot(ax, canvas, [], [], label_x, label_y)
```

Next, modify **digital_oscilloscope_mcc.py**'s `acquire_waveform()` function to become the following definition of a function called `generate_waveform()`. Here, the code will be familiar from **simple_ao_waveform_mcc.py**.

```
def generate_waveform():

    num, samp_freq = gc.get_dig_entries(var_num,
                                        var_samp_freq)
    a, f, shape = gc.get_wav_entries(var_a, var_f,
                                     var_shape)

    t = wc.create_time_data(num, samp_freq)
    y = wc.create_waveform_data(t, a, f, shape)

    pc.update_plot(ax, canvas, 1000*t, y, label_x, label_y)

    num_points, rate = num, int(samp_freq)

    memhandle, ctypes_array = mo.create_buffer(num_points)

    for i in range(num_points):
        ctypes_array[i] = y[i]

    mo.write_waveform(board_num, low_chan, high_chan,
                      num_points, rate, ao_range,
                      memhandle)

    mo.free_buffer(memhandle)

    return
```

Finally, revise **digital_oscilloscope_mcc.py**'s `toggle()` function as shown next. Here, pressing the **Start** button will spawn a thread named `t` upon which the function `generate_waveform()` executes. Through its execution, `generate_waveform()` loads the desired waveform data into the DAQ device's on-board buffer and starts the continuous analog output operation. As the AO operation continues, a `while` loop keeps the DAQ-related thread `t` active until the **Stop** button is pressed, at which time the waveform generation is stopped, the allocated memory is released, and the button label is morphed back to **Start**.

With its `return` statement, `generate_waveform()` has then completed its execution and so the thread `t` terminates. If the user wishes, the control parameters can be changed and, with a press of the **Start** button, the process is repeated. Throughout, the GUI code runs on the main thread initiated when the program begins its execution.

```
def toggle():

    if btn.cget('text') == 'Start':
        btn.config(text='Stop')
        t = threading.Thread(target=generate_waveform)
        t.start()
    elif btn.cget('text') == 'Stop':
        btn.config(text='Start')
        mo.stop_waveform(board_num)

    return
```

Save your work. The complete **waveform_generator_mcc.py** should appear as follows:

```
import Utilities.waveform_creator as wc
import Utilities.gui_creator as gc
import Utilities.plot_creator as pc
import Utilities.mccdaq_operations as mo
import threading

board_num, low_chan, high_chan = 1, 0, 0
ao_range = mo.chan_range('bipolar_10')

label_x,label_y = 'Time (ms)', 'Voltage (V)'

root = gc.create_window(title='Waveform Generator')
control_box, right_frame = gc.create_frames(root)

var_num, var_samp_freq = gc.digitizing_control(control_box)
var_a, var_f, var_shape = gc.waveform_control(control_box)

ax,canvas = pc.create_plot(right_frame)
pc.update_plot(ax, canvas, [], [], label_x, label_y)


def generate_waveform():

    num, samp_freq = gc.get_dig_entries(var_num,
                                        var_samp_freq)
    a, f, shape = gc.get_wav_entries(var_a, var_f,
                                     var_shape)

    t = wc.create_time_data(num, samp_freq)
    y = wc.create_waveform_data(t, a, f, shape)
```

```
    pc.update_plot(ax, canvas, 1000*t, y, label_x, label_y)

    num_points, rate = num, int(samp_freq)

    memhandle, ctypes_array = mo.create_buffer(num_points)

    for i in range(num_points):
        ctypes_array[i] = y[i]

    mo.write_waveform(board_num, low_chan, high_chan,
                      num_points, rate, ao_range,
                      memhandle)

    mo.free_buffer(memhandle)

    return


def toggle():

    if btn.cget('text') == 'Start':
        btn.config(text='Stop')
        t = threading.Thread(target=generate_waveform)
        t.start()
    elif btn.cget('text') == 'Stop':
        btn.config(text='Start')
        mo.stop_waveform(board_num)

    return


btn = gc.action_button(control_box, label='Start',
                       action=toggle)

root.mainloop()
```

Connect the positive and negative (ground) inputs of an oscilloscope to the AOUT0 and AGND pins of your DAQ device, respectively. Set your scope to the **Normal** or **Auto** triggering mode and the **Trigger Level** to somewhere within the range between −5 and +5 V).

Test-drive **waveform_generator_mcc.py** by using it to output a 50-Hz sine wave of amplitude 5 V. First, in the **Waveform Parameters** control, set **Amplitude**, **Frequency (Hz)**, and **Shape** to be *5*, *50*, and *Sine*, respectively. Next, assuming 100 points are required to well define each cycle of the sine wave, the sampling frequency should be 100 times larger than the sine wave's frequency. Thus, for **waveform_generator_mcc.py** to create five complete cycles, in the **Digitizing Parameters** control, set **Number of Samples** and **Sampling Frequency (Hz)** be *500* and *5000*, respectively.

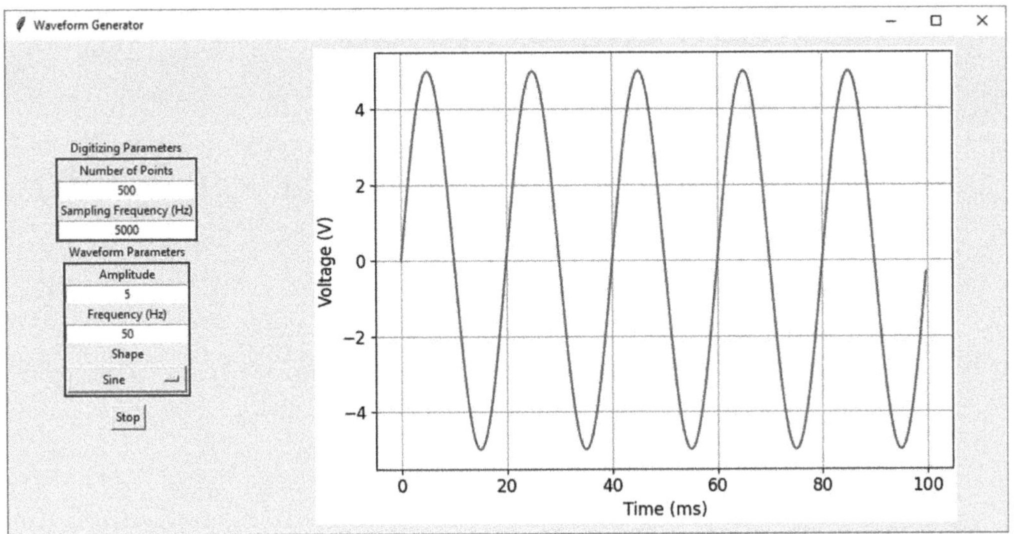

Press the **Start** button. The software waveform produced by `create_waveform_data()` will be displayed on the GUI's plot for you to view. Using your oscilloscope, observe the voltage waveform output at AO channel 0. Is it a (quasi-) continuous sine wave with a 50-Hz frequency and 5-V amplitude?

Press the **Stop** button. Then, keeping all other parameters the same, change **Frequency (Hz)** (in the **Waveform Parameters** control) to *75*. From the GUI's plot, you will find that `create_waveform_data()` produces 1½ cycles of this waveform. Using your knowledge of the circular fashion in which the DAQ device sequences through the memory buffer values, can you explain the discontinuities that you observe in the voltage waveform being output at AO channel 0?

With all other settings held the same, try generating sine waves with **Frequency (Hz)** set to *100*, *200*, *250*, and *500*. With each of these choices, `create_waveform_data()` will produce an integer number of complete cycles of the waveform and so produce voltage waveform outputs with no discontinuities. Why does the waveform output at AO channel 0 become less well defined for the higher frequencies? Also, you might enjoy changing **Shape** and **Amplitude** and viewing the resulting waveforms on the oscilloscope.

A final reminder: When program execution is stopped (by closing the **Waveform Generator** window), your MCC DAQ device is designed to maintain the last voltage value it generated. Thus, before stopping the program, press **Start** one more time with **Amplitude** set to *0* (and all other parameters set to nonzero values), so the AO channel 0 is left in the zero-voltage state.

Do It Yourself

Lighting LED Using DO Line In this exercise, you will write a program that alternately lights one LED and then another. The signals that toggle these LEDs on and off will be provided by digital output (DO) pins on a DAQ device.

(a) First, light a single LED under computer control as follows: Using the DIO0 line on your DAQ device, construct the hardware circuit shown in Fig. 8.1. In this circuit, when the line goes HIGH, it supplies a DAQ device-dependent voltage difference (common values are 3.3 and 5 V) across the resistor–LED series combination. Assuming this voltage difference is 3 V, the predicted current through the LED will be $3\text{ V}/220\ \Omega \approx 14$ mA (for a more precise prediction of the current, the voltage drop across the forward-biased LED should be taken into account; for a red LED, this voltage drop is about 1.8 V). If this predicted current is larger than the current-sourcing capability of your DAQ device's digital output line, the current will top out at the maximum allowed value. This maximum value can be found on your DAQ device's datasheet (the USB-202 is able to supply currents up to 24 mA).

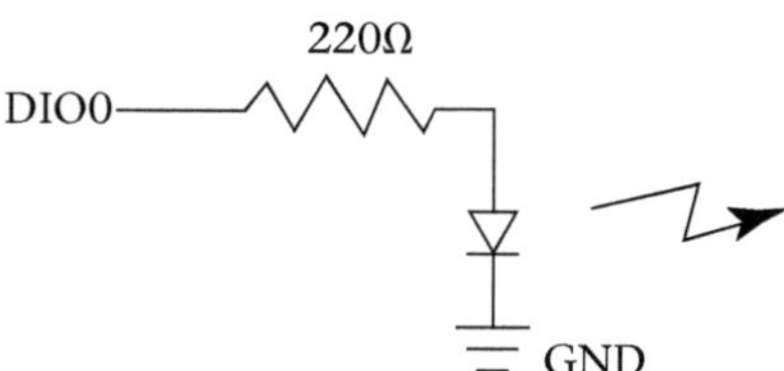

Fig. 8.1 Hardware circuit for Do It Yourself project. The digital I/O line DIO0 of the data-acquisition device is connected to the (current-limiting) resistor in series with LED.

Write a GUI-equipped program called **single_led_mcc.py** that can be used to illuminate this single LED via a GUI button. For use in your program, add the following two functions to **mccdaq_operations.py**. Here, `config_do_port()` configures the pins labeled DIO0 through DIO7 (i.e., the 8-bit DIO port) as digital output lines, and `write_do_port()` sets each of the eight pins in their LOW or HIGH state as determined by the decimal integer `data_value`. `data_value` ranges from 0 to 255, which corresponds to the 8-bit binary numbers 00000000_2 to 11111111_2. When the bit corresponding to a particular pin is *0* (*1*), that pin is set LOW (HIGH):

```
def config_do_port(board_num):

    port_type = me.DigitalPortType.AUXPORT
    direction = me.DigitalIODirection.OUT

    mu.d_config_port(board_num, port_type, direction)

    return

def write_do_port(board_num, data_value):

    port_type = me.DigitalPortType.AUXPORT

    mu.d_out(board_num, port_type, data_value)

    return
```

To write **single_led_mcc.py**, you may wish to start with the code in **toggle_button.py**. Then, add coding to configure the DIO port for DO operations. Within the `toggle()` function, switch the line between its LOW and HIGH states using `write_do_port()` with `data_value=0` and `data_value=1`, respectively.

Run **single_led_mcc.py** several times and show that you can toggle the LED in your hardware circuit on and off using the GUI button.

(b) Next, using the DIO0 and DIO1 lines on your DAQ device, construct the hardware circuit shown in Fig. 8.2 with two LEDs, which we will call A and B, respectively.

Fig. 8.2 Hardware circuits controlled by alternating LEDs VI. In this figure, digital I/O lines DIO0 and DIO1 are used.

Then, write a GUI-equipped `while`-loop-based program named **toggling_leds_mcc.py** such that, when a GUI **Start** button is pressed, a `while` loop commences execution, where each loop iteration lasts 2 seconds. During each loop iteration make A lit and B dark for 1 second, followed by A dark and B lit for 1 second.

To write **toggling_leds_mcc.py**, you may wish to start with the code in **dc_voltmeter_mcc_live.py**. Then, add coding to configure the DIO port for DO operations. Replace `acquire_sample()` from **dc_voltmeter_mcc_live.py** with a new function called `toggle_leds()`. In using `write_do_port()`, to make A lit and B dark set `data_value=1`, while to make A dark and B lit set `data_value=2`, since the decimal integers 1 and 2 correspond to the 8-bit binary numbers 00000001_2 to 00000010_2, respectively.

Use It!

Anti-aliasing Filter By passing an analog signal through a lowpass filter prior to inputting it to a digitizer, the aliasing effect can be suppressed. When used in this manner, the lowpass filter is called an *anti-aliasing filter*. The gain of a filter is defined as its output voltage divided by its input voltage: $G \equiv V_{out}/V_{in}$. An ideal anti-aliasing filter will transmit all signals with frequencies less than the Nyquist frequency to the digitizer at unity gain ($V_{out} = V_{in}$), but will completely block—that is, attenuate to zero gain—all signals with frequencies greater than the Nyquist frequency. The two frequency ranges $f < f_{nyquist}$ and $f > f_{nyquist}$ are called the *passband* and *stopband*, respectively. For real filters, the shift from the passband to the stopband is not abrupt, but takes place over an intermediate range of frequencies called the *transition band*; for frequencies within the transition band, gains are in the range $0 < G < 1$.

Butterworth lowpass filters of various orders n are commonly used as anti-aliasing filters. This class of filter provides "maximally flat" signal transmission (i.e., close to constant unity gain) up to a given cutoff frequency f_c. In the transition region above f_c, an nth-order filter's gain changes by (approximately) a factor of $(1/2)^n$ for each doubling of signal frequency. A doubling of frequency is called an *octave*. For example, if $f_c = 100$ Hz, 200-Hz and 400-Hz signals are one and two octaves above the cutoff frequency, respectively. Thus, for a first-order ($n = 1$) Butterworth filter with $f_c = 100$ Hz (assuming $G \approx 1$ at the cutoff frequency), the gain at 200 and 400 Hz will be $(1/2)^1 = 0.5$ and $(1/2)^1(1/2)^1 = 0.25$; for an eighth-order ($n = 8$)

Butterworth filter with f_c = 100 Hz, the gain at 200 and 400 Hz will be $(1/2)^8 = 4 \times 10^{-3}$ and $(1/2)^8(1/2)^8 = 2 \times 10^{-5}$, respectively.

To demonstrate the design of an anti-aliasing filter, construct the op-amp circuit shown in Fig. 8.3, which is a first-order Butterworth lowpass filter. The cutoff frequency (also known as the *3 dB frequency*) for this circuit is given by $f_c = 1/2\pi RC$; for your circuit, use R = 10 kΩ and C = 0.1 μF so that $f_c \approx$ 150 Hz. Attach a function generator to the filter's input and connect its output to the AI channel pins programmed into **digital_oscilloscope_mcc.py**. Also, attach the function generator's sync output to the proper DAQ device pins to facilitate digital triggering.

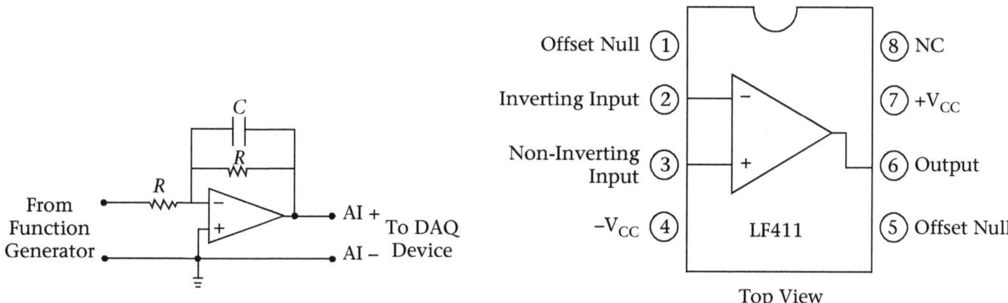

Fig. 8.3 First-order Butterworth low-pass filter. Suggested op-amp is LF411.

Run **digital_oscilloscope_mcc.py** and set **Number of Samples** and **Sampling Frequency (Hz)** equal to *200* and *2000*, respectively, making $f_{nyquist}$ = 1000 Hz. Configure your function generator to input a (smaller than Nyquist frequency) 50-Hz sine wave with amplitude 1 V to the low-pass filter, then press the **Start** button. Since the sine-wave frequency is less than the filter's cutoff frequency, you will see a plot of about five cycles of the unattenuated ($G \approx 1$) 50-Hz sine wave.

Next, input a 1200-Hz sine wave of amplitude 1 V to the filter. This frequency is greater than the Nyquist frequency as well as the filter's cutoff frequency and so, instead of a 1200-Hz waveform, your data-acquisition system will produce an aliased waveform of 800 Hz, but with an amplitude attenuated from the 1-V input value. Since 1200 Hz is three octaves above $f_c \approx$ 150 Hz, we expect the gain due to this first-order filter to be approximately $(1/2)^1(1/2)^1(1/2)^1 = 0.125$. Thus, this simple filter suppresses aliasing, but not to the level that we might hope. Since 1200 Hz exceeds $f_{nyquist}$, this input sine wave is in the stopband and so the gain would ideally be much closer to zero.

To be optimally effective, an anti-aliasing filter will receive a large-amplitude signal within the stopband and attenuate it to below the DAQ system's smallest resolvable voltage level. Thus, for a 12-bit digitizer with a range of ±10 V and voltage resolution ΔV = 5 mV, the optimal gain for this filter within its stopband will be on the order of $G \equiv V_{out}V_{in}$ = 0.005 V/10 V = 5×10^{-4}. From our above calculations, we see that an eighth-order Butterworth filter can achieve this order of attenuation for frequencies greater than two octaves above f_c. Such high-order Butterworth filters are available as low-cost integrated circuits. For example, the Maxim MAX291 chip is an eighth-order Butterworth low-pass filter, whose cutoff frequency can be chosen anywhere within the range from 0.1 Hz to 25 kHz. When placed in the circuit shown in Fig. 8.4, the cutoff frequency (in Hertz) is given by $f_c = 333/C$, where C is in nF. Build this circuit with C = 1 nF so that $f_c \approx$ 300 Hz and demonstrate its ability to completely suppress the aliasing of a 1200-Hz input sine wave (1200 Hz is two octaves above 300 Hz).

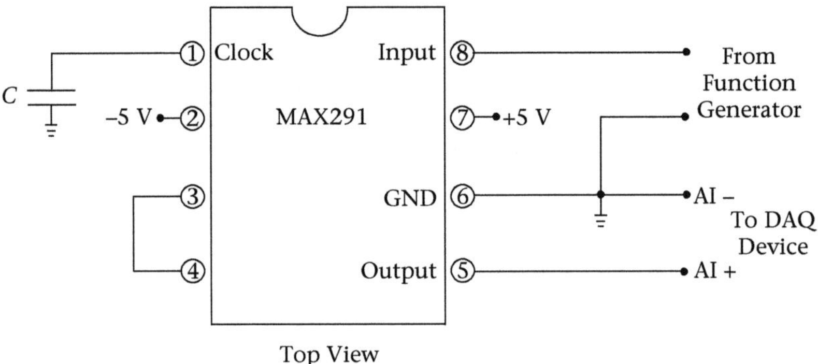

Fig. 8.4 MAX291 eighth-order Butterworth low-pass filter. Capacitor *C* determines cutoff frequency.

Problems

1. **dual_channel_digital_oscilloscope_mcc.py** Build a two-channel digital oscilloscope, which acquires and plots digitalized waveforms on both AI channels CH0 and CH1. With **digital_oscilloscope_mcc.py** open, use **File>>Save As...** to create the new program and save it in **YourNamePythonWork**. The changes required are the following:
   - To specify that sampling is to be done on the two channels CH0 and CH1, use the assignments `board_num, low_chan, high_chan = 0, 0, 1`. Also, within `acquire_waveform()`, make `num_points = num*(high_chan-low_chan + 1)`.
   - In `acquire_waveform()`, the command `mo.read_waveform()` remains as is. Now, however, instead of the buffer being an array with `num` elements, consisting of the `num` samples acquired on a single channel, the buffer has an array with 2×`num` elements, where the zeroth and first elements are the first samples on CH0 and CH1, the second and third elements are the second samples on CH0 and CH1, and so on. Thus, the following two list comprehensions will create the `num`-element lists `y` and `z`, which contain the `num` samples from CH0 and CH1, respectively:

```
    y = [ctypes_array[2*i] for i in range(num)]
    z = [ctypes_array[2*i+1] for i in range(num)]
```

   - For the display, a plot of two curves is needed. Because this plot is rapidly updated within the while Loop, it's best to write a dedicated plotting routine. In **plot_creator.py**, add the following function and then use it in your program:

```
def update_two_curve_plot(ax, canvas, x, y, z,
                           label_x, label_y,
                           marker='-'):

    size = 12

    ax.clear()
    ax.tick_params(axis='both', labelsize=size)
    ax.set_xlabel(label_x, fontsize=size)
    ax.set_ylabel(label_y, fontsize=size)
    ax.grid()
```

```
    ax.plot(x, y, marker)
    ax.plot(x, z, marker)
    canvas.draw()

    return
```

On your DAQ device, attach a voltage waveform from a function generator to AI channel CH0 and the function generator's sync output digital signal to the AI channel CH1. Additionally, attach the sync's positive and negative (ground) terminals to the TRIG and GND pins of your DAQ device. Run your program and verify that it simultaneously displays traces of the two inputs.

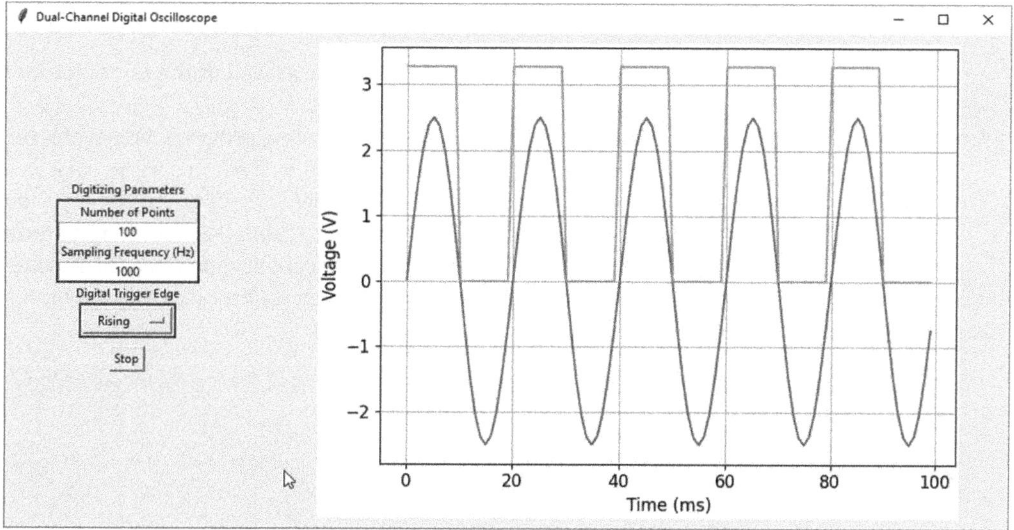

2. **tmp36_thermometer_mcc.py** The Analog Devices TMP36 temperature sensor (available from https://www.adafruit.com and https://www.sparkfun.com) is a low-cost three-terminal semiconductor device, which can be used to measure temperatures in the range from -40 to 125°C with accuracy of ± 1°C. As shown in Fig. 8.5, to operate the TMP36, simply connect a power supply voltage to pin 1 (this voltage can be anything in the range from +2.7 to 5.5 V) and ground to pin 3. Then measure the voltage difference V between pins 2 and 3. The temperature T in degrees Celsius is then given by $T = 100 \times V - 50$.

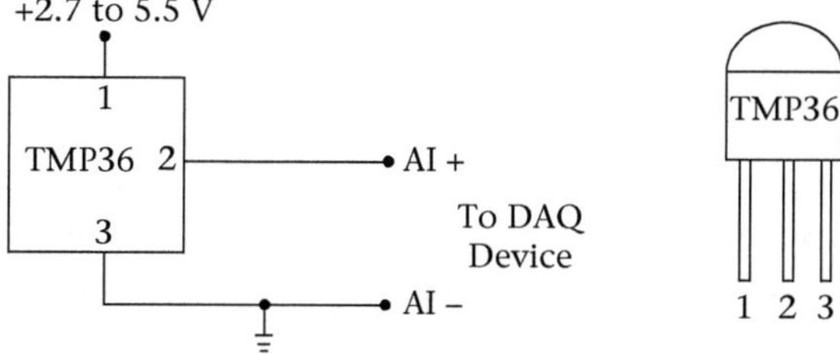

Fig. 8.5 TMP36 temperature sensor.

Connect a powered TMP36 device to the AI channel CH0 and AGND pins of your DAQ device. Using **dc_voltmeter_mcc_live.py** as your guide, write a GUI-equipped program named **tmp36_thermometer_mcc.py** that, every 250 ms until a **Stop Button** is pressed, reads the voltage V, converts this value to the corresponding temperature in Celsius, and then displays this temperature in a labeled GUI indicator.

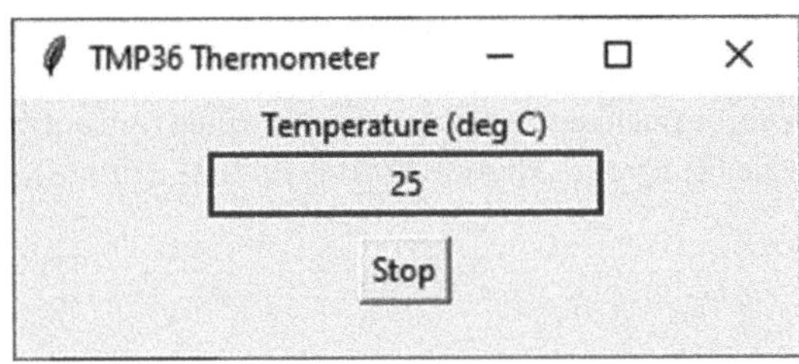

Run your program and use it to measure room temperature as well as the temperature of your skin.

3. **sinewave_generator_mcc_software_timed.py** Build the following program, which creates a sine-wave voltage output using the software timing provided by Python's `time.sleep()` function (this exercise is possible with the MCC USB-202 DAQ device). As the `for` loop iterates, the code the function $y(x) = 2 + \sin x$ is evaluated at N equally spaced arguments x in each sine-wave cycle. The spacing in radians between adjacent arguments is $2\pi/N$, where N is the number of samples per cycle. Since each `for`-loop iteration takes 0.01 second, its 1000 iterations will take place over 10 seconds.

```
import Utilities.mccdaq_operations as mo
import math
import time

board_num, channel = 0, 0
ao_range = mo.chan_range('unipolar_5')

n=10

for i in range(1000):
    x = 2*math.pi*i/n
    y = 2 + math.sin(x)
    mo.write_sample(board_num, channel, ao_range, y)
    i +=1
    time.sleep(0.01)
```

Connect the AOUT0 and AGND pins of your DAQ device to the positive and negative (ground) inputs of an oscilloscope, respectively. Then run your program with samples per cycle set to `n=10` so that each sine cycle is represented by ten voltage levels. Given that each `for`-loop iteration takes 10 ms, the period of the observed sine wave should be (10 ms/update)× (10 updates) = 100 ms, making its frequency 10 Hz. You will find that, although the underlying sine pattern is apparent, the ten voltage levels used to form this waveform are too well resolved to give the appearance of a continuous function.

Try increasing samples per cycle to `n=20`, then `n=30`, and so on. Since the `for`-loop iteration time remains constant, as you increase the samples per cycle, the frequency of the output waveform will decrease. You will need to adjust your oscilloscope's *Time/Div* control appropriately to view a complete cycle of the waveform. At what value for samples per cycle does the output waveform start to take on a continuous appearance?

You will find the generated sine wave has a frequency a bit smaller than expected due to the inherent accuracy of your computer's clock, which is the time base for Python's `time.sleep()` function as well as the time taken to execute the `write_sample()` command in each `for` loop iteration).

4. **frequency_meter_mcc.py** Build a frequency meter program that determines the frequency of a digital signal input to your DAQ device as follows.
   (a) Add the two functions given below to **mccdaq_operation.py**, which clears the counter on your DAQ device to zero counts and reads the current count on a 32-bit counter (the type on the USB-202 DAQ device). The particular counter is specified by `counter_num`:

```
def clear_counter(board_num, counter_num):

    mu.c_clear(board_num, counter_num)

    return

def read_counter(board_num, counter_num):

    count = mu.c_in_32(board_num, counter_num)

    return count
```

   (b) Using **dc_voltmeter_mcc_live.py** as your model, write a GUI-equipped program called **frequency_meter_mcc.py** that displays the frequency of a digital waveform input at the counter's input pin. Measure the frequency as follows: Within each iteration of a `while` loop, clear the counter, use `time.sleep()` to define the counting interval to be 1.0 second, read the number of counts N accumulated during the 1.0-second interval, then display N in a GUI indicator. This process is coded as follows:

```
        mo.clear_counter (board_num, counter_num)
        time.sleep(1.0)
        count = mo.read_counter(board_num, counter_num)
        string_data = str(count)

        lbl_2.config(text=string_data)
```

   (c) Configure the program's GUI so that a **Start** button initiates the `while` loop as it morphs into a **Stop** button. Pressing the **Stop** button ceases the while Loop's operation.

After you have completed your program, program a function generator to output a certain frequency, say, 1000 Hz. Then, connect the sync output of the function generator to the input pin of counter CTR (and GND) on your DAQ device, and run **frequency_meter_mcc.py** to measure the frequency of the sync output's digital signal. You may find a slight error in the measured frequency due to the inherent accuracy of your computer's clock, which is the time base for Python's `time.sleep()` function.

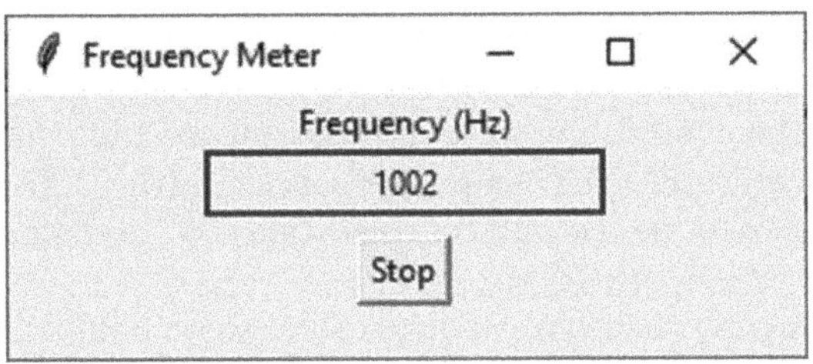

5. **data_logger_mcc.py** Starting with Chapter 4's **sinewave_animation_blit_gui.py,** use **File>>Save As**... to create **data_logger_mcc.py,** then modify the function `animate()` so that the quantity `y_i` is a single voltage sample obtained from your MCC DAQ device, rather than the calculated value `y_i = a*math.sin(2.0*math.pi* f*t_i)`. Configure the program to acquire one voltage sample every 0.5 seconds. Input a low-frequency (e.g., 0.1 Hz) waveform (e.g., a triangle wave as shown below) from a function generator to AI channel 0 of your DAQ device to demonstrate real-time plot of this input signal over a time span of 50 seconds.

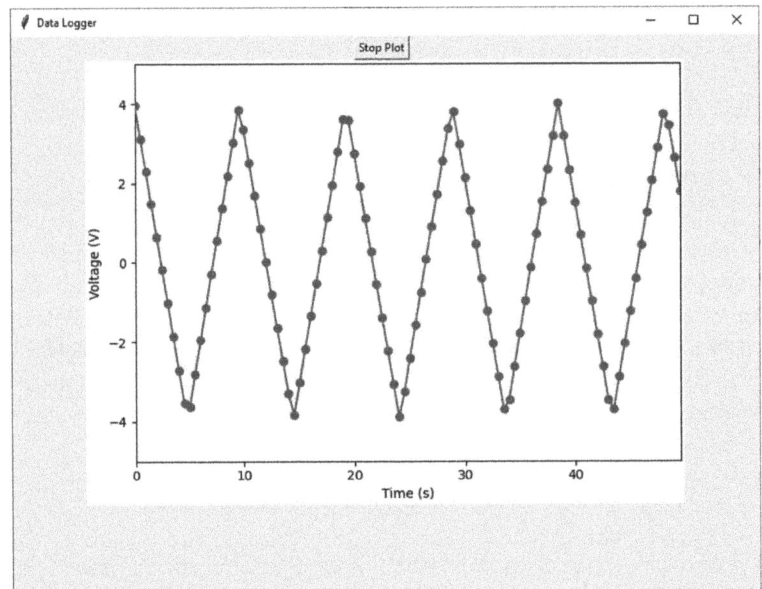

CHAPTER 9

Data Files and Character Strings

9.1 Text and Binary Data Files

The outcome of any experiment is embodied in the data it generates. In computer-controlled experimentation, then, it is important to have the ability to save experimental data in computer files for future analysis. Ideally, these data are stored in a convenient format so that they can be easily read by a user-written program or a commercially available software package that analyzes and displays the experimental results.

Subsequent to its definition in 1963, computers commonly communicated alphanumeric data using the *American Standard Code for Information Interchange* (ASCII). In this coding scheme, seven bits of a byte are used to represent a character, while the byte's eighth "parity" bit is used for error checking when the data are received by a reader. The ASCII set of $2^7 = 128$ distinct states is used to represent all of the English-language keyboard's alphanumeric characters as well as nondisplayable control characters such as *carriage return* (CR) and *line feed* (LF); LF is also commonly called *newline*. In "extended" ASCII standards, notably *Latin-1* (formerly, ISO 8859-1) and *Windows-1252*, all eight bits of the byte are utilized for character coding, yielding up to 256 distinct states (with the first 128 states assigned to the same characters as 7-bit ASCII).

Since the early 1990s, a new international encoding standard called *Unicode* has been widely adopted for electronic communication, whose ongoing goal is to define a unique numeric code for every character in every one of the world's languages, along with numeric codes for symbols from mathematics and other fields (and emojis!). The most common implementation of Unicode (and the one adopted by Python) is called *UTF-8*, which is short for *Unicode Transformation Format–8-bit*. UTF-8 is capable of encoding over 1.1 million unique characters, where (in its quite involved scheme) some simple characters are represented by a single byte, while other more complex characters require representation by up to four bytes. Thankfully, UTF-8 was designed to be backward compatible with the ASCII standard, so each of its first 128 characters are represented by a single byte and they have a one-to-one correspondence to the ASCII standard (as well as Latin-1 and Windows-1252). In this chapter we will be investigating how to store (and retrieve) experimental data in a computer file, which simply involves the English-based alphanumeric characters in the ASCII standard or, equivalently, the first 128 single-byte UTF-8 characters. Thus, on a practical level, the characters being processed in our Python programs will be equivalent in all the above-mentioned coding standards.

A sequence of characters is called a *string*. Character strings, of course, can be used to represent text messages. However, strings also are useful for passing commands and data to and from the computer and stand-alone instruments. Additionally, as you will find in this chapter, strings can be used to store numeric data in files on a computer *drive* (e.g., hard drive, solid-state drive, flash drive).

The Python-Based Laboratory. John Essick, Oxford University Press. © John Essick (2024). DOI: 10.1093/9780191998478.003.0009

The closest approximation to a universally readable data file format is the *text file*, i.e., a file where data are represented by characters rendered using, say, the Unicode standard. Storing data in this manner has the following advantages. Your files will be read accurately by computers of all manufacturers. Additionally, the files can be viewed with a word processor and, if desired, easily cut and pasted into a document for report generation. Finally, your data will be easy to import into commercially available data analysis software packages. A large number of these application programs prefer that your file is formatted as a *spreadsheet*, in which data values are arranged in the rows and columns of a grid. One common spreadsheet configuration is the *comma-separated values (CSV) format*, where commas separate ("delimit") columns and end-of-line (EOL) characters separate rows as shown next, where the *paragraph symbol* (¶) is the EOL character.

```
0.00,·0.4258¶
1.00,·0.3073¶
2.00,·0.9453¶
3.00,·0.9640¶
4.00,·0.9517¶
```

Opening this file using a spreadsheet program such as Microsoft Excel yields the following:

	A	B	C
1	0	0.4258	
2	1	0.3073	
3	2	0.9453	
4	3	0.964	
5	4	0.9517	
6			

Alternatively, numeric data may be stored as a *binary file*. This file format is simply a bit-for-bit image of the data that reside in your computer's memory. Thus, when data are exchanged between a binary file and computer memory, little or no data conversion is required, so you get maximum performance. Also, binary data files provide the most memory-efficient storage method for numeric data. To demonstrate this fact, consider the number of bytes necessary to store the integer *54321*. Since this number is less than 2^{16} = 65636, it will require only two bytes of memory in a binary file. However, in a text file, this number will occupy five bytes of memory, one byte for each of the characters *5*, *4*, *3*, *2*, and *1*. The disadvantage of binary files is their lack of portability. They cannot be viewed by a word processor, and they cannot be read by any program without a detailed knowledge of the file's format.

Thus, if ease of use and compatibility with a spreadsheet application program are of most concern, use text file storage. If, on the other hand, efficient memory use and high speed are desired, a binary file is the best choice.

The Python Standard Library contains built-in functions that facilitate data storage and retrieval using either the text or binary file format. These functions are general purpose in that they can deal with files containing any type of alphanumeric data, including primarily word-based data (e.g., a company's product information or a vacation itinerary) as well as numeric-based data related to a laboratory experiment. For this latter case, which is our chief concern, the NumPy library offers alternative functions tailor-made for interacting with files containing numbers placed in the spreadsheet format.

In this chapter, you will learn how to use the NumPy file-related functions to store and retrieve data in a spreadsheet file. In the following sections, you will write programs that generate one- and two-dimensional numeric data arrays and store these data in a spreadsheet file. Then, you'll open the spreadsheet file and read its contents into computer memory. These data file management skills play a key role in the process of analyzing and presenting the data acquired in your experiments.

9.2 NumPy Text File Functions

To begin, here's an introduction to the two file-related functions in the NumPy library that we will utilize in our upcoming work. Both of these functions can be programmed to operate in a number of optional ways by assigning desired values to their arguments. Here, the form of each function is given, along with the default values of its arguments. We'll learn the meaning of each option as we work through examples throughout the chapter.

First, the function `savetxt()` is designed to store a NumPy array of numeric values in a text file on your computer's drive. The form of this function is as follows:

```
numpy.savetxt(fname, X, fmt='%.18e', delimiter=' ',
              newline='\n', header='', footer='',
              comments='# ')
```

In using `savetxt()` to store data in a spreadsheet-formatted file, `delimiter` is the character placed between each data value within a row (thus defining the columns), while `newline` will be the character placed at the end of a row (i.e., it is the EOL character).

For now, let's assume we are satisfied with all of the default values provided for the last six arguments, so that we can focus our attention on how to assign the `fname` and `X` arguments. `fname` specifies the file path (i.e., the directory in which the file resides, along with the file's name) into which the data are to be written, while `X` is the "array-like" arrangement of data to be written in the file. In NumPy, the term "array-like" means any data structure that can be converted to a NumPy array and so includes Python lists and tuples. Thus, `savetxt()` can flexibly adapt to a range of ways that `X` is constructed. By default, the characters representing each digit of the data are rendered using the Latin-1 coding scheme, which is equivalent to Unicode for the English-language alphanumeric characters involved in this process.

Second, the function `loadtxt()` is used to retrieve the numeric data saved in a text file on your computer's drive and place it in a NumPy array within a program. The form of this function is as follows (where a few arguments not pertinent to our study are omitted):

```
a = numpy.loadtxt(fname, dtype=<class 'float'>,
                  comments='#', delimiter=None, skiprows=0,
                  usecols=None, unpack=False)
```

If one is satisfied with all of the default values provided for the last six arguments, only `fname` need be supplied, i.e., the path of the data file to be read. The retrieved data are placed in the NumPy array `a`, where each element in the array is (by default) a floating-point number. In converting the text representation of the data within the file to numeric float values within the NumPy array, the Latin-1 coding scheme is used by default.

9.3 Storing One-Dimensional Data Array

Let's start by developing a program to store a one-dimensional array of numbers in a file with the comma-delimited spreadsheet format. Here's the plan for our data-storage program: Generate a 1D array of data, and then store these data in a spreadsheet file, which is saved on your computer's drive.

To aid in the creation of the required file, we'll use the following handy function in the `filedialog` module of `tkinter` that, when executed, pops up a dialog window in which a user can navigate to a desired directory and create a file there:

```
fname = tkinter.filedialog.asksaveasfilename()
```

The name assigned to the file within the dialog window become the value of the object `fname`.

Start by creating a new program in the Editor. Using **File≫New File**, create the program and then save it under the name **1d_spreadsheet_storage.py** in **YourNamePythonWork**. Begin coding this program as shown below. Here, `np.linspace(start,stop,num)` generates a one-dimensional NumPy array of ten values called `t`, and `asksaveas-filename()` opens a dialog window where a user creates a data file named `fname`. Note that, by default, 1D NumPy array behaves mathematically like a row vector.

```
import numpy as np
import tkinter.filedialog as fd

start, stop, num = 0, 1, 10

t = np.linspace(start, stop, num)

fname = fd.asksaveasfilename()
```

To complete the program, we'll use `savetxt(fname,X,...)` to convert the numeric values in `t` into character strings and store them in `fname`, where this file is of the form of a comma-delimited spreadsheet. As mentioned previously, the argument `X` can receive the data `t` in a variety of ways. For consistency with the manner in which we will input two-dimensional arrays later in the chapter, we'll make `t` the sole element of a singleton tuplet, i.e., `X = (t,)`; remember the comma that trails `t` denotes this data structure is a tuple. Additionally, the default delimiter for `savetxt()` is a space, but for the comma-delimited spreadsheet format we must choose `delimiter=','`. Add the final line to your program, which will store the data as desired.

```
import numpy as np
import tkinter.filedialog as fd

start, stop, num = 0, 1, 10

t = np.linspace(start, stop, num)

fname = fd.asksaveasfilename()

np.savetxt(fname, (t,), delimiter=',')
```

Note to Windows users: If, in the following exercises, your results differ from those shown, modify the last line as given below; the reason is explained is Section 9.4:

```
np.savetxt(fname, (t,), delimiter=',', newline='\r\n')
```

Save your work, then run **1d_spreadsheet_storage.py**. A dialog window titled **Save** will appear. Navigate to **YourNamePythonWork**, and enter **1d_spreadsheet_data.csv** for the name of the to-be-created text file in the **File name:** box, and then click on the **Save** button.

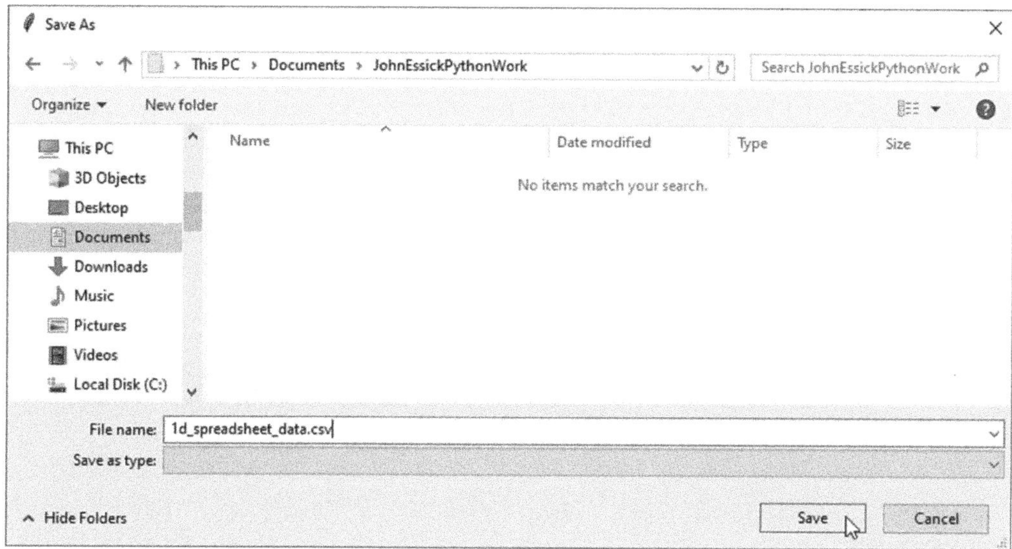

The program will then complete execution by creating the requested spreadsheet file in **YourNamePythonWork**.

Use a word-processing program to open the text-based **1d_spreadsheet_data.csv** spreadsheet file; in the **Open** dialog window, you may have to choose the file type as **All Files** to see **1d_spreadsheet_data.csv** listed and you may have to pass through a pop-up file conversion window. Shown next is what I found by viewing this data file in Microsoft Word (the appearance of your file may vary slightly because of the particular tab settings of your word processor). Here, I have activated Word's **Show ¶** command, which allows one to see the file's nondisplayable characters such as the EOL character. If your word processor has a similar option, use it to view these usually hidden characters.

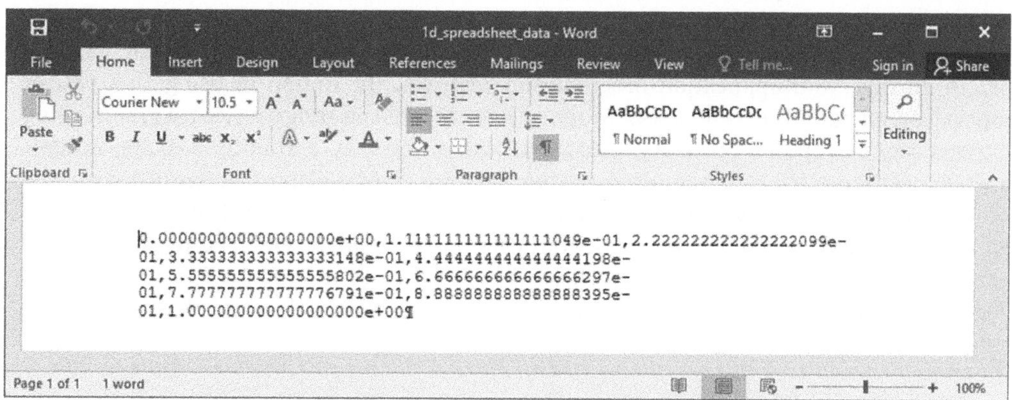

The ten values of `t` have been stored in this file with each value separated from its neighbor by a comma. However, these values are written in scientific notation with many digits of precision causing this fact to be a bit obscured. Close **1d_spreadsheet_data.csv** with the word processor, but don't save it (or else the program will possibly embed its own characteristic formatting statements within the file).

For the sake of clarity, let's modify **1d_spreadsheet_storage.py** so that the formatting of the stored values is something more readable. This change is accomplished by adding the following number-formatting argument `fmt='%.3f'` to `savetxt()`. For now, simply make this change. We'll explain its meaning in a few moments:

```
np.savetxt(fname, (t,), fmt='%.3f', delimiter=',')
```

Save your work. With this change to the `fmt` argument, rerun **1d_spreadsheet_storage.py** to create a new spreadsheet file of ten values, reusing the **1d_spreadsheet_data.csv** file name (by selecting this name in the **Save** dialog window, assuming the original file is initially closed, and pressing subsequent **Save** and/or **Replace** buttons). Use a word-processing program to open the newly created **1d _spreadsheet_data.csv** file, which will now appear in the more easy-to-decipher form shown here:

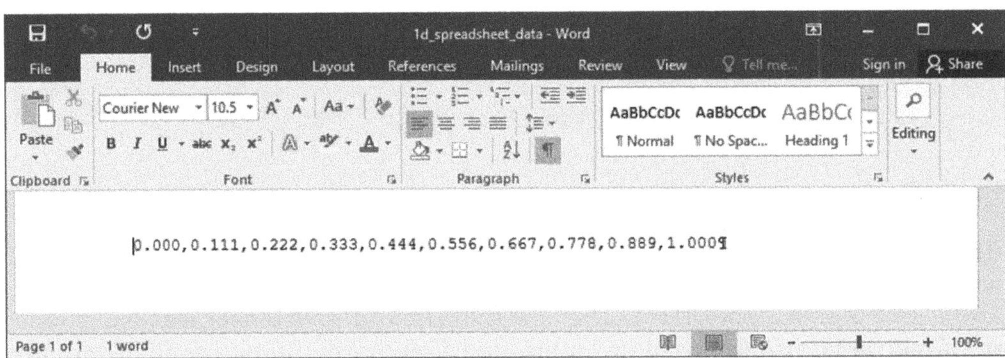

We now see clearly that the ten sine-wave values are delimited by commas and that the sequence concludes with an EOL character (on some computing systems, `savetxt` also places an additional EOL character at the very end of the file). From our knowledge of the spreadsheet formatting convention, we conclude that a spreadsheet application program will interpret this collection of values as a row of data, placing these ten numeric values in a sequence of columns contained within a single row. Close **1d_spreadsheet_data.csv** with the word processor (but don't save it).

If available, use a spreadsheet application program to read your **1d_spreadsheet_data.csv** file. To do this, your application program may give you the option of using an **Open** or **Import** command. Either will work. You may have to tell the program that you're reading in a text file (as opposed to binary, Excel, etc.). Also, you may encounter a dialog window in which you must tell the program that your data are comma-delimited. Next, I show the result of reading the **1d_spreadsheet_data.csv** file using Microsoft Excel. As expected, the ten numeric values appear in ten sequential columns of a single row. The reason that the ten values are distributed along a row can be traced back to `linspace()`. This function created these values and placed them in a 1D NumPy array, which by default behaves like a row vector.

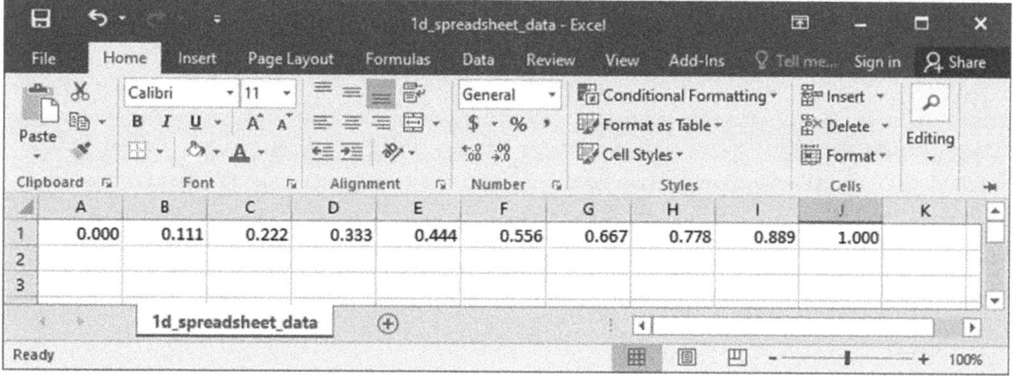

If you have a bit of experience with spreadsheet applications, the above illustration will cause some concern. In using such programs for generating plots, performing curve fitting, and other useful data analysis operations, the array of values for a particular quantity (e.g., time sequence, waveform displacement) is expected to reside in a single column, with the array's elements indexed by the row numbers. This organizational scheme for data is often called *column-major order*. In providing the argument `X` as the tuple `(t,)`, we have found that `savetxt()` does not function in a manner consistent with the column-major order convention. However, once this problem has been identified, it is simple to remedy by using NumPy's `transpose()` function, a function that switches rows into columns. That is, by making `X=np.transpose((t,))`, the file will be recorded in a "column-like" manner, rather than "row-like."

Modify **1d_spreadsheet_storage.py** so that its final line becomes the following:

```
np.savetxt(fname, np.transpose((t,)), fmt='%.3f',
           delimiter=',')
```

Save your work. Then, rerun **1d_spreadsheet_storage.py**, reusing the **1d_spreadsheet_data.csv** file name. As shown next (using Microsoft Word), each data value in **1d_spreadsheet_data.csv** is now separated from its neighbor by an EOL character. Thus, when read by a spreadsheet application, this array of values will be placed in a single column.

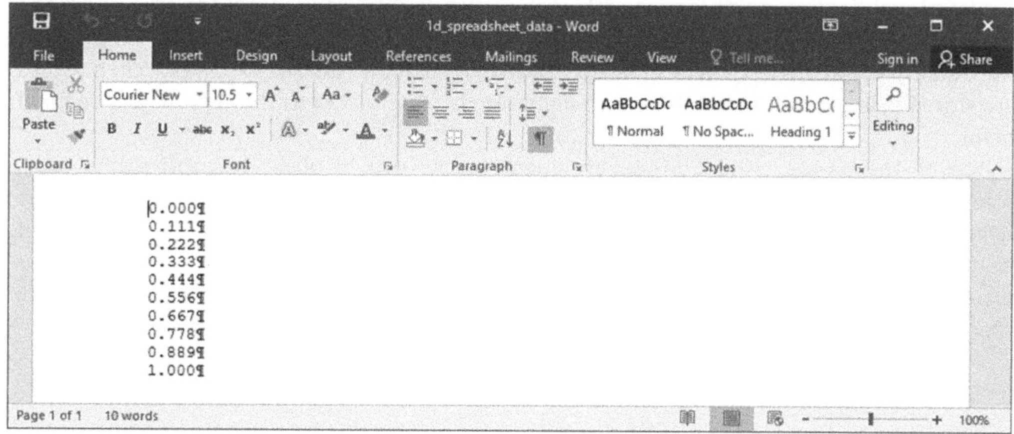

If available, open the data file in a spreadsheet application. In the next illustration (using Microsoft Excel), we see that our expectation of column-like data is fulfilled.

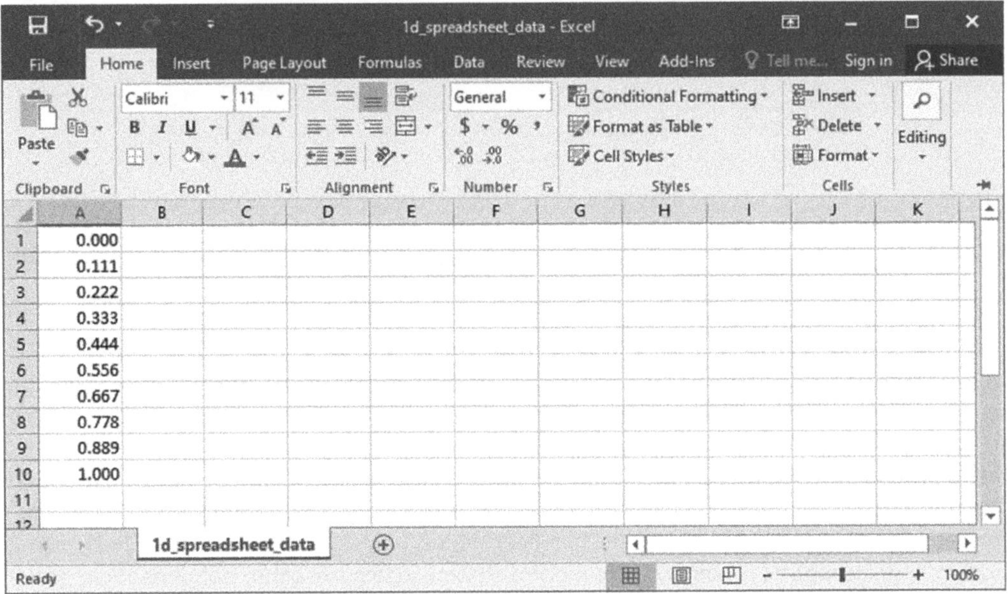

Finally, let's circle back and discuss the number-formatting argument `fmt`. We found that the instruction `fmt='%.3f'` directs that the values stored in the spreadsheet file are truncated at the thousandths decimal place. To understand this instruction, we need to know that the value assigned to `fmt` obeys the syntax

```
%[WidthString] [.PrecisionString] ConversionCharacter
```

Here, optional features are enclosed in square brackets. Each syntax element, and how it affects the form of the stored number within the data file, is explained in Table 9.1.

From Table 9.1, we see that setting `fmt='%.3f'` instructs `savetext()` to store numbers in the fractional floating-point format with each number's total width auto-adjusted under the restriction of maintaining three digits to the right of the decimal point. In our initial work, we found that when `fmt` is unspecified, it takes on its default value of `'%.18e'`. Then values are stored using scientific notation with eighteen decimal places to the right of the decimal point.

Finally, the `header` and `footer` arguments allow you to place descriptive text (e.g., column headings, record of experimental parameters) at the top and bottom, respectively, of the spreadsheet file. The `comments` argument selects the character(s) that will be prepended to the `header` and `footer` text to mark them as non-numerical comments. The default value for `comments` is `'#  '`, i.e., the hash symbol followed by a space.

Table 9.1 Syntax Elements of Format String

%	Character that denotes the beginning of a format specification
WidthString (Optional)	Integer specifying the total number of characters to be used to represent the stored number. If your specified value of **WidthString** exceeds the number of characters actually required to represent a particular number, the excess portion of the string will be padded with spaces. If **WidthString** is not specified (or the stored number requires more characters than the specified **WidthString**), the string will expand to as long as necessary to represent the stored number.
.PrecisionString (Optional)	Period (.) followed by an integer specifying the number of digits to the right of the decimal point in the stored number. If **.PrecisionString** is not specified, six digits to the right of the decimal point are stored. If only the period (.) appears and **PrecisionString** is missing or zero, no digits to the right of the decimal point are stored.
ConversionCharacter	Single character that specifies in what manner the number is to be stored with the following code: d decimal integer x hex integer o octal integer f floating point with fractional format e floating point with scientific notation

The final version of **1d_spreadsheet_storage.py** is as follows:

```
import numpy as np
import tkinter.filedialog as fd

start, stop, num = 0, 1, 10

t = np.linspace(start, stop, num)

fname = fd.asksaveasfilename()

header_text='This is the header.'
footer_text='This is the footer.'

np.savetxt(fname, np.transpose((t,)), fmt='%.3f',
           delimiter=',', header=header_text,
           footer=footer_text)
```

Running this program produces the following **1d_spreadsheet_data.csv** file, when viewed in a word processor:

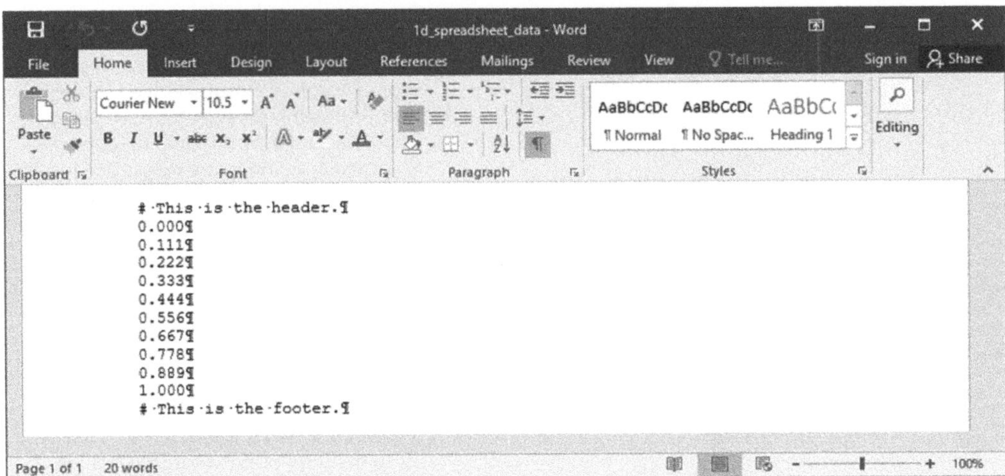

9.4 Escape Sequences and End-of-Line Markers

Among the first 128 Unicode characters are several nonprintable characters, which were part of the ASCII standard. These characters are used by word processors and other applications to format the appearance of a document or display. A few of the commonly used such characters are horizontal tab, line feed (also called newline), carriage return, and space. In Python, these nonprintable characters are coded by *escape sequences*, each of which consists of a backslash (\) immediately followed by a lowercase letter as listed in Table 9.2. When used in this manner, the backslash is called the *escape character*. As shown in the table, an escape sequence can also be used to represent a hexadecimal-encoded byte (e.g., \xFF is equivalent to the binary number 11111111_2) or to write a Unicode character, if you know its code number (e.g., the Greek letter beta can be printed in the Shell by typing `print('\u03B2')` and then pressing <*Enter*>).

Table 9.2 Commonly Used Escape Sequences

Escape sequence	Meaning
\t	Tab (HT, Unicode \u0009)
\n	Line Feed* (LF, Unicode \u000A)
\r	Carriage Return (CR, Unicode \u000D)
\s	Space (Unicode \u0020)
\\	Backslash (\, Unicode \u005C)
\u	Unicode Character (append hexadecimal character code to \u)
\x00–\xFF	Hexadecimal value of an eight-bit character

*Commonly called Newline.

For text files, the EOL character is constructed from escape sequences and, unfortunately, is platform-dependent. The definitions for EOL peculiar to each computing systems are as follows:

Windows: Carriage Return, then Line Feed (\r\n)

Macintosh: Line Feed (\n)

Linux: Line Feed (\n)

The EOL character used by `savetxt()` is set by its `newline` argument, which by default is `'\n'`. This default is consistent with the Mac and Linux computing systems, but `savetxt()` is designed to place the correct EOL character on a Windows system even under the default setting `newline='\n'`. However, if you are a Windows user and find that `savetxt()` is not creating new lines at the proper locations in your spreadsheet file, try setting `newline='\r\n'`.

9.5 Storing Two-Dimensional Data Array

A much-used design for a laboratory experiment is the following: A stimulus of given intensity (the independent variable) is applied to an experimental setup and the intensity of the setup's associated response (the dependent variable) is measured. By repeating this procedure for varying intensities of the stimulus, a data table of the dependent variable as a function of the independent variable is generated. Stated another way, the experiment produces two data arrays: The array `x_data`, whose elements are the varying stimulus intensities, and the array `y_data`, whose elements are the associated response intensities. To store the acquired data in such an experiment in a spreadsheet-formatted file, the experimenter will need to write the `x_data` and `y_data` arrays into two spreadsheet columns. Two simple modifications to **1d_spreadsheet_storage.py** will provide the experimenter with that skill.

Here are the simple changes needed: To simulate the independent-variable and dependent-variable data arrays produced in an experiment, we'll evaluate a 10-Hz sine function over a range of times as shown here:

```
start, stop, num = 0, 1, 10
A, f = 5, 10

t = np.linspace(start, stop, num)
y = A*np.sin(2*np.pi*f*t)
```

Then, to store these two arrays in a spreadsheet using `savetxt(fname,X, delimiter=',')`, we simply set `X=np.transpose((t,y))`, i.e., change the tuple `(t,)` in **1d_spreadsheet_storage.py** to `(t,y)`.

With **1d_spreadsheet_storage.py** open in the Editor, use **File≫Save As...** to create a program named **2d_spreadsheet_storage.py** and save it in **YourNamePythonWork**. Change the coding as shown next:

```
import numpy as np
import tkinter.filedialog as fd

start, stop, num = 0, 1, 10
A, f = 5, 10

t = np.linspace(start, stop, num)
y = A*np.sin(2*np.pi*f*t)
```

```
fname = fd.asksaveasfilename()

header_text='This is the header.'
footer_text='This is the footer.'

np.savetxt(fname, np.transpose((t, y)), fmt='%.3f',
           delimiter=',', header=header_text,
           footer=footer_text)
```

Save your work. Run **2d_spreadsheet_storage.py** and use it to create a spreadsheet named **2d_spreadsheet_data.csv** in **YourNamePythonWork**. Then view your resulting data file using a word processor. It should appear like this.

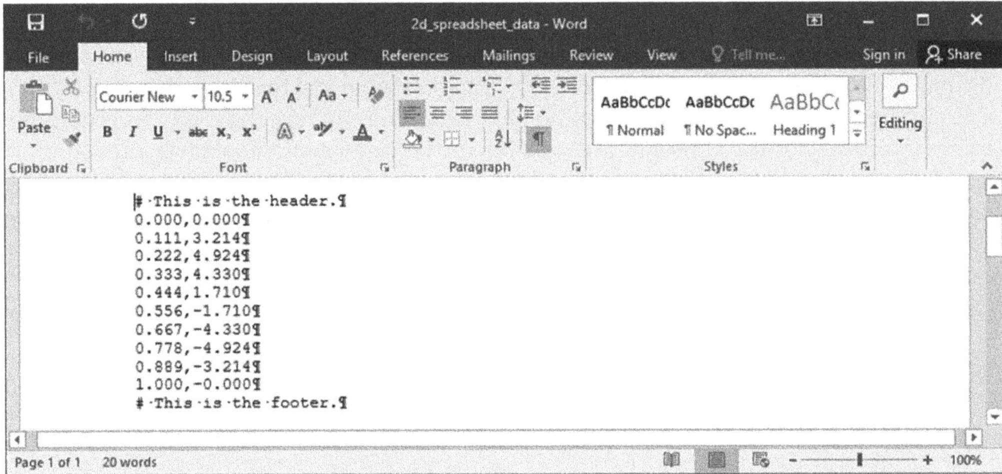

This comma-delimited formatting appears correct for importing the two data arrays into two adjacent columns of a spreadsheet. To verify this expectation, open your file with a spreadsheet application program. It should appear as below.

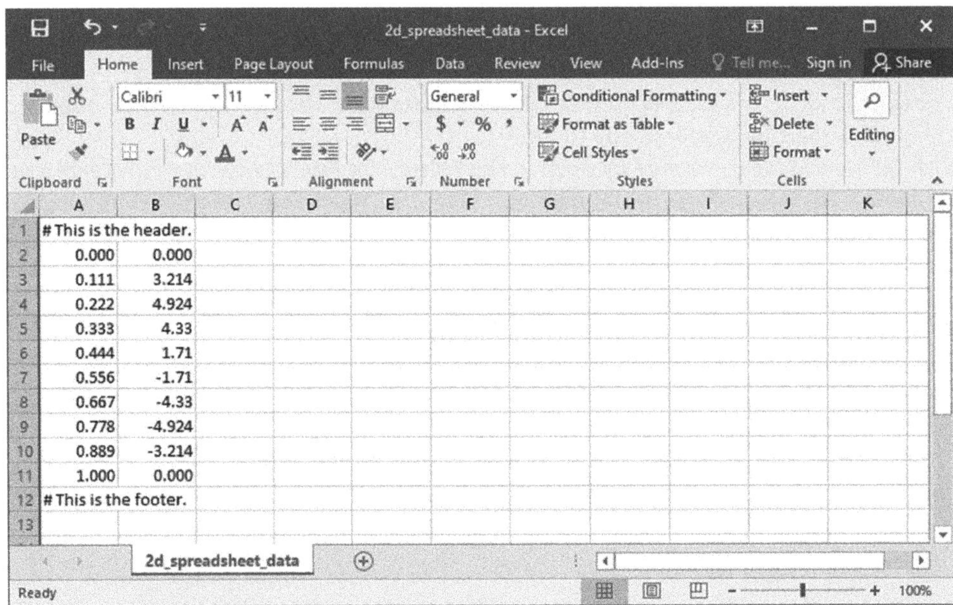

9.6 Reading One-Column Spreadsheet File

In this section, we will learn how to read the data values stored in a single-column spreadsheet.

First, we note that, regardless of the number of columns, to retrieve numeric data written into a spreadsheet file by savetxt() we will use the NumPy function loadtxt(). The form of this function is as follows, where only the argument fname is required and the default values of optional arguments are shown:

```
a = numpy.loadtxt(fname, dtype=<class 'float'>,
                  comments='#', delimiter=None,
                  skiprows=0, usecols=None, unpack=False)
```

As long as the spreadsheet file's data are formatted as floating-point numbers (i.e., f or e in Table 9.1), the default value for dtype is appropriate (and thus the argument dtype does not need to appear in your coding). The argument comments='#' instructs loadtxt() to ignore any line that begins with '#' (in our case, the header and footer) as these lines are comments, not numeric data. Again, this default option is appropriate for our savetxt()-created files. If a nonzero integer is assigned to skiprows, that number of rows will be skipped at the start of the file before reading begins; we will not need this option, but it may be of use to you in the future. The usecols and unpack arguments pertain to multicolumn spreadsheets. We will discuss these two arguments in the Section 9.7, where we will need to opt for values other than the defaults.

To aid in finding the path of the spreadsheet file we intend to read, we'll use the following handy function in the filedialog module of tkinter that, when executed, pops up a dialog window in which a user can navigate to a desired directory and there select the name of the file to be read:

```
fname = tkinter.filedialog.askopenfilename()
```

The name of the file selected within the dialog window become the value of the object fname.

To read a single-column spreadsheet file, use **File≫New File** in the Editor to create a new program called **1d_spreadsheet_read.py** in **YourNamePythonWork**. Then, enter the code

```
import numpy as np
import tkinter.filedialog as fd

fname = fd.askopenfilename()

a = np.loadtxt(fname, delimiter=',')

print(a)
```

Save your work, then run **1d_spreadsheet_read.py**. In the dialog window that pops up, select the file named **1d_spreadsheet_data.csv** in **YourNamePythonWork**, then press the **Open** button. If all goes well, the one column from **1d_spreadsheet_data.csv**, which is composed of ten data values, will be loaded into the 1D NumPy array named a and then the ten elements of a will be printed in the Shell. Compare the ten values you find printed in the Shell with the ten values stored in **1d_spreadsheet_data.csv** to confirm that the read process was carried out correctly.

9.7 Reading Two-Column Spreadsheet File

To read a two-column spreadsheet, we just need to write the `loadtxt()` command with values assigned to its `usecols` and `unpack` arguments as follows:

```
t, y = np.loadtxt(fname, delimiter=',', usecols=(0,1),
                  unpack=True)
```

Here, `usecols=(0,1)` provides the indices of the columns to be read, in this case the first (column 0) and second (column 1). The default value `usecols=None` instructs `loadtxt()` to read all of the columns within the spreadsheet and, for a two-column spreadsheet, is equivalent to our explicit choice of `usecols=(0,1)`. Setting `unpack=True` instructs the spreadsheet to be unpacked in the manner of a tuple. Thus, column 0 and column 1 will be assigned to the first element (`t`) and second element (`y`) of the tuple on the left-hand side of the equal sign, respectively.

With **1d_spreadsheet_read.py** open in the Editor, use **File≫Save As**... to create a program named **2d_spreadsheet_read.py** and save it in **YourNamePythonWork**. Change the coding as shown next:

```
import numpy as np
import tkinter.filedialog as fd

fname = fd.askopenfilename()

t, y = np.loadtxt(fname, delimiter=',', usecols=(0,1),
                  unpack=True)

print(t)
print(y)
```

Save your work, then run **2d_spreadsheet_read.py**. In the dialog window that pops up, select the file named **2d_spreadsheet_data.csv** in **YourNamePythonWork**, then press the **Open** button. If your program is coded correctly, the two columns from **2d_spreadsheet_data.csv**, which are composed of ten data values each, will be loaded into the two 1D NumPy arrays named `t` and `y`, then these arrays will be printed in the Shell. Compare the array values you find printed in the Shell with the values stored in **2d_spreadsheet_data.csv** to confirm that the read process was carried out correctly.

For a more visual comparison, modify **2d_spreadsheet_storage.py** so that it samples a 10-Hz sine-wave at 100 equally spaced times over a one-second time interval. That is, change 10 to 500 in the third line as shown here:

```
import numpy as np
import tkinter.filedialog as fd

start, stop, num = 0, 1, 500
A, f = 5, 10
```

Save you work, then run **2d_spreadsheet_storage.py** to create a new 500-row spreadsheet named **2d_spreadsheet_data.csv**.

Next, modify **2d_spreadsheet_read.py** so that, rather than printing the two arrays in the Shell, they are plotted by Matplotlib.

```
import numpy as np
import tkinter.filedialog as fd
import matplotlib.pyplot as plt

fname = fd.askopenfilename()

t, y = np.loadtxt(fname, delimiter=',', usecols=(0,1),
                  unpack=True)

plt.plot(t, y)
plt.show()
```

Save your work, then run **2d_spreadsheet_read.py**. In the dialog window that pops up, select the file named **2d_spreadsheet_data.csv** in **YourNamePythonWork**, then press the **Open** button. Matplotlib will pop up a plot of ten cycles of the 100-Hz sine wave, with fifty samples taken during each cycle as shown next.

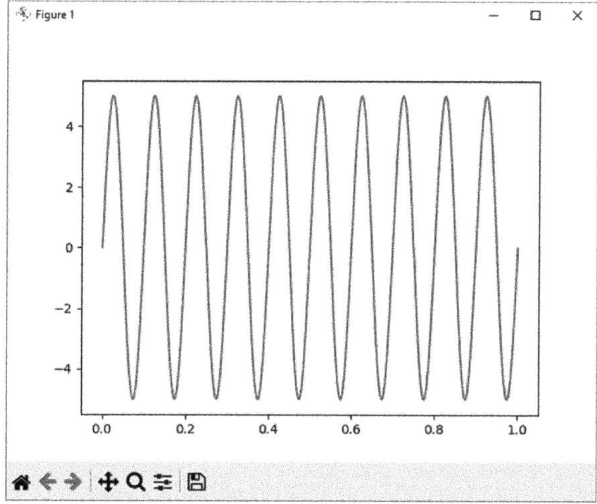

9.8 File Paths

In some cases, it may be desirable to hardwire the name of a data file into your program code, rather than acquire the file path using an interactive dialog window as we have done in this chapter. Thus, let's look more closely at the act of naming the data file to understand the alternate approach of explicitly including a specific file name within a program.

As you probably know, the *path* to a particular file in your computer system is specified by a hierarchical directory structure. The path is typically denoted by a **drivename**, followed by **directory names** (commonly called **folders**), followed by the **filename** itself. In Windows, these various levels of the hierarchy are separated by backslashes (\); on the Macintosh and in Linux, forward slashes (/) are used. Thus, on a Windows machine with a drive named **C:**, the file

2d_spreadsheet_data.csv within the directory **Users\Account\Desktop\YouNamePythonWork** has the path **C:\Users\Account\Desktop\YouNamePythonWork\2d_spreadsheet_data.csv**. On a Mac, the root of the system is "/", so a similarly named file on the **Desktop** has the path **/Users/Account/Desktop/YouNamePythonWork/2d_spreadsheet_data.csv**.

Thus, in this alternate approach on a Windows machine, **2d_spreadsheet_read.py** is written as follows. Here, a method called `join()` in the `path` submodule of the Python Standard Library's `os` module is used to properly concatenate the folder and file names in order to form the desired path. In general, `join()` allows one to construct a path from two or more components; note `import os` has been added at the top of the program.

```
import numpy as np
import tkinter.filedialog as fd
import matplotlib.pyplot as plt
import os

folder = 'C:\\Users\\Account\\Desktop\\YourNamePythonWork'
file = '2d_spreadsheet_data.csv'
fname = os.path.join(folder, file)

t, y = np.loadtxt(fname, delimiter=',', usecols=(0,1),
                  unpack=True)

plt.plot(t, y)
plt.show()
```

Note that, since backslash (\) signals an escape sequence, the backslash involved in a Windows path needs to be written using the escape sequence \\. Another option is to write the path as a *raw string literal* as shown below. Here the "`r`" prefix instructs Python to not treat any backslashes in the string as escape characters.

```
folder = r'C:\Users\Account\Desktop\YourNamePythonWork'
file = '2d_spreadsheet_data.csv'
fname = os.path.join(folder, file)
```

Since Mac and Linux path use forward slashes, no complications related to implied escape sequences arise.

The two modules within the Python Standard Library called `os.path` and `pathlib` contain many functions and classes for creating paths programmatically.

9.9 Unicode Character Coding Scheme

In the Unicode coding scheme, each character is assigned a unique *code point*, where the code point is represented by a *U+* followed by a four-digit hexadecimal index number. For example, the code point for the uppercase English-alphabet letter *A* is U+0041. Unicode's code points are distributed in seventeen *planes*, where each plane can accommodate up to $2^{16} = 65,536$

code points. For a given code point U+XXXX, the first two hexadecimal digits denote the index of the plane in which the code point is found and the last two hexadecimal digits are the index of the code point within that plane. Plane 00 (i.e., the plane with index 00) includes the ASCII character set and, for the ASCII characters within Plane 00, the indexing from the ASCII standard is maintained. So, within Plane 00, for example, the index number for uppercase A is hexadecimal 0041 (or, equivalently, decimal 65), which is the same index for uppercase A in the ASCII standard.

In Python, a Unicode character can be accessed by the escape sequence *u*, followed by the four-digit hexadecimal index number of the character's code point, surrounded by single (or double) quotation marks. To demonstrate this syntax, enter the coding shown below in the Shell:

```
>>> character = '\u0041'
>>> print(character)
```

When you press <*Enter*>, the uppercase letter `A` will be printed.

In working with Unicode characters, two handy Python functions are `ord(character)` and `chr(index_number)`. Given a (string-formatted) Unicode `character` as input, the `ord()` function returns Unicode's index number for that character, stated as a decimal integer. Test-drive the `ord()` function in the Shell as follows:

```
>>> index = ord('A')
>>> print(index)
    65

>>> print(type(index))
    <class 'int'>

>>> hex_index = hex(index)
>>> print(hex_index)
    0x41
```

The argument `character` can also be entered in the form of a `'\uXXXX'` escape sequence:

```
>>> index = ord('\u0041')
>>> print(index)
    65
```

The `chr(index_number)` function performs the opposite operation of `ord()`. Given a character's Unicode (decimal-formatted) `index_number` as input, `chr()` returns the associated (string-formatted) character. Test-drive the `chr()` function in the Shell as follows:

```
>>> character = chr(65)
>>> print(character)
    A
>>> print(type(character))
    <class 'str'>
```

Do It Yourself

1. **Binary Data Files** (a) Write the following program named **2d_binary_storage.py**, which stores a 2D array of time and 1-Hz sine-wave displacement values in a binary-formatted file. Much of the program you will write is similar to **2d_spreadsheet_storage.py**, so you may want to begin with that program open, then create the new program using **File≫Save As...**, and save it in **YourNamePythonWork**.

```
import numpy as np
import tkinter.filedialog as fd

start, stop, num = 0, 1, 5
A, f = 5, 1

t = np.linspace(start, stop, num)
y = A*np.sin(2*np.pi*f*t)
print(t)
print(y)

fname = fd.asksaveasfilename()

row = np.hstack((t, y))
row.tofile(fname)
```

Save your work. Here, the NumPy function `hstack()` concatenates the two five-element `t` and `y` 1D NumPy arrays into one ten-element 1D NumPy array `row`, which has the characteristics of a row vector. By default, the data elements in the arrays are formatted as eight-byte floats. The method `tofile()` associated with a NumPy array is then used to write the array `row` to a binary-formatted file named `fname`. The `t` and `y` arrays are printed in the Shell, for future reference.

(b) Run **2d_binary_storage.py** and store the binary-formatted data in a file named **2d_binary_data.dat** in **YourNamePythonWork**. Determine of the size of this file (in bytes). In Windows, the file's exact size is found by right-clicking on the icon and selecting **Properties**. Given that each element in the array being stored is represented by eight bytes, explain why **2d_binary_data.dat** has the size it does.

(c) Open **2d_binary_data.dat** with a word processor. Offer a qualitative explanation for what you observe. Close **2d_binary_data.dat**, without saving it.

(d) Write the following program named **2d_binary_read.py**, which reads a binary-formatted file in which a 2D array of floats is stored. Much of the program you will write is similar to **2d_spreadsheet_read.py**, so you may want to begin with that program open, then create the new program using **File≫Save As...**, and save it in **YourNamePythonWork**.

```
import numpy as np
import tkinter.filedialog as fd

fname = fd.askopenfilename()
```

```
row = np.fromfile(fname, dtype='f8')

num = int(len(row)/2)

t = row[0:num]
y = row[num:2*num]

print(t)
print(y)
```

Save your work. Here, the NumPy function `fromfile()` decodes the binary-formatted data in the file `fname` into the 1D NumPy array `row`. To accomplish this decoding process, the argument `dtype='f8'` alerts `fromfile()` that each element of `row` is an eight-byte float. The first `num` elements of `row` are the `t` array, while the second `num` elements of `row` are the `y` array. The Python function `len(row)` gives the number of elements in `row` and so the integer `num` is found by `int(len(row)/2)`. The slicing operations `row[0:num]` and `row[num:2*num]` yield `t` and `y`, respectively.

(e) Run **2d_binary_read.py** and use it to read and decode **2d_binary_data.dat**. How do the decoded `t` and `y` arrays printed in the Shell compare with the original arrays found when running **2d_binary_storage.py**?

Use It!

2d_spreadsheet_runtime_storage.py It is often an advantage to "stream data to disk" during a data-acquisition program's runtime, i.e., add each data sample to an open computer file at the instant it is acquired, rather than store the entire collection of data samples at the close of the program's execution (as would be done when using the NumPy function `savetxt()`). With this approach, if your computing system crashes before the data-acquisition program has completed its full execution, the data up to the time of the crash have been saved, a feature especially valuable for long data runs.

Write the following program and store it in **YourNamePythonWork** under the name **2d_spreadsheet_runtime_storage.py**.

```
import numpy as np
import tkinter.filedialog as fd
import time

delta_t, num = 0.1, 10
A, f = 5, 1

fname = fd.asksaveasfilename()

fobject = open(fname, mode='w')
```

```
for i in range(num):

    t = i*delta_t
    y = A*np.sin(2*np.pi*f*t)
    print(t, y)

    row = f'{t:.3f}' + ',' + f'{y:.3f}' + '\n'
    time.sleep(1.)
    fobject.write(row)

fobject.close()
```

This program simulates an experiment in which sine-wave data is taken once a second for a total runtime of 10 seconds. Python's `open()` command opens the file `fname` and returns a file object named `fobject`. The argument `mode='w'` configures the file so that character strings can be written into it. If the file `fname` does not exist, it will be created; if it does exist, any existing content will be overwritten. The `for` loop then simulates the time and sine-wave displacement data being acquired one sample at a time at a rate of one sample per second. In line with the CSV spreadsheet-format, the string `row` is created with the time and sine-wave displacement values separated by a comma and the string terminating with the EOL character `\n` (Windows users may need to use `\r\n`). The float-formatted sample values are converted to characters using f-string commands and the four component strings comprised in a row are concatenated using plus signs (+). Once constructed, the string row is written to the file `fname` using the `write()` method associated with the object `fobject`. Once the `for` loop completes its ten-iteration execution, the file is closed using the `close()` method also associated with `fobject`.

Save your work, then run name **2d_spreadsheet_runtime_storage.py**, using it to save the acquired data in a file named **2d_runtime_data.csv** in **YourNamePythonWork**. After **2d_spreadsheet_runtime_storage.py** completes its 10-second execution, opening **2d_runtime_data.csv** in a word processor or spreadsheet program such as Microsoft Excel yields the following:

	A	B	C	D	E	F	G	H	I	J	K
1	0.0	0.000									
2	0.1	2.939									
3	0.2	4.755									
4	0.3	4.755									
5	0.4	2.939									
6	0.5	0.000									
7	0.6	-2.939									
8	0.7	-4.755									
9	0.8	-4.755									
10	0.9	-2.939									
11											

2d_runtime_data

An advantage of using Python commands, rather than NumPy's `savetxt()` command as in **2d_spreadsheet_storage.py**, is that different digits of precision can be chosen for the time and sine-wave displacement columns. Let's see whether our intended benefit of data storage prior to a crash is also present. Introduce a runtime error into **2d_spreadsheet_runtime_storage.py** by placing a `ZeroDivisionError` when `i = 5` within the `for` loop as shown below:

```
for i in range(num):

    if i == 5: i/0

    t = i*delta_t
    y = A*np.sin(2*np.pi*f*t)
    print(t, y)

    row = f'{t:.3f}' + ',' + f'{y:.3f}' + '\n'
    time.sleep(1.)
    fobject.write(row)

fobject.close()
```

Run **2d_spreadsheet_runtime_storage.py**, choosing **2d_runtime_data.csv** again as the storage file. You will find that the `ZeroDivisionError` exception aborts program execution when `i = 5`. Our hope, though, is that the data for the `i=0` through `i=4` iteration was saved in **2d_runtime_data.csv**. Unfortunately, when you open **2d_runtime_data.csv** in a word processor program, you will find that it is an empty file. The reason that our hopes are dashed is that the changes within the file do not go into effect until the file is properly closed by the `fobject.close()` command and, given our program structure, this command isn't executed until the `for` loop has completed all of its iterations.

The solution to this problem is to use Python's `with` statement to control file operations. The syntax for this command is as follows:

```
    with open(fname, mode='w') as fobject:
```

The `with` statement will take care to open and close the file at the proper times. Modify **2d_spreadsheet_runtime_storage.py** as shown below, which includes the runtime-exception-producing *if* statement. Note that the `fobject.close()` command is no longer needed:

```
import numpy as np
import tkinter.filedialog as fd
import time

delta_t, num = 0.1, 10
A, f = 5, 1

fname = fd.asksaveasfilename()

with open(fname, mode='w') as fobject:

    for i in range(num):

        if i == 5: i/0
```

```
t = i*delta_t
y = A*np.sin(2*np.pi*f*t)
print(t, y)

row = f'{t:.3f}' + ',' + f'{y:.3f}' + '\n'
time.sleep(1.)
fobject.write(row)
```

Save your work. Run **2d_spreadsheet_runtime_storage.py**, choosing **2d_runtime_data.csv** again as the storage file. Program execution will again be halted by a `ZeroDivisionError` exception when `i = 5`. Open **2d_runtime_data.csv** in a word processor program. I think you will be pleased with what you find.

Problems

1. **Text Versus Binary File Size** (a) Open **2d_spreadsheet_storage.py** and choose the parameters

```
start, stop, num = 0, 1, 10
A, f = 5, 10
```

Also, make the header and footer empty strings (there is no space between the single quotation marks):

```
header_text=''
footer_text=''
```

Finally, make sure that `fmt='%.3f'` in the `savetxt()` function.

Run **2d_spreadsheet_storage.py** and store the resulting spreadsheet-formatted text file in **YourNamePythonWork** under the name **2d_spreadsheet_data.csv**.

Next, open **2d_binary_storage.py** and choose the same parameters as above:

```
start, stop, num = 0, 1, 10
A, f = 5, 10
```

Run **2d_binary_storage.py** and store the resulting binary-formatted file in **YourNamePythonWork** under the name **2d_binary_data.dat**.

(b) Determine the size (in bytes) of the spreadsheet-formatted file **2d_spreadsheet_data.csv** and the binary-formatted file **2d_binary_data.dat**. Open **2d_spreadsheet_data.csv** in a word processor program and count the number of characters within the file (including EOL characters). Then explain why **2d_spreadsheet_data.csv** and **2d_binary_data.dat** have the sizes they do and why the former file is (maybe surprisingly) smaller.

(c) By changing a particular argument in **2d_spreadsheet_storage.py**, you can cause the spreadsheet-formatted text file **2d_spreadsheet_data.csv** to be approximately twice as large as the binary-formatted file **2d_binary_data.dat**. Predict the change needed to result in this file-size doubling. Make this change and verify your prediction.

2. **print_spreadsheet_byte_stream.py** A spreadsheet file is stored as a single string ("byte stream") that contains the contents of the spreadsheet-formatted text file. Write a program that opens a spreadsheet file, reads its byte stream, and then prints the byte stream in the Shell. To write this program, first create a `fname` object by asking for a file name in a dialog window. Then, the Python `open()` command can be used to create a file object `fobject` in the "read" mode via the statement

```
fobject = open(fname, mode='r')
```

Next, the byte stream (call it `byte_stream`) can be read using the `read()` method of `fobject`. Finally, since `byte_stream` contains nonprintable characters such as newline, convert it to a printable representation, using the Python function `repr(byte_stream)`. The string object created by this command can be printed by the `print()` command. By way of explanation, when printing a string object in a printable representation, a newline character will be printed as `'\n'` in the Shell, rather than causing the printing to skip to a new line.

After you complete **print_spreadsheet_byte_stream.py**, run **2d_spreadsheet_storage.py** and use it to create a spreadsheet-formatted **2d_spreadsheet_data.csv**. Next, run **print_spreadsheet_byte_stream.py** and use it to open **2d_spreadsheet_data.csv** and read its byte stream. This byte stream will be displayed in the Shell and you will see something similar to this:

```
'0.000,0.000\n0.111,3.214\n0.222,4.924\n0.333,4.330\
n0.444,1.710\n0.556,-1.710\n0.667,-4.330\n0.778,-4.924\
n0.889,-3.214\n1.000,-0.000\n'
```

Explain the pattern of characters that you see.

To better understand the functioning of `repr()`, run your program with `print(byte_stream)` rather than `print(repr(byte_stream))`.

3. **2d_spreadsheet_storage_labels.py** In comparison with the fully defined NumPy file function `savetxt()`, the use of Python file functions makes it possible to customize the features of a data file. Demonstrate this advantage by writing a spreadsheet-storage program that includes labels at the top of the columns as shown here:

	A	B	C	D	E	F	G	H	I	J	K
1	Time (s)	Displacement									
2	0.0	0.000									
3	0.1	2.939									
4	0.2	4.755									
5	0.3	4.755									
6	0.4	2.939									
7	0.5	0.000									
8	0.6	-2.939									
9	0.7	-4.755									
10	0.8	-4.755									
11	0.9	-2.939									
12											

2d_spreadsheet_data_labels

Programming tip: Use the coding of **2d_spreadsheet_runtime_storage.py** as your guide.

4. **alphabet_string.py** (a) Using a `for` loop, create a program that displays a twenty-six-character string in the Shell, where the string consists of the uppercase letters of the English alphabet in proper order. When run, the display in the Shell should appear as

```
ABCDEFGHIJKLMNOPQRSTUVWXYZ
```

Programming tip: Make sure your alphabet string includes the last letter Z.

(b) Produce a lowercase Greek alphabet string. The Unicode code points for lowercase alpha through omega runs from U+03B1 through U+03C9.

5. **palindrome_detector.py** Write a program that determines whether an input word, phrase, or sentence is a palindrome, that is, reads the same backwards as forwards. As shown below, a message **Entry is Palindrome** is displayed when the input text is determined to be a palindrome; otherwise, the message is **Entry is Not Palindrome**. When completed, test your VI with the following inputs: *radar*, *Able was I ere I saw Elba.*

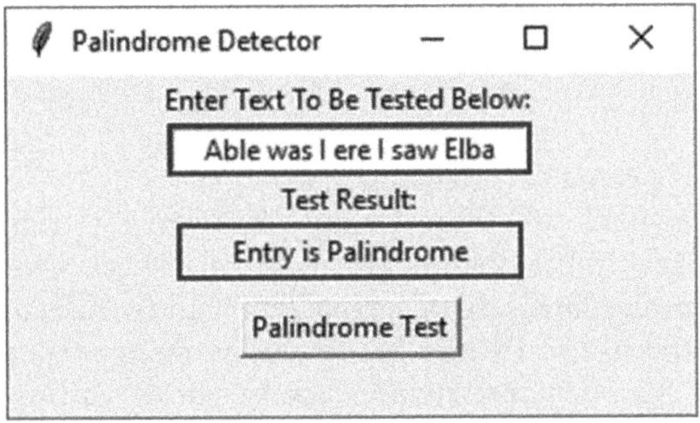

Programming tips: Prior to the palindrome test, change all characters to lowercase and remove all spaces from the entered text. The string method `string.lower()` will return a fully lowercase version of `string`, while the string method `string.replace(old,new)` returns a version of string with every `old` character replaced by a `new` character. To change the text in a `label` object, use `label.config(text=message)`.

6. **parse_and_convert_multimeter_string.py** Computer-controlled scientific instruments often report their measurement results in the form of a character string. As an example, assume a multimeter sends the string *VOLTS DC +1.345e+01* in which *VOLTS DC* identifies

the type of measurement being reported and *+1.345e+01* is the actual measured value. For this measured value to be useful within a program (e.g., plotted on a chart, input into a calculation), the string received from the multimeter must be "parsed" (i.e., the identifier portion separated for the measured value portion), and the measured value portion converted from its character string representation to a numeric format.

Write a program in which the first line is `string = 'VOLTS DC +1.345e+01'` (note there is a space between `VOLTS` and `DC` as well as between `DC` and `+1.345e+01`). Then, in the program, parse `string` appropriately, convert the measured value portion to a float, and display this float in the Shell (with three digits to the right of the decimal point). Programming tip: Review the Do It Yourself project in Chapter 1.

CHAPTER 10

Data Analysis: Curve Fitting

In this chapter's exercises, you will explore the use of a *thermistor* as a temperature sensor. A thermistor has the useful characteristic that its resistance versus temperature (R–t) curve can be fit to a well-known analytic function, facilitating its use as a calibrated thermometer. In Table 10.1, the resistance at particular values of temperature for a thermistor (Epcos Model B57863S0103F040) are given. This data set is called the thermistor's *calibration data.* If you will be using a different thermistor in the Do It Yourself project at the end of this chapter, please use the calibration data for the thermistor that will be employed in your particular experimental setup.

Throughout the chapter, lowercase t will indicate temperature on the Celsius scale, while uppercase T will be temperature on the Kelvin scale. A one-degree temperature change is exactly equal on both scales, while the value of absolute zero differs (0 K versus −273.15°C). Thus, the conversion relation between these two temperature scales is given by $T = t + 273.15$.

10.1 Thermistor Resistance–Temperature Data File

Over the course of this chapter, you will learn how to calibrate a thermistor by fitting its calibration data to several functional forms, including the widely used *Steinhart–Hart equation*. As the first step in accomplishing this feat, you will need to read the calibration data into your fitting program. In Chapter 9, we found that the NumPy function `loadtxt()` makes this task easy, as long as the calibration data are stored in a comma-delimited spreadsheet file. Thus, prior to working on this chapter's exercises, please create and save a spreadsheet-formatted file containing a thermistor's calibration data using either the data provided in Table 10.1 or the data for your own device. In Chapter 9, you learned that in the comma-separated-values (CSV) spreadsheet format, commas separate columns and end-of-line (EOL) characters separate rows. You can use either a spreadsheet data analysis program or a word processing program to create the required file and store it under a descriptive name like **thermistor_calibration_data.csv**. For example, I used Microsoft Word to construct the spreadsheet file shown on page 278 of the thermistor resistance (in kilohms) versus temperature (in degrees Celsius) data given in Table 10.1. I put in blank spaces around each comma to make the document more decipherable; these blank spaces will be skipped over

The Python-Based Laboratory. John Essick, Oxford University Press. © John Essick (2024). DOI: 10.1093/9780191998478.003.0010

by `loadtxt()`. When saving the file using a word processing program, it is important to use the **Save** or **Save As...** command, and then select **Plain Text (.txt)**, as opposed to, for example, **Word Document (.docx)**, in the **File Format**: selection. This procedure creates a Unicode text file devoid of any word-processor formatting characters. For negative temperatures, make sure to use minus signs (i.e., keyboard hyphen, Unicode \u002D), not en or em dashes.

Table 10.1 Calibration Data for Epcos Model B57863S0103F040 Thermistor

Temperature (°C)	Resistance (kΩ)
-50.00	670.1
-45.00	471.7
-40.00	336.5
-35.00	242.6
-30.00	177.0
-25.00	130.4
-20.00	97.07
-15.00	72.93
-10.00	55.33
-5.00	42.32
0.00	32.65
5.00	25.39
10.00	19.90
15.00	15.71
20.00	12.49
25.00	10.00
30.00	8.057
35.00	6.531
40.00	5.327
45.00	4.369
50.00	3.603
55.00	2.986
60.00	2.488
65.00	2.083
70.00	1.752
75.00	1.481
80.00	1.258
85.00	1.072
90.00	0.9177
95.00	0.7885
100.00	0.6800
105.00	0.5886
110.00	0.5112

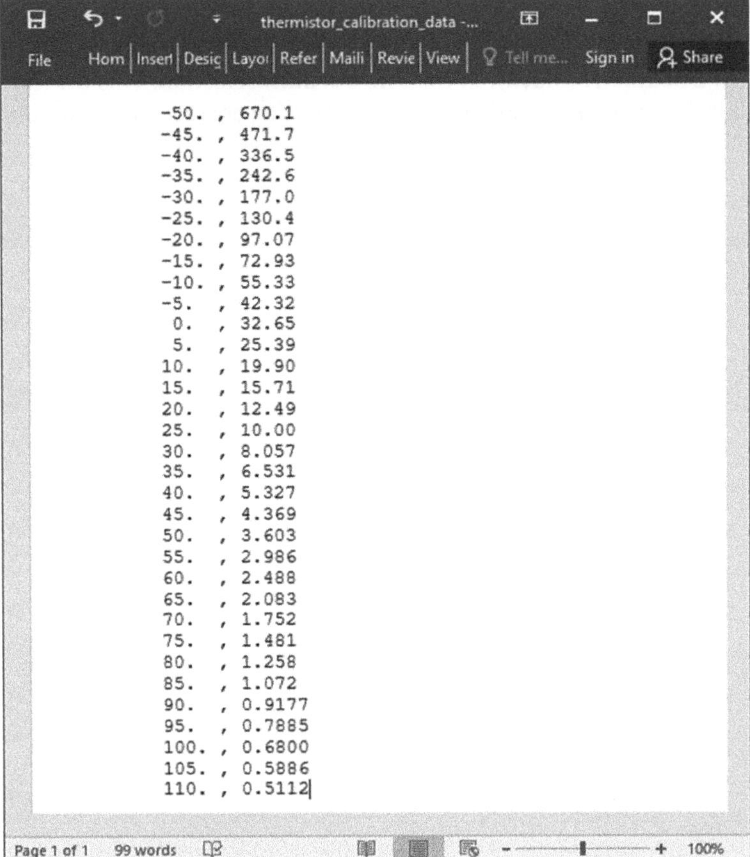

```
-50. , 670.1
-45. , 471.7
-40. , 336.5
-35. , 242.6
-30. , 177.0
-25. , 130.4
-20. , 97.07
-15. , 72.93
-10. , 55.33
-5.  , 42.32
 0.  , 32.65
 5.  , 25.39
10.  , 19.90
15.  , 15.71
20.  , 12.49
25.  , 10.00
30.  , 8.057
35.  , 6.531
40.  , 5.327
45.  , 4.369
50.  , 3.603
55.  , 2.986
60.  , 2.488
65.  , 2.083
70.  , 1.752
75.  , 1.481
80.  , 1.258
85.  , 1.072
90.  , 0.9177
95.  , 0.7885
100. , 0.6800
105. , 0.5886
110. , 0.5112
```

Alternatively, if you use a spreadsheet-based program to create the file, you will want to save it as a **CSV UTF-8 (Comma delimited) (.csv)** file. This option may be listed under a **Save**, **Save As...**, or **Export** menu item, depending on the program you use. Make sure that you save this file in the correct manner to avoid including a bunch of program-dependent formatting characters in the file. Perhaps the best way to check that you've produced the desired file is to open it using a word processing program to see whether it appears as above.

Take a minute to create the spreadsheet file of the calibration data and name it **thermistor_calibration_data.csv.** You will need this file during your work in this chapter.

10.2 Temperature Measurement Using Thermistors

Among the many electronic temperature sensors available—including platinum resistance thermometers, thermocouples, and semiconductor devices such as the Analog Devices AD590—thermistors are used in applications that require high sensitivity, small size, ruggedness, fast response time, and low cost. When utilized with the proper support circuitry, a thermistor can measure temperatures to an accuracy of 0.01°C over a fairly wide range (about 100°C). The only price to be paid for the thermistor's high sensitivity is that its resistance is a nonlinear function of temperature. In the past, the nonlinearity of thermistors was a distinct disadvantage, causing users to employ resistor-based linearizing networks or less-than-accurate resistance-to-temperature conversion methods such as the Beta formula (see Eq. (10.9)). Fortunately, the widespread availability of microprocessors has enabled the use of accurate (and somewhat complex) temperature-calibration models, greatly simplifying the use of thermistors in accurate temperature measurement.

The active layer of most thermistors is composed of a semiconducting metal-oxide alloy. As is characteristic of a semiconductor, when the thermistor's temperature is elevated, loosely bound (valence) electrons within its active layer are thermally released from their respective binding sites to become conduction electrons. By this mechanism, the thermistor's resistance decreases with temperature and results in the utility of this device as a temperature sensor. Since the bound electronic state is of lower energy, an activation energy barrier E separates the conducting state from the insulating state. Thus, at a given temperature T (in degrees Kelvin), we expect that the probability of such activation, and therefore the density n of conduction electrons, is proportional to a Boltzmann factor. That is,

$$n = n_0 \exp\left[-\frac{E}{kT}\right], \tag{10.1}$$

where n_0 is the density of conduction electrons at very high temperatures, and k is Boltzmann's constant. It is easily shown that these conduction electrons imbue the material with a resistivity ρ given by

$$\rho = \frac{m}{ne^2\tau}, \tag{10.2}$$

where e and m are the charge and mass of an electron, respectively. The scattering time τ is the average time traveled by a conduction electron until it is scattered by a lattice vibration, an impurity, or some other scattering mechanism peculiar to the material.

Thus, assuming that e, m, and τ are independent of temperature, we expect that the temperature dependence of ρ is entirely caused by the Boltzmann factor related to n. Putting Eq. (10.1) into Eq. (10.2), we predict that

$$\rho = \rho_0 \exp\left[\frac{E}{kT}\right], \tag{10.3}$$

where ρ_0 is the temperature-independent constant given by

$$\rho_0 = \frac{m}{n_0e^2\tau}. \tag{10.4}$$

The macroscopic resistance R of a sample is determined by

$$R = \rho\frac{L}{A}, \tag{10.5}$$

where L and A are the sample's length and cross-sectional area, respectively.

Because a solid sample's geometric shape is little changed with temperature, we expect that the temperature dependence of R will simply mimic that of ρ. Thus,

$$R = R_0 \exp\left[\frac{E}{kT}\right], \tag{10.6}$$

where R_0 is the (assumed) temperature-independent constant

$$R_0 = \frac{mL}{An_0e^2\tau}. \tag{10.7}$$

If all of our logic is correct, we predict that the temperature dependence of a thermistor's resistance should be described by Eq. (10.6). By taking the natural logarithm of both sides, this relation can be rewritten as the linear relation

$$\frac{1}{T} = -\frac{k}{E}\ln R_0 + \frac{k}{E}\ln R \tag{10.8}$$

or

$$\frac{1}{T} = A + B\ln R, \tag{10.9}$$

where A and B are constants defined to be $A \equiv -k\ln R_0/E$ and $B \equiv k/E$. Equation (10.9) is the so-called *Beta formula* (B is sometimes written as β) and it predicts that when a thermistor's temperature-dependent resistance data are plotted as $1/T$ (y-axis) versus $\ln R$ (x-axis), a straight line will result with the y-intercept and slope equal to A and B, respectively.

Despite this expectation, it is found experimentally that a real thermistor's $R - T$ data deviates somewhat from a straight line when plotted as $1/T$ versus $\ln R$. Rather than being constant, the slope B decreases with decreasing temperature. Over a narrow temperature range (say, of 10°C), B varies so slightly that the Beta formula can predict the thermistor's behavior with a temperature uncertainty of about 0.01°C. However, for the much larger temperature span over which a thermistor is useful, the Beta formula inadequately models the data.

Something, then, must be missing in our theory. In reviewing the above derivation, one might suspect it is our assumption of a temperature-independent scattering time. It is easy to imagine that as temperature increases, the conduction electrons move faster and/or lattice vibrations become more pronounced and consequently shorten the scattering time τ. While the exponential Boltzmann factor activation of conduction electron density will still dominate the resistance's temperature dependence, the milder temperature-dependent scattering time can possibly account for the observed slight nonlinearity of a $1/T$ versus $\ln R$ plot.

How, then, can we more accurately model the $R-T$ data? Unfortunately, the detailed theory for electrical conduction in metal-oxide thermistors, which attempts to account for deviations from the predictions of the semiconductor energy band model advocated above, is quite complex and does not offer us a more refined analytic relation to replace Eq. (10.9). In light of this theoretical vacuum, those using thermistors as a temperature sensor take an empirical approach, where the (slight) nonlinearity of the $1/T$ versus $\ln R$ curve is described via the standard curve-fitting technique of considering $1/T$ to be a polynomial of $\ln R$. That is, the inverse temperature is written as the nth-order polynomial

$$\frac{1}{T} = A + B\ \ln R + C\ (\ln R)^2 + \ \dots\ + Q\ (\ln R)^n. \tag{10.10}$$

From this point of view, Eq. (10.9) is such a polynomial that has been truncated at the first-order term. One then is led to the conclusion that by retaining higher-order terms, the equation will remain valid over a wider range of temperatures.

Starting with this idea, researchers have found that the 100°C temperature span over which a typical thermistor is active can be accurately modeled by a third-order polynomial:

$$\frac{1}{T} = A + B\ \ln R + C\ (\ln R)^2 + D\ (\ln R)^3. \tag{10.11}$$

This relation can be used to convert resistance to temperature with a precision equal to that of the calibration data used to determine the constant A, B, C, and D (typically, 0.005 to 0.01°C).

In 1968, the two oceanographers Steinhart and Hart announced their following discovery: When modeling up to 200° spans contained within the temperature range of −70 to 135°C (they were especially interested in the oceanographic range of −2 to +30°C), the second-order term in the above relation could be neglected without significant loss of accuracy. Then Eq. (10.11) reduces to

$$\frac{1}{T} = A + B \ \ln R + D \ (\ln R)^3. \tag{10.12}$$

Equation (10.12) is called the *Steinhart–Hart equation* and is the conventional relation used in thermistor calibration. For a typical thermistor, the constants A, B, and D have order of magnitude values of 10^{-3}, 10^{-4}, and 10^{-7}, respectively, when resistance is measured in ohms. The temperature T is, of course, measured on the Kelvin scale.

10.3 Least-Squares Curve-Fitting Method

The process of scientific inquiry many times proceeds as follows: Researchers perform an experiment to investigate a particular physical phenomenon and to obtain a set of N data samples (x_j, y_j). After shutting off their instruments, the investigators apply appropriate curve-fitting techniques to determine the functional relationship $y = f(x)$ manifest in their data. Once armed with this $f(x)$, the scientific community can then judge the veracity of theoretical models posited to explain the phenomenon under investigation by testing each model's capacity for correctly predicting the experimentally observed $f(x)$. In this section, we will explore Python's curve-fitting functions that allows one to extract an accurate $y = f(x)$ from a given data set (x_j, y_j), a skill that plays a crucial role in the above-described process. We will put this skill to immediate use in calibrating a temperature-sensing thermistor.

Quite often, it is convenient to adopt a *linear model* for a functional relationship $y = f(x)$ in which the relation is represented by a linear combination of a set of *basis functions* b_k. The general form of a linear model is

$$y = f(x) = \sum_{k=0}^{M-1} a_k b_k(x) = a_0 b_0 + a_1 b_1 + \ldots + a_{M-1} b_{M-1}, \tag{10.13}$$

where the basis functions $b_k(x)$ are known functions of x and the a_k are the *linear expansion coefficients*. For example, if $b_k = x^k$, then $f(x) = a_0 + a_1 x + a_2 x^2 + \ldots + a_{M-1} x^{M-1}$, which is a polynomial of order $M - 1$. Alternately, if $b_k = \cos(kx)$, then $f(x) = a_0 + a_1 \cos(x) + a_2 \cos(2x) + \ldots$, which is a Fourier series. Note that, in the linear model, the basis functions $b_k(x)$ can be nonlinear functions of x such as trigonometric, exponential, or logarithmic functions. However, the coefficients a_k appear in a linear fashion in Eq. (10.13).

Once the decision is made to model $f(x)$ as a linear combination of a set of basis functions, one needs some method to determine the appropriate choice of the a_k in Eq. (10.13) so that a given data set is accurately described by Eq. (10.13). The theory of data analysis provides just such a method. Because all data-taking processes are subject to random errors, the repeated acquisition of a particular experimental quantity produces an array of values that is distributed as a Gaussian ("bell curve") distribution. By carefully considering the ramification of this fact, it can be shown that Eq. (10.13) will best describe a given N data samples (x_j, y_j) when the a_k

are chosen such that they minimize an error function e defined as

$$e \equiv \sum_{j=0}^{N-1} \left[y_j - f(x_j)\right]^2 = \sum_{j=0}^{N-1} \left[y_j - \sum_{k=0}^{M-1} a_k b_k(x_j)\right]^2. \tag{10.14}$$

Note that e is a measure of the deviation between the experimentally determined y_j and the y obtained from the function $y = f(x_j)$, which depends on the values chosen for the coefficients a_k. Various algorithms have been proposed to identify the a_k that minimize e for a given data set, all of which are classified under the rubric of *linear least-squares* curve-fitting methods. Here, "linear" means that the function $f(x)$ being fit to a given data set (termed the *model function*) has the form of Eq. (10.13); i.e., it is a linear combination of some chosen basis functions.

The `polynomial` module of NumPy offers several linear least-squares curve-fitting functions. All of these curve-fitting functions implement a model function that consists of a linear combination of basis functions and differ from each other only through their choices of the basis functions. Then, given a (x, y) data set and their particular choice for basis functions, each of these curve-fitting functions will obtain the values for the linear expansion coefficients that produce a function $f(x)$ that optimally agrees with the data. This process is implemented through a linear-algebra technique called *singular value decomposition* (SVD). The curve-fitting function `polyfit()` uses $b_k = x^k$ as its basis functions and so its model function is the familiar "standard" polynomial (i.e., a sum of terms with increasing powers of an argument `x`). On the other hand, the linear curve-fitting functions `chebfit()`, `legfit()`, `lagfit()`, and `hermfit()` use model functions that are linear combinations of the specialized polynomials of Chebyshev, Legendre, Laguerre, and Hermite, respectively. These specialized polynomials appear in the solutions to many advanced applications including electronic filters and vibrational modes of spherical objects as well as quantum-mechanical wavefunctions of atoms and simple harmonic oscillators.

For our thermistor work, our model function is a standard polynomial as given in Eq. (10.11), so will employ the `polyfit()` curve-fitting function. This NumPy function is (confusingly) found in the `polynomial` submodule of the `polynomial` module (the `polynomial` module contains curve-fitting functions with all types of polynomial-based model functions, while the `polynomial` submodule pertains just to the "standard" polynomial model function). The form of `polyfit()` is as follows (a few arguments not relevant to our study are omitted):

```
coef = numpy.polynomial.polynomial.polyfit(x, y, deg,
                                           w=None)
```

Here, `x` and `y` are independent- and dependent-variable arrays, respectively, of input data to which the polynomial of degree `deg` is to be fit (`deg` is the highest power in the polynomial). The argument `w` selects whether a weighted fit is be carried out. In the default case of `w=None`, the fit is unweighted. For a weight fit, `w` equals an array of weighting factors. The function returns `coef`, which is a NumPy array of the optimized linear expansion coefficients, ordered from lowest-power to highest-power terms.

For some applications, the required model function is nonlinear; that is, the function is not of the form given in Eq. (10.13). For example, the Gaussian function appears often as the model function in statistical analysis:

$$y(x) = A \exp\left[-\frac{(x-B)^2}{2C^2}\right].$$

This function does not follow the form of Eq. (10.13) in that two of the "to-be-optimized" parameters B and C are within the exponential function. For this type of model function, SciPy offers the curve-fitting function `curve_fit()`, which can fit data to a given nonlinear model function. To carry out this process, `curve_fit()` implements (by default) the *Levenberg–Marquardt* (L–M) algorithm, a quite general method that can handle both linear and nonlinear model functions. `curve_fit()` is found in the `optimize` module of SciPy and has the following form (arguments not relevant to our study, including those for a weighted fit, are omitted):

```
popt, pcov = scipy.optimize.curve_fit(f, x, y, p0)
```

Again, `x` and `y` are independent- and dependent-variable arrays, respectively, of input data to which the model function is to be fit. `f` is the name of the model function, whose definition must be in the form `f(x, A, B, C, ...)`; i.e., its first parameter is the independent variable `x` and the remaining parameters are the fit parameters to be optimized. `curve_fit()` requires an initial guess for the fit parameters. This guess is provided through the "array-like" argument `p0`; thus, `p0` can, for instance, be a NumPy array, a Python list, or Python tuple. The optimized values of the fit parameters determined by `curve_fit()` are returned as the NumPy array `popt`. Also returned is `pcov`, a 2D *covariance matrix*, which provides a quantitative measure of the goodness of fit. The diagonal elements of `pcov` are the variance σ^2 of each fit parameter, so the square root of each diagonal element yields the standard deviation σ (or uncertainty) of each optimized fit parameter.

In the following pages, you will learn how to implement both `polyfit()` and `curve_fit()` in programs that fit a thermistor's calibration data to different model functions, including the Steinhart–Hart equation. Through this work, we will be able to judge the validity of Steinhart–Hart's "discovery" that the second-order term in Eq. (10.11) be neglected without significant loss of accuracy.

10.4 Plotting Thermistor Calibration Data

For our curve-fitting program, an important initial operation will be inputting the given array of thermistor calibration data into a Python program, an operation for which we have already developed all of the needed tools. Let's then begin by writing a program that can receive and plot the thermistor spreadsheet data that you stored in a computer file in Section 10.1. We'll equip this program with a GUI and write it generically so that it can be easily reused in the future to read other spreadsheet data. If you haven't created the thermistor data file yet, now is the time to do it.

Open IDLE and in the Editor, create a new file using **File≫New File**, and then save it in **YourNamePythonWork** under the name **r-t_spreadsheet_plot.py**. Enter the code for this program as shown below. In it, the two-column spreadsheet data is read into two NumPy arrays, then an operating system window named `root` is created and a plot is placed in it. Finally, the calibration data are plotted, with each R–t data point represented by a filled circle marker. This program uses coding components from **2d_spreadsheet_read.py** and **waveform_simulator.py**. Consider copying-and-pasting from those programs to expedite your coding here.

```
import Utilities.gui_creator as gc
import Utilities.plot_creator as pc
import numpy as np
```

```
import tkinter.filedialog as fd


fname = fd.askopenfilename()
t, r = np.loadtxt(fname, delimiter=',', usecols=(0,1),
                  unpack=True)

root = gc.create_window(title='Thermistor R-t Data Plot',
                        size='700x500')
ax, canvas = pc.create_plot(root)


pc.update_plot(ax, canvas, t, r,
               label_x='Temperature (deg Celsius)',
               label_y='Resistance (kOhm)',
               marker='o')
```

Save your work, then run **r-t_spreadsheet_plot.py**. In the dialog window that pops up, open the **thermistor_calibration_data.csv** file in **YourNamePythonWork**. Once the data file is read, the GUI window will appear, presenting you with a graph of your thermistor's resistance (y-axis) versus temperature (x-axis). Note the exponential-looking decay of resistance with increasing temperature. This is a signature of the thermally activated process taking place within the semiconducting material of which the thermistor is composed.

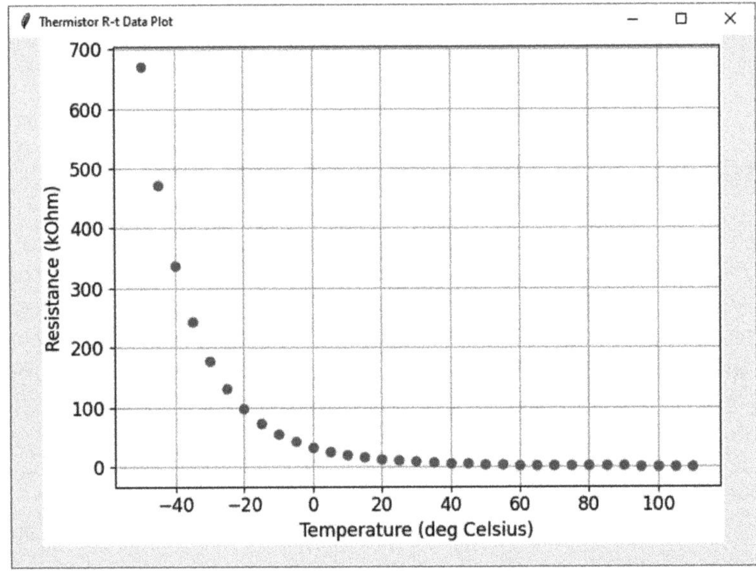

If you encounter an error when your program tries to open the data file, carefully check to make sure that your data file is correctly constructed. An extraneous character as the last element in an array is a common problem. Also make sure your negative temperatures have minus signs and not, say, en dashes.

10.5 Linear Least-Squares Curve Fitting Using polyfit()

Now that you know how to input your thermistor calibration data into a program, let's write a program that uses the NumPy function `polyfit()` to carry out a linear least-squares fit to these data. The model function assumed by `polyfit()` is a polynomial of degree `deg`. As mentioned in Section 10.2, researchers found that a typical thermistor's calibration data can be accurately modeled by the function

$$\frac{1}{T} = A + B\ \ln R + C\ (\ln R)^2 + D\ (\ln R)^3. \tag{10.15}$$

At first glance, this relation does not appear to follow the form assumed by the linear least-squares theory (Eq. (10.13)). However, if we define $y \equiv 1/T$ and $x \equiv \ln(R)$, then Eq. (10.11) becomes the third-order polynomial

$$y = A + Bx + Cx^2 + Dx^3, \tag{10.16}$$

which is of the form given in Eq. (10.13) with the basis functions $b_0 = 1$, $b_1 = x$, $b_2 = x^2$, and $b_3 = x^3$ and the linear expansion coefficients $a_0 = A$, $a_1 = B$, $a_2 = C$, and $a_3 = D$. Equation (10.16) is the *lineriazed* version of Eq. (10.15). The strategy for our curve-fitting program is this: Input the thermistor's calibration data, recast these data as $x = \ln R$ and $y = 1/T$, and then use the `polyfit()` function to perform a linear least-squares fit to obtain the values of A, B, C, and D. Note that we will not be able to fit the data to the Steinhart–Hart equation (which omits the second-order term in Eq. (10.16)), because `polyfit()` assumes all terms are present in the polynomial model function up to order `deg`.

In the Editor, select **File≫New File**, and then save it in **YourNamePythonWork** under the name **r-t_polyfit.py**. Enter the code for this program as shown below. In it, the resistance array, which is in units of kiloohms in the original calibration data, is multiplied by 1000 in order to convert these values to ohms, then the natural logarithm of this value is taken. In addition, since the temperature array in the spreadsheet is in units of degrees Celsius, 273.15 is added to each value in order to convert to temperature on the Kelvin scale. (Remember that the degree intervals on the Kelvin and Celsius scales are defined to be the same. Because absolute zero on the Kelvin scale is define as zero degrees and the ice point is 273.15 K, temperatures on the two scales simply differ by the constant value of 273.15.)

```
import numpy as np
import tkinter.filedialog as fd
import numpy.polynomial.polynomial as npp

fname = fd.askopenfilename()
t,r = np.loadtxt(fname, delimiter=',', usecols=(0,1),
                 unpack=True)

x = np.log(r*1000)
y = 1/(t+273.15)

coef = npp.polyfit(x, y, 3)

print(coef)
```

Save your work, then run **r-t_polyfit.py**. In the dialog window that pops up, select the **thermistor_calibation_data.csv** file in **YourNamePythonWork**. After the spreadsheet file is read, the optimized values for linear expansion coefficients A, B, C, and D will be displayed in the Shell, packaged in a NumPy array. Do your values for A, B, and D have expected order of magnitude values of 10^{-3}, 10^{-4}, and 10^{-7}, respectively?

Next, explore the Steinhart–Hart "discovery." At the end of your program, add the following `for` loop, which prints the contribution of each term of Eq. (10.16) in the sum that calculates `y`. The statement `a, b, c, d = coef` unpacks the four parameters from the array `coef`.

```
a, b, c, d = coef

for value in x:
    print(a, b*value, c*value**2, d*value**3)
```

Save your work, then rerun **r-t_polyfit.py**. The values of the four terms for each element of the array `x` (i.e., corresponding to each temperature in the original calibration data) will be printed row by row in the Shell.

To make the output in the Shell more readable, consider using the f-string operation to limit the number of digits in each value and place tabs between neighboring value as shown here:

```
a, b, c, d = coef

for value in x:
    print(f'{a:.3e}', '\t', f'{b*value:.3e}', '\t',
          f'{c*value**2:.3e}', '\t', f'{d*value**3:.3e}')
```

By comparing the magnitudes of the four contributions in each row, does the Steinhart–Hart "discovery" (that the second-order term can be ignored) seem valid?

10.6 GUI-Equipped Program for polyfit()

Let's make a GUI-equipped program that compares the original thermistor calibration data with the fitted curve derived by the function `polyfit()`. In the Editor, begin by opening **r-t_spreadsheet_plot.py** and then selecting **File≫Save As...** to create a new file named **r-t_plotfit_and_plot.py** in **YourNamePythonWork**. Then, add `create_frames()` from **gui_creator.py** and place the plot on the right frame and increase the size of the `root` window as shown next:

```
import Utilities.gui_creator as gc
import Utilities.plot_creator as pc
import numpy as np
import tkinter.filedialog as fd


fname = fd.askopenfilename()
t, r = np.loadtxt(fname, delimiter=',', usecols=(0,1),
                  unpack=True)
```

```
root = gc.create_window(title='Thermistor R-t Data Plot',
                        size='1000x500')
control_box, right_frame = gc.create_frames(root)
ax, canvas = pc.create_plot(right_frame)

pc.update_plot(ax, canvas, t, r,
               label_x='Temperature (deg Celsius)',
               label_y='Resistance (kOhm)',
               marker='o')
```

Save your work, then run the program. It will produce a window as shown here.

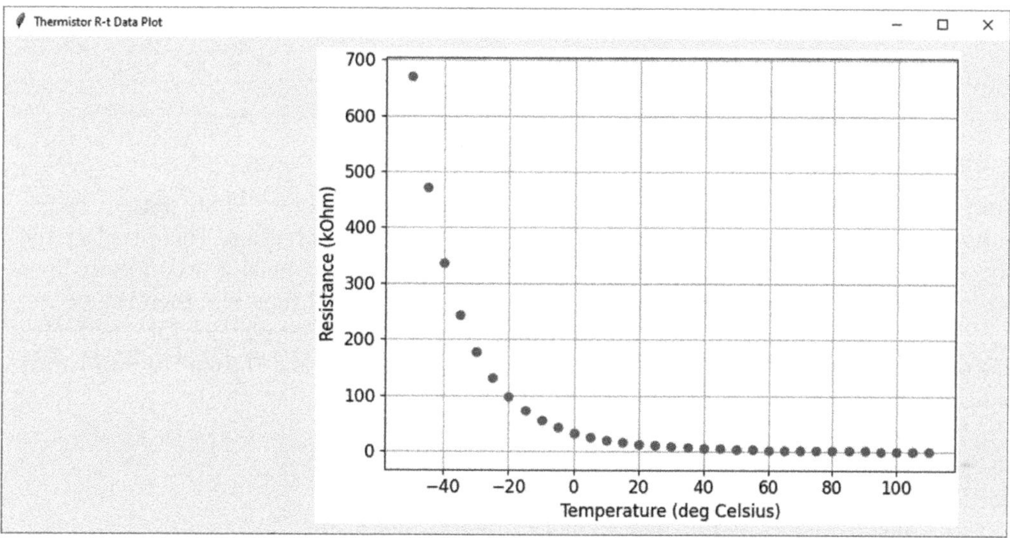

Next, at the bottom of the program, add the following code. Here, the function `create_indictor()` from **gui_creator.py** (see Chapter 5's Use It! project) is used to produce four indicators in the `control_box` frame in order to display the optimized linear expansion coefficients *A*, *B*, *C*, and *D*. For now, zeros are displayed in these indicators.

```
a, b, c, d = 0, 0, 0, 0

gc.create_indicator(control_box, label_1='A',
                    label_2=f'{a:.8e}')
gc.create_indicator(control_box, label_1='B',
                    label_2=f'{b:.8e}')
gc.create_indicator(control_box, label_1='C',
                    label_2=f'{c:.8e}')
gc.create_indicator(control_box, label_1='D',
                    label_2=f'{d:.8e}')
```

Save your work, then run the program. The window will now appear as shown.

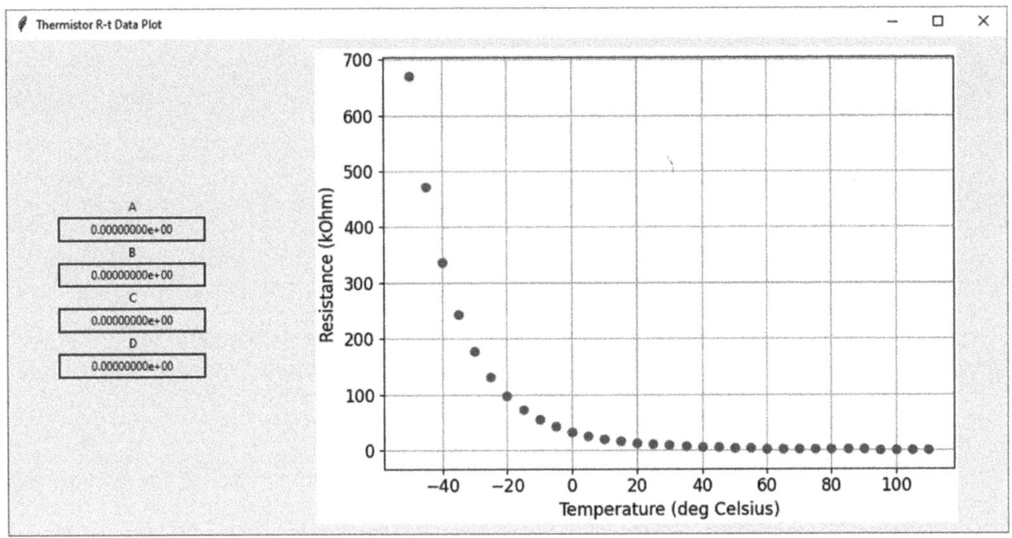

Now, add the curve-fitting component to the program as follows: First, place `import numpy.polynomial.polynomial as npp` at the top of the program. Then, replace the line `a,b,c,d = 0,0,0,0` with the following four lines of code. The first two of these lines carry out the change in variables required to linearize our model function, then the `polyfit()` function performs the linear least-square fit to a third-order polynomial. The fourth line unpacks the optimized linear expansion coefficients A, B, C, and D from the `coef` array produced by `polyfit()`:

```
x = np.log(r*1000)
y = 1/(t + 273.15)

coef = npp.polyfit(x, y, 3)
a, b, c, d = coef
```

Lastly, at the bottom of the program, add the following lines. The first of these lines calculates the values of the `y`-variable on the fitted curve at each of the resistances in the original calibration data (represented by the values of `x`). Remembering that $y = 1/T$, where T is temperature in degrees Kelvin, the second line calculates the temperatures on the fitted curve in the units of degrees Celsius. The final line involving the `update_plot()` function adds a plot of the fitted curve to the existing plot of the calibration data. Setting `clear_flag=False` ensures that the first plot of the calibration data is not cleared as this second plot is made:

```
y_fit = a + b*x + c*x**2 + d*x**3
t_fit = 1/y_fit - 273.15

pc.update_plot(ax, canvas, t_fit, r,
               label_x='Temperature (deg Celsius)',
               label_y='Resistance (kOhm)',
               clear_flag=False)
```

Save your work. The finalized **r-t_plotfit_and_plot.py** program will now appear as shown here. Remember to place `import numpy.polynomial.polynomial as npp` at the top of the program:

```
import Utilities.gui_creator as gc
import Utilities.plot_creator as pc
import numpy as np
import tkinter.filedialog as fd
import numpy.polynomial.polynomial as npp


fname = fd.askopenfilename()
t, r = np.loadtxt(fname, delimiter=',', usecols=(0,1),
                  unpack=True)

root = gc.create_window(title='Thermistor R-t Data Plot',
                        size='1000x500')
control_box, right_frame = gc.create_frames(root)
ax, canvas = pc.create_plot(right_frame)

pc.update_plot(ax, canvas, t, r,
               label_x='Temperature (deg Celsius)',
               label_y='Resistance (kOhm)',
               marker='o')

x = np.log(r*1000)
y = 1/(t + 273.15)

coef = npp.polyfit(x, y, 3)
a, b, c, d = coef

gc.create_indicator(control_box, label_1='A',
                    label_2=f'{a:.8e}')
gc.create_indicator(control_box, label_1='B',
                    label_2=f'{b:.8e}')
gc.create_indicator(control_box, label_1='C',
                    label_2=f'{c:.8e}')
gc.create_indicator(control_box, label_1='D',
                    label_2=f'{d:.8e}')

y_fit = a + b*x + c*x**2 + d*x**3
t_fit = 1/y_fit - 273.15

pc.update_plot(ax, canvas, t_fit, r,
               label_x='Temperature (deg Celsius)',
               label_y='Resistance (kOhm)',
               clear_flag=False)
```

Save your work, then run **r-t_plotfit_and_plot.py**. The plot of the original data from **thermistor_calibration_data.csv** is plotted alongside the fitted third-order polynomial curve determined by `polyfit()`. The optimized values of the linear expansion coefficient are displayed in the four indicators. As you can see, the fitted curve agrees very well with the data as shown here:

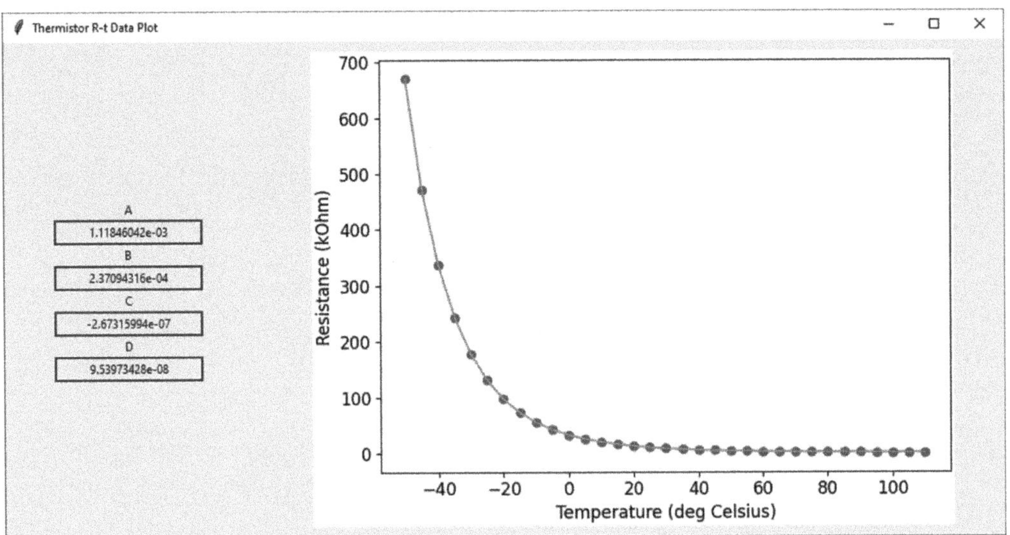

10.7 Nonlinear Least-Squares Curve Fitting Using curve_fit()

We'll now explore the use of the SciPy curve-fitting function `curve_fit()`. This function employs the powerful Levenberg–Marquardt minimization algorithm, which allows least-squares curve fitting of data to both linear and nonlinear model functions. Thus, almost any function that depends on one or more *fit parameters* (i.e., adjustable constants within the expression) can be used as the model function in the curve-fit process.

Although `curve_fit()` is much more generally applicable than the polynomial-only function `polyfit()`, with this added power comes a drawback. When using `curve_fit()`, one must provide fairly accurate initial guesses for the fit parameters, introducing some difficulty in acquiring final optimized values as well as the possibility that these final values may not be the best solution to be found. To recognize the source of this difficulty, it's important to understand that optimization algorithms seek to minimize (or maximize) a particular quantity through a search of *parameter space*, where parameter space comprises all possible combinations for the values of the fit parameters. In the case of least-squares curve-fitting, the quantity being minimized is the error function e given by Eq. (10.14) and this minimum is found by checking the value of e as the values of the fit parameters are being systematically changed through excursions in parameter space. An analogy to this search process may help to understand how it can go awry. Image a house with an adjoining cement backyard patio. A few meters above ground level, a clogged rain gutter is attached to the edge ("eave") of the house's downward-sloping roof, while (due to poor-quality construction) the patio is concave downward, with all of its sides raised and sloping down toward a depression at the patio's center. Now imagine a rainy day at this house. Raindrops that fall anywhere on the patio will flow down to the depression at its center and form a puddle there. However, raindrops that fall on the roof will flow down and be trapped in the clogged gutter. Stated in technical terms, with height being the quantity to be minimized in this problem, the patio's central depression is the *global minimum*, while the clogged gutter is a *local minimum*. As we reasoned, the initial location where a raindrop alights—patio or roof—determines the minimum at which it ends its journey—global or local, respectively. In an analogous way, in the L–M curve-fitting process, as fit parameters are varied throughout parameter space, the error function associated with a particular model function will most likely have a unique global minimum, but it may also have one or (possibly, many) more local minima. For complex model functions, especially those with many fit parameters, it's almost guaranteed there are many local minima. Thus, the downside of the L–M algorithm's ability to accept any form of model function is that there is no general way for it to start its search in parameter space in a region that will guarantee it ends up at the global minimum of the error function. So, you are required to provide initial guesses of the fit parameters, which are sufficiently close to the optimized values so that the search ends successfully (at the global minimum) and not unsuccessfully (at a local minimum). This requirement can be thorny for problems where there is no way to provide initial "nearly optimal" fit-parameter values, often calling for a laborious iterative process (i.e., taking output of initial run and feeding it back as input to subsequent run) until convergence to a final result is achieved. Take note that no initial guesswork was required when we implemented `ployfit()`. It's maybe now easier to appreciate that the restricted form of the model function in linear least-squares curve fitting (Eq. (10.13)) allows the global minimum of the error function to be found without need of initial fit-parameter guesses by the user.

Given the ability of `curve_fit()` to handle any form of the model function, we will be able to use `curve_fit()` to fit our thermistor data to the nonlinear Steinhart–Hart equation:

$$\frac{1}{T} = A + B\ \ln R + D\ (\ln R)^3. \tag{10.17}$$

As we will show, `curve_fit()` can indeed perform a nonlinear fit to this nonlinear equation. However, it is often possible to algebraically morph a nonlinear equation into a linear form (in a process called *linearization*), thereby lending a helping hand to the fitting process. At times, taking the logarithm of both sides of an equation can produce a linearized form but, this time, the change of variables we implemented earlier in the chapter does the trick. Again, defining $y \equiv 1/T$ and $x \equiv \ln(R)$, the Steinhart–Hart equation becomes

$$y = A + Bx + Dx^3. \tag{10.18}$$

This *linearized Steinhart–Hart equation* is a linear combination of the form given in Eq. (10.13), with basis functions $b_0 = 1$, $b_1 = x$, and $b_2 = x^3$ and the linear expansion coefficients $a_0 = A$, $a_1 = B$, and $a_2 = D$.

The program we will write is similar to **r-t_polyfit.py**. So, in the Editor, open **r-t_polyfit.py** and select **File≫Save As..**, creating a new program named **r-t_curve_fit.py** in **YourNamePythonWork**. In the top several lines of the program, simply replace the `import numpy.polynomial.polynomial as npp` statement with `import scipy.optimize as opt` as shown here:

```
import numpy as np
import tkinter.filedialog as fd
import scipy.optimize as opt

fname = fd.askopenfilename()
t, r = np.loadtxt(fname, delimiter=',', usecols=(0,1),
                   unpack=True)

x = np.log(r*1000)
y = 1/(t + 273.15)
```

Then, replace the rest of the program with the following code. The linearized Steinhart–Hart equation is defined as the function `linearized_sh(x,a,b,d)`. By assigning the name of this function as the first argument of `curve_fit()`, it becomes the model function used for the curve-fitting process. `p0` is the "array-like" object (in this case, it's a tuple) that contains the initial guess of the three fit parameters *A*, *B*, and *D*. These guesses follow from the knowledge that, for a typical thermistor, *A*, *B*, and *D* have expected order of magnitude values of 10^{-3}, 10^{-4}, and 10^{-7}, respectively. The function `curve_fit()` returns the two quantities `popt` and `pcov`. For now, we are just printing `popt`.

```
def linearized_sh(x, a, b, d):
    return a + b*x + d*x**3


p0 = 1.e-3, 1.e-4, 1e-7

popt, pcov = opt.curve_fit(linearized_sh, x, y, p0)

print(popt)
```

The complete **r-t_curvefit.py** program appear as follows:

```
import numpy as np
import tkinter.filedialog as fd
import scipy.optimize as opt

fname = fd.askopenfilename()
t, r = np.loadtxt(fname, delimiter=',', usecols=(0,1),
                   unpack=True)

x = np.log(r*1000)
y = 1/(t + 273.15)

def linearized_sh(x, a, b,d):
    return a + b*x + d*x**3


p0 = 1.e-3, 1.e-4, 1e-7

popt, pcov = opt.curve_fit(linearized_sh, x, y, p0)

print(popt)
```

Save your work. Then, run **r-t_curve_fit.py**, using the dialog window that pops up to select the file **thermistor_calibration_data.csv** in **YourNamePythonWork**. After the spreadsheet file is read, the optimized values for fit parameters A, B, and D will be displayed in the Shell, packaged in a NumPy array. Do your values for A, B, and D have expected order of magnitude values of 10^{-3}, 10^{-4}, and 10^{-7}, respectively?

Add the statement `print(pcov)` to the end of your program and then rerun it. In the Shell, the 2D covariance matrix returned by `curve_fit()` will be displayed, which provides a quantitative measure of the goodness of fit. The three diagonal elements of this matrix, from top to bottom, are the variance σ^2 of the fit parameters A, B, and D, respectively. Thus, the square root of each of these elements yields the standard deviation σ (taken as the measure of uncertainty) of each optimized fit parameter. For my program, the uncertainties σ were on the order of 10^{-7}, 10^{-8}, and 10^{-10} for A, B, and D, respectively. Thus, given that my program outputs the optimized value for A as 0.0011262203706949162, only the digits up to the seventh decimal point are significant; i.e., the proper value to report is $A = 0.0011262$.

Check the robustness of the curve-fitting procedure in regard to the choice of `p0` by making initial fit-parameter guesses that highly diverge from the optimal values. Try changing your initial guesses to `p0=1,1,1`, and rerun **r-t_curve_fit.py**. Do the output values for A, B, and D deviate from those you obtained with the previous initial guesses (that were closer to the optimized values)? What happens when `p0=100,100,100`? Remember that our model function has been linearized.

Now, let's explore how `curve_fit()` performs with a nonlinear model function. Again, we will fit our thermistor data to the Steinhart–Hart equation (Eq. (10.17)), but this time written as

$$T = \frac{1}{A + Bx + Dx^3}, \tag{10.19}$$

where $x = \ln R$. Remember, temperature T is in degrees Kelvin, so $T = t + 273.15$, where temperature t is in degrees Celsius. In this form, the constants A, B, and D do not appear in the manner dictated by Eq. (10.13), and so this is a nonlinear model function.

Modify **r-t_curvefit.py** as follows in order to carry out a curve fit of our thermistor data using the nonlinear Eq. (10.19) as the model function. Here, we have added `z` to convert the Celsius temperatures in the calibration data set to the Kelvin scale, defined the model function `nonlinear_sh(x,a,b,d)`, and changed the arguments of the curve-fitting function to `curve_fit(nonlinear_sh,x,z,p0)`; note `y` has been changed to `z`:

```
import numpy as np
import tkinter.filedialog as fd
import scipy.optimize as opt

fname = fd.askopenfilename()
t, r = np.loadtxt(fname, delimiter=',', usecols=(0,1),
                  unpack=True)

x = np.log(r*1000)
y = 1/(t+273.15)
z = t + 273.15


def linearized_sh(x, a, b, d):
    return a + b*x + d*x**3
```

```
def nonlinear_sh(x, a, b, d):
    return 1/(a + b*x + d*x**3)


p0 = 1.e-3, 1.e-4, 1e-7

popt, pcov = opt.curve_fit(nonlinear_sh, x, z, p0)

print(popt)
print(pcov)
```

Save your work. Then, run **r-t_curve_fit.py**, using the dialog window that pops up to select the file **thermistor_calibration_data.csv** in **YourNamePythonWork**. Are the optimized values you obtain for A, B, and D consistent with those you obtained when using the linearized Steinhart–Hart equation as the model function? How about the values for uncertainties σ obtained from the square root of the diagonal elements of the 2D covariance matrix?

Check the robustness of the curve-fitting procedure in regard to the choice of `p0` when using a nonlinear model function. Try changing your initial guesses to `p0=1,1,1`, and rerun **r-t_curve_fit.py**. Do the output values for A, B, and D deviate from those you obtained with the previous initial guesses that were closer to the optimized values? That is, for the region of parameter space in which you started it, did the optimization algorithm get stuck in a local minimum of the error function? What happens when `p0=100,100,100`?

Finally, try using the same third-order polynomial model function that we used when working with `polyfit()` by adding the following definition for the function `polynomial(x,a,b,c,d)` to **r-t_curve_fit.py**. Also, now `p0=1.e-3,1.e-4,1e-7,1e-7` and the arguments of the curve-fitting function are `curve_fit(polynomial,x,y,p0)`.

```
import numpy as np
import tkinter.filedialog as fd
import scipy.optimize as opt

fname = fd.askopenfilename()
t, r = np.loadtxt(fname, delimiter=',', usecols=(0,1),
                  unpack=True)

x = np.log(r*1000)
y = 1/(t+273.15)
z = t + 273.15

def linearized_sh(x, a, b, d):
    return a + b*x + d*x**3

def nonlinear_sh(x, a, b, d):
    return 1/(a + b*x + d*x**3)

def polynomial(x, a, b, c, d):
    return a + b*x + c*x**2 + d*x**3


p0 = 1.e-3, 1.e-4, 1e-7, 1e-7
```

```
popt, pcov = opt.curve_fit(polynomial, x, y, p0)

print(popt)
print(pcov)
```

Save your work. Then, run **r-t_curve_fit.py**, using the file **thermistor_calibration_data.csv** to input the thermistor data. Do the optimized values you obtain for A, B, C, and D agree with those you obtained from **r-t_polyfit.py**?

10.8 GUI-Equipped Program for curve-fit()

With just a few changes to **r-t_plotfit_and_plot.py**, we can construct a GUI-equipped program that compares the original thermistor calibration data with the curve derived by the curve-fitting function `curve_fit()`. In the Editor, open **r-t_plotfit_and_plot.py** and then select **File≫Save As...** to create a new file named **r-t_curve_fit_and_plot.py** in **YourNamePython-Work**. First, replace `import numpy.polynomial.polynomial as npp` with `import scipy.optimize as opt`. Then, in the portion of the code that is `polyfit()`-related, as shown here:

```
x = np.log(r*1000)
y = 1/(t + 273.15)

coef = npp.polyfit(x, y, 3)
a, b, c, d = coef
```

Substitute the `curve_fit()`-related coding

```
x = np.log(r*1000)
y = 1/(t + 273.15)
z = t + 273.15

def linearized_sh(x, a, b, d):
    return a + b*x + d*x**3

def nonlinear_sh(x, a, b, d):
    return 1/(a + b*x + d*x**3)

def polynomial(x, a, b, c, d):
    return a + b*x + c*x**2 + d*x**3


p0 = 1.e-3, 1.e-4, 1e-7

popt, pcov = opt.curve_fit(linearized_sh, x, y, p0)

c = 0
a, b, d = popt
```

The finalized **r-t_curve_fit_and_plot.py** program will now appear as shown here:

```
import Utilities.gui_creator as gc
import Utilities.plot_creator as pc
import numpy as np
import tkinter.filedialog as fd
import scipy.optimize as opt

fname = fd.askopenfilename()
t, r = np.loadtxt(fname, delimiter=',', usecols=(0,1),
                  unpack=True)

root = gc.create_window(title='Thermistor R-t Data Plot',
                        size='1000x500')
control_box, right_frame = gc.create_frames(root)
ax, canvas = pc.create_plot(right_frame)

pc.update_plot(ax, canvas, t, r,
               label_x='Temperature (deg Celsius)',
               label_y='Resistance (kOhm)',
               marker='o')

x = np.log(r*1000)
y = 1/(t + 273.15)
z = t + 273.15

def linearized_sh(x, a, b, d):
    return a + b*x + d*x**3

def nonlinear_sh(x, a, b, d):
    return 1/(a + b*x + d*x**3)

def polynomial(x, a, b, c, d):
    return a + b*x + c*x**2 + d*x**3


p0 = 1.e-3, 1.e-4, 1e-7

popt, pcov = opt.curve_fit(linearized_sh, x, y, p0)

c = 0
a, b, d = popt

gc.create_indicator(control_box, label_1='A',
                    label_2=f'{a:.8e}')
gc.create_indicator(control_box, label_1='B',
                    label_2=f'{b:.8e}')
gc.create_indicator(control_box, label_1='C',
                    label_2=f'{c:.8e}')
gc.create_indicator(control_box, label_1='D',
                    label_2=f'{d:.8e}')
```

```
y_fit = a + b*x + c*x**2 + d*x**3
t_fit = 1/y_fit - 273.15

pc.update_plot(ax, canvas, t_fit, r,
               label_x='Temperature (deg Celsius)',
               label_y='Resistance (kOhm)',
               clear_flag=False)
```

Save your work, then run **r-t_curve_fit_and_plot.py**. The plot of the original data from **thermistor_calibration_data.csv** is plotted alongside the curve determined by `curve_fit()` when the linearized Steinhart–Hart equation is used as the model function. As you can see, the fitted curve agrees very well with the data as shown here:

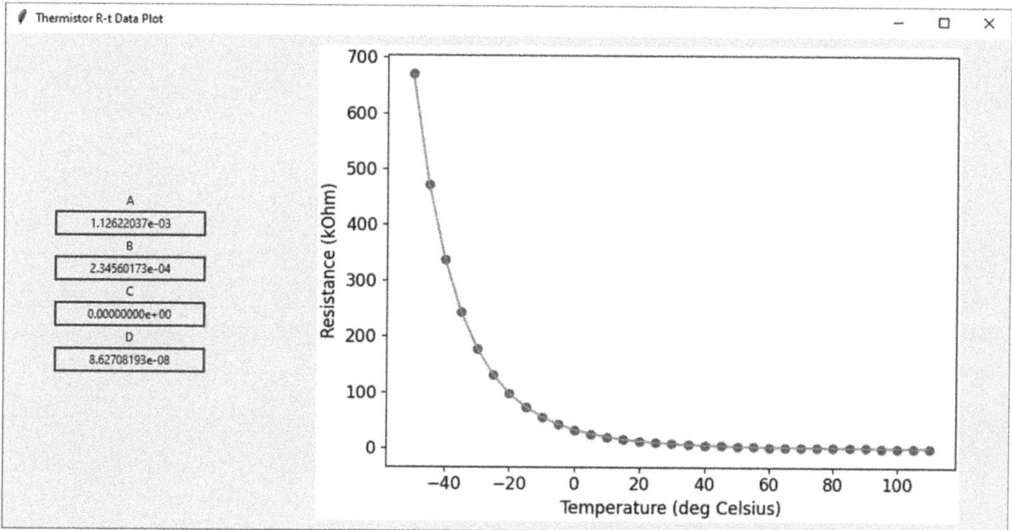

Try modifying the program so that it runs with the third-order polynomial and then the nonlinear Steinhart–Hart equation as the model functions.

10.9 Residual Plot

Is there a quantitative method for judging how well the Steinhart–Hart equation describes the thermistor's calibration data? Yes, there is, and it's called the *residual plot*. At the conclusion of the curve-fitting procedure, the model function can be used to predict the array of "fitted" temperatures t_{fit} (in degrees Celsius) associated with each of the resistance values in the original calibration data set. Then, the deviation of these fitted temperatures t_{fit} from the actual Celsius temperatures `t` at each resistance value is found by calculating the array of residuals R, where a residual is defined to be $R \equiv t - t_{fit}$. Thus, the residuals provide a quantitative measure of the "goodness" of the fitting procedure.

To write a program that plots the residuals for a given choice of model function, in **r-t_curve_fit_and_plot.py** and then select **File≫Save As...** to create a new file named **r-t_curve_fit_residuals.py** in **YourNamePythonWork**. Then, modify the program so that it appears as shown next. Here, the `update_plot()` function near the middle of the program has been deleted, a few titles and labels have been changed, and the residuals are calculated and plotted at the end. As shown here, the program is configured to display the residuals for the case of the linearized Steinhart–Hart model function.

```
import Utilities.gui_creator as gc
import Utilities.plot_creator as pc
import numpy as np
import tkinter.filedialog as fd
import scipy.optimize as opt

fname = fd.askopenfilename()
t, r = np.loadtxt(fname, delimiter=',', usecols=(0,1),
                  unpack=True)

root = gc.create_window(title='Thermistor Residuals Plot',
                        size='1000x500')
control_box, right_frame = gc.create_frames(root)
ax, canvas = pc.create_plot(right_frame)

x = np.log(r*1000)
y = 1/(t + 273.15)
z = t + 273.15

def linearized_sh(x, a, b, d):
    return a + b*x + d*x**3

def nonlinear_sh(x, a, b, d):
    return 1/(a + b*x + d*x**3)

def polynomial(x, a, b, c, d):
    return a + b*x + c*x**2 + d*x**3


p0 = 1.e-3, 1.e-4, 1e-7

popt, pcov = opt.curve_fit(linearized_sh, x, y, p0)

c = 0
a, b, d = popt

gc.create_indicator(control_box, label_1='A',
                    label_2=f'{a:.8e}')
gc.create_indicator(control_box, label_1='B',
                    label_2=f'{b:.8e}')
gc.create_indicator(control_box, label_1='C',
                    label_2=f'{c:.8e}')
gc.create_indicator(control_box, label_1='D',
                    label_2=f'{d:.8e}')
```

```
y_fit = a + b*x + c*x**2 + d*x**3
t_fit = 1/y_fit - 273.15

residual = t - t_fit

pc.update_plot(ax, canvas, t, residual,
               label_x='Temperature (deg Celsius)',
               label_y='Residual (deg Celsius)',
               marker='o')
```

Save your work, then run **r-t_curve_fit_residuals.py** using **thermistor_calibration_data.csv** as your input data set. As seen in the plot, the fitted curve agrees very well with the calibration data, with the deviation typically on the order of 0.01°C. However, there does seem to be more pronounced (although still small) deviation at the highest temperatures as well as near 70°C.

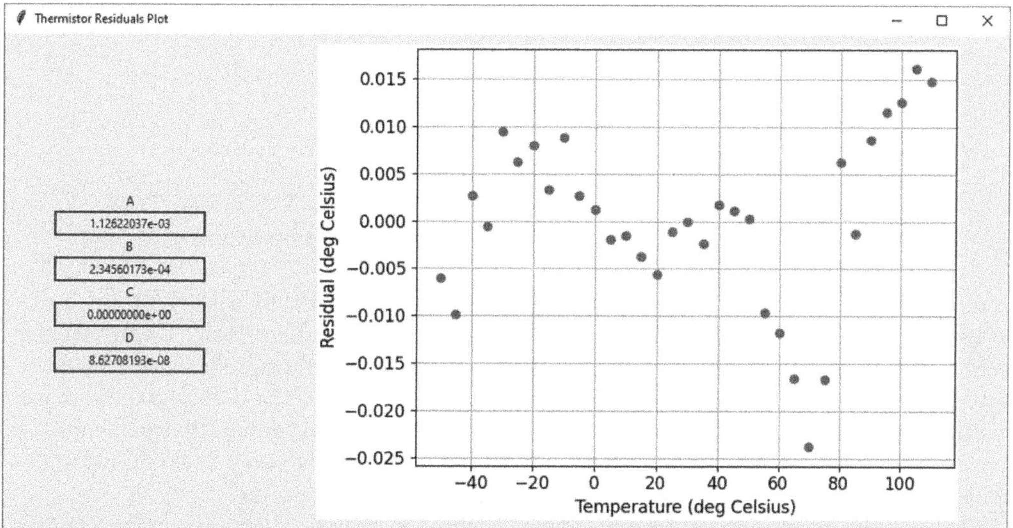

The deviation at around 70°C is surprising and may point to some novel effect within the semiconducting material of the thermistor or, perhaps, a problem in the calibration data. From this plot, one can also consider the following: Is there a temperature range within which the Steinhart–Hart consistently overestimates the temperature value ($R < 0$)? Is there a range within which the temperature is consistently underestimated ($R > 0$)? If you need your thermistor to be calibrated with an accuracy of ±0.05°C (i.e., the Steinhart–Hart predicted value falls within ±0.05°C of the actual temperature), within what temperature range can you operate your thermistor?

When curve fitting, one expects that a better fit can be achieved if the number of basis functions used in the fitting algorithm is increased. Thus, we would expect that a third-order polynomial (Eq. (10.11)), which involves the four basis functions $b_0 = 1$, $b_1 = x$, $b_2 = x^2$, and $b_3 = x^3$, should fit our thermistor data with more accuracy than the Steinhart–Hart equation (Eq. (10.12)), which deletes any contribution from the second-order basis function.

Let's see if this conjecture is true. Configure **r-t_curve_fit_residuals.py** to make a third-order polynomial fit to the thermistor data. The needed changes are here:

```
p0 = 1.e-3, 1.e-4, 1e-7, 1e-7

popt, pcov = opt.curve_fit(polynomial, x, y, p0)

a, b, c, d = popt
```

Save your work. Upon running **r-t_curve_fit_residual.py** using **thermistor_calibration_data.csv** as the input data set, here is my result:

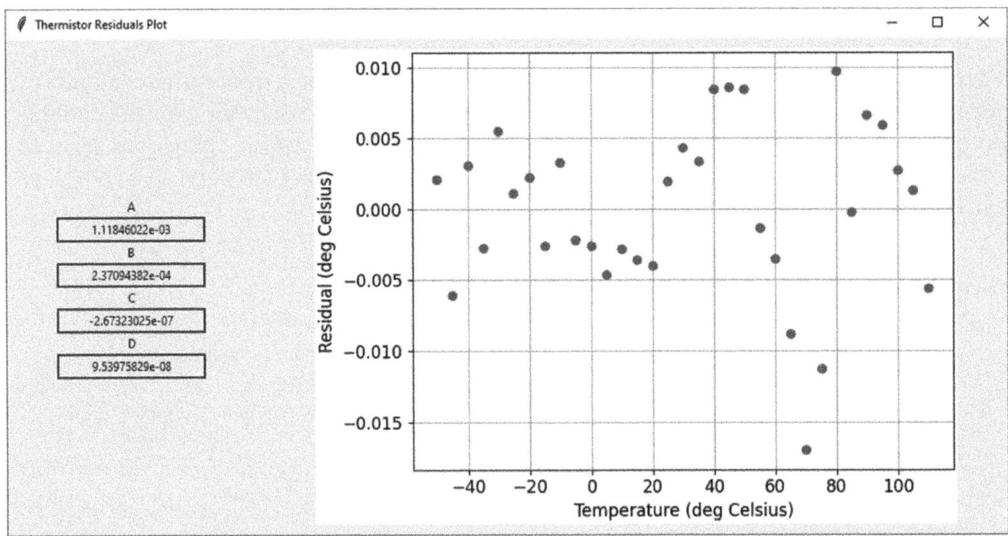

In comparing the residuals plots for the linearized Steinhart–Hart fit with the third-order polynomial fit, do you come to a conclusion consistent with Steinhart–Hart's claim that the second-order term can be deleted from the fitting equation without significant loss of accuracy? Remember that Steinhart and Hart were most interested in the oceanographic range and that this claim comes to us from an era when each additional term in the temperature-conversion equation would have been calculated manually and possibly necessitated additional analog circuitry, so any labor-saving tip would have been welcome.

Note that the third-order polynomial fit adds the additional $(\ln R)^2$ term to the model function used in the linearized Steinhart–Hart fit. Maybe somewhat surprisingly, the new optimized values for A, B, and D from the polynomial fit become different from those derived in the Steinhart–Hart case. This phenomenon results from the fact that the functions x^k that we are using as a basis are not *orthogonal*. Take a linear algebra course for more details.

Next, produce the residuals plot when the nonlinear Steinhart–Hart equation is used as the model function. The changes required this time are as follows:

```
p0 = 1.e-3, 1.e-4, 1e-7

popt,pcov = opt.curve_fit(nonlinear_sh, x, z, p0)

c=0
a, b, d = popt
```

Save your work. This time after running **r-t_curve_fit_residual.py** using **thermistor_calibration_data.csv** as the input data set, my result is shown. Is this plot comparable to that of the linearized Steinhart–Hart fit?

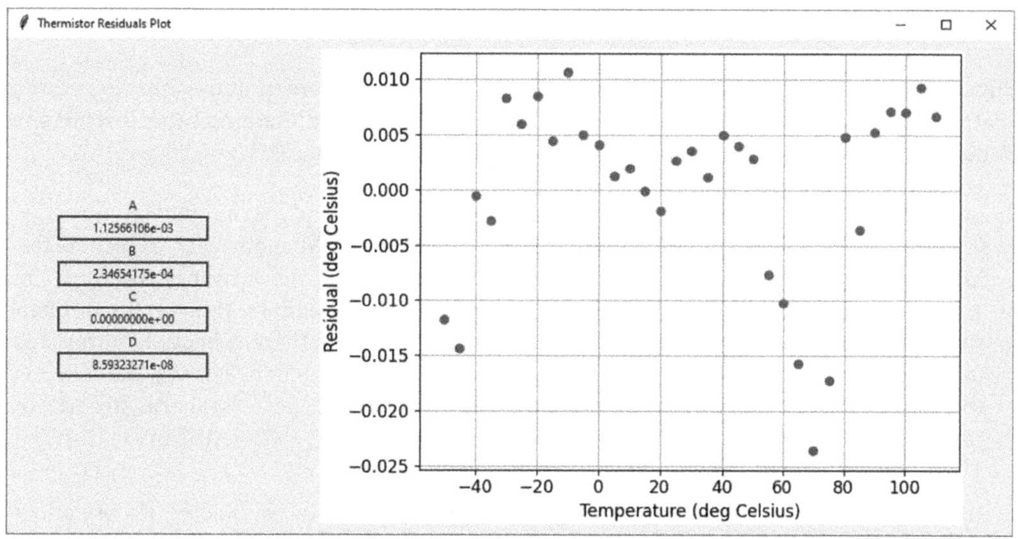

Finally, try fitting the thermistor calibration data to the Beta formula ((Eq. (10.9)). In this first-order polynomial fit, only two basis functions are required: $b_0 = 1$ and $b_1 = x$. Define the function `beta_function()`, which carries out the calculation `a + b*x`. Then make the following changes to running **r-t_curve_fit_residual.py**.

```
p0 = 1.e-3, 1.e-4

popt, pcov = opt.curve_fit(beta_function, x, y, p0)

c, d = 0,0
a, b = popt
```

Save your work, then run **r-t_curve_fit_residual.py** using **thermistor_calibration_data.csv** as the input data set. From the resulting plot (shown below), comment on the accuracy achieved by the Beta formula in comparison to the Steinhart–Hart equation. The impact of one or two higher-order terms in the model function is pretty remarkable, don't you agree?

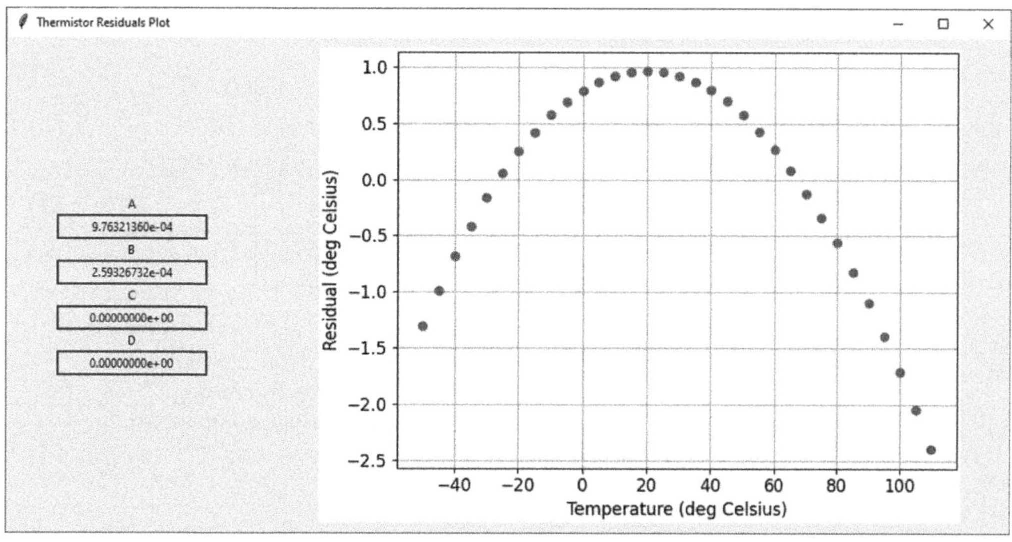

Do It Yourself

digital_thermometer.py Build a computer-based digital thermometer that uses the thermistor that you have calibrated in this chapter as its temperature sensor. Construct this instrument in the following steps:

1. Build the constant-current circuit shown in Fig. 10.1, which produces a constant current of 0.1 mA through the thermistor independent of its resistance. You may wish to include the high input-impedance unity-gain op-amp buffer to isolate the thermistor circuit from the voltage-sensing circuitry. If available, use an ammeter to determine the current through your thermistor precisely. It probably will differ slightly from 0.1 mA because of resistor tolerances. With a voltmeter, check that the voltage difference between V_{out} and GND is within the expected range. For example, the voltage difference across a thermistor whose resistance is about 10 kΩ near room temperature should be $V_{out} = IR \approx (0.1\ \text{mA})\ (10\ \text{k}\Omega) = 1\ \text{V}$.
2. In preparation for measuring the voltage difference across your thermistor, choose an AI channel on your DAQ device and connect its positive and negative (or AGND) pins to V_{out} and GND, respectively, in your circuit.

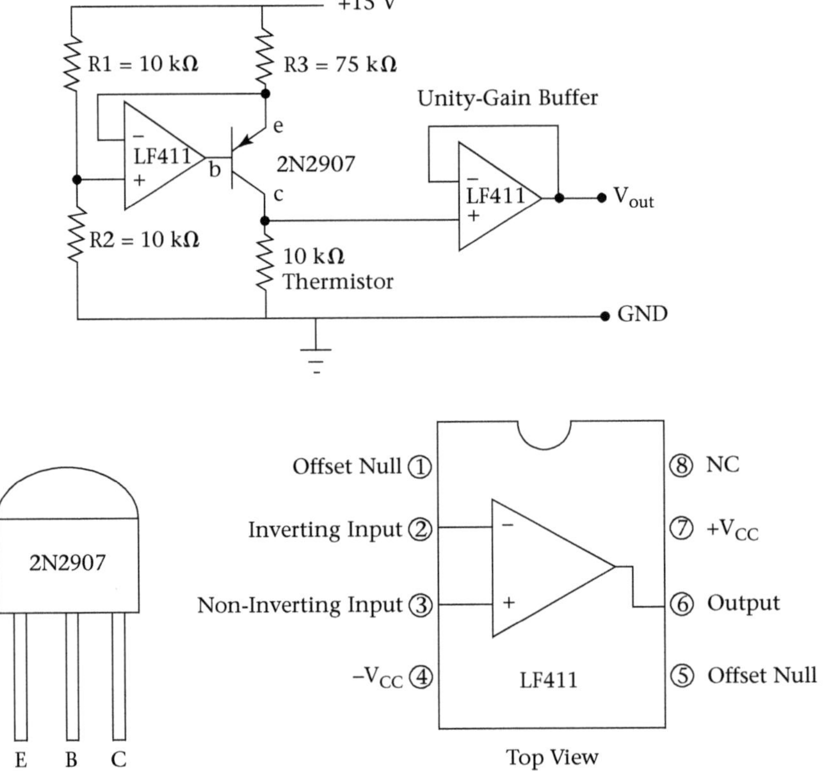

Fig. 10.1 Hardware circuit for the Do It Yourself project. Constant current of 0.1 mA flows through the thermistor independent of its resistance. The op amp depicted here is the LF 411, but any general-purpose op amp with compatible pin layout (e.g., 741) will work as well.

3. Write a GUI-equipped program with an indicator labeled **Temperature (deg Celsius)** and a toggling **Start/Stop Button** as shown here. Save your program in **YourNamePythonWork**.

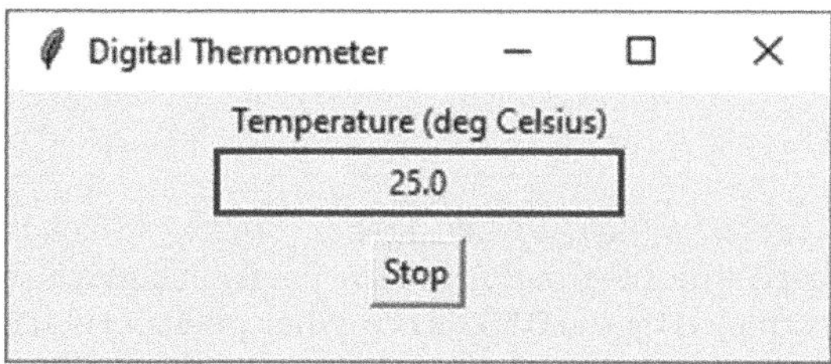

Your coding will include many of the features of **dc_voltmeter_ni_live.py** (or **dc_voltmeter_mcc_live.py**), so you may wish to open that program and create **digital_thermometer.py** using **File≫Save As...** Then, incorporate the following process in your program: When the **Start** button is pressed (and it toggles to a **Stop** button), initiate execution of a data-acquisition `while` loop, whose time per iteration is specified (say, 0.5 or 1 second). With each iteration of the loop, instruct your DAQ device to sample the voltage difference across the thermistor. Using this acquired voltage difference and the known constant current flowing through the thermistor (use your measured value for the current, if available), apply Ohm's Law to determine the thermistor's resistance. Next, with the three Steinhart–Hart coefficients A, B, and D you determined in this chapter, utilize the Steinhart–Hart equation to calculate the thermistor's temperature on the Celsius scale. Finally, display this temperature in the GUI indicator, so that it updated every 0.5 or 1 second. Have your program repeat this process continually until the **Stop** button is pressed.

Once completed, run **digital_thermometer.py** and use it to measure the room's temperature and the temperature of your skin.

Use It!

Signal Conditioning for Thermocouple Temperature Measurement Because of their many favorable properties—including wide operating range and low cost—thermocouples are the most widely used temperature sensors. A thermocouple is constructed by joining the ends of two dissimilar metals and the resulting junction produces a well-characterized temperature-dependent voltage. However, thermocouples also have several unfavorable properties that pose challenges for the user. First, the voltage produced by a thermocouple is in the millivolt range and typically requires amplification before being digitized in order to resolve small temperature changes. Second, a thermocouple's voltage versus temperature (V–t) curve is not a straight line and is not predicted by a simple theoretical expression. As a result, the thermocouple's temperature conversion equation $t(V)$—that is, temperature as a function of voltage—comes in the form of a high-order polynomial. As explored in Problem 2 of this chapter, this temperature conversion equation has been established by the National Institute of Standards and Technology (NIST) through a curve-fitting procedure applied to the empirical V–t calibration data. Finally, a process called *cold-junction compensation* is required to account for the dissimilar-metal junctions formed at the points where the thermocouple wire connects to the supporting measurement circuitry.

Signal conditioning using a low-cost integrated circuit removes many of the difficulties of thermocouple measurement. For example, the Analog Devices AD8495 chip both amplifies the small thermocouple voltage produced by a type K thermocouple and performs cold-junction

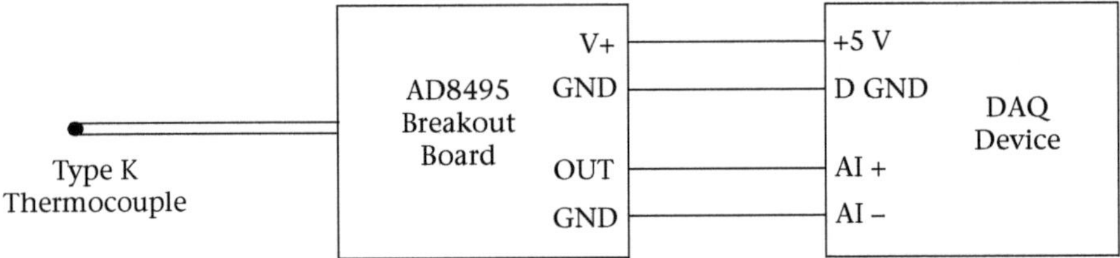

Fig. 10.2 Hardware circuit for the Use It! example consisting of the Adafruit breakout board with the attached type K thermocouple. The breakout board is powered by, and its output voltage is sampled by, the DAQ device. When powered by +5 V, this setup is designed to measure temperatures accurately in the range from −25 to +400°C.

compensation. In addition, the chip produces a linear 5 mV /°C output, which provides a fairly good approximation (±1°C near room temperature) to the true nonlinear temperature conversion curve. The AD8495 chip, conveniently mounted on a small printed circuit board along with its required supporting circuitry, can be purchased from Adafruit Industries (Product ID 1778). This "breakout board," when powered by the +5 V and D GND pins of a DAQ device, shifts the type K thermocouple voltage by +1.25 V so that both positive and negative Celsius temperatures can be measured. The required connections between the breakout board and the DAQ device are shown in Fig. 10.2.

The hardware setup shown in Fig. 10.2 can be used to measure the temperature of a type K thermocouple under Python control. Starting with the **digital_thermometer.py** program developed in this chapter's Do It Yourself project, make this modification: In each iteration of the data-acquisition `while` loop, the output voltage V produced by the Adafruit breakout board will be sampled by the DAQ device. The thermocouple's temperature t (in Celsius) is then determined by the "linearized" approximation of the calibration equation $t = (V - 1.25\ \text{V})/(0.005\ \text{V}/°\text{C})$. To suppress noise, rather than a single voltage sample, one can opt to read an array of, say, 100 voltage samples over a time interval of 0.1 second, and the mean V of this array can then be used in the temperature calculation.

Alternatively, the actual thermocouple voltage V_{therm} can be calculated from $V_{therm} = (V - 1.25\ \text{V} - 0.00125\ \text{V})\ /122.4$ (see Analog Devices Application Note 1087). Then, using the NIST temperature conversion equation determined in Problem 2, V_{therm} can produce a more accurate determination of the temperature.

Problems

1. **Energy Barrier Calculation** According to the theoretical considerations discussed at the beginning of this chapter, the value of B determined from fitting a thermistor's calibration data to the Steinhart–Hart equation is related to the activation energy barrier E for electronic conduction in the thermistor. Use the value you obtained for B to obtain E for your thermistor in the units of electronvolts (eV). The value for Boltzmann's constant is $k = 8.6 \times 10^{-5}\text{eV/K}$.
2. **thermocouple_fit.py** A type K thermocouple is constructed by joining the ends of two metallic alloy (named chromel and alumel) wires. This junction between the two dissimilar metals produces a temperature-dependent voltage, which can be used as a temperature sensor. The type K thermocouple's empirically determined voltage (in millivolts) as a function of temperature (in degrees Celsius) produced the following calibration data:

Temperature t (°C)	Voltage V (mV)
−25	−0.968
0	0.000
50	2.023
100	4.096
150	6.138
200	8.138
250	10.153
300	12.209
350	14.293
400	16.397

The temperature conversion equation for the type K thermocouple has been established by NIST by fitting its calibration data to the following equation, where V and t are in mV and °C, respectively:

$$t = a_1 V + a_2 V^2 + a_3 V^3 + a_4 V^4 + a_5 V^5 + a_6 V^6 + a_7 V^7 + a_8 V^8 + a_9 V^9. \tag{10.20}$$

Write a GUI-equipped program that uses `polyfit()` and the calibration data given in the above table to determine the coefficients a_1 through a_9 in Eq. (10.20). Then, on a GUI graph, plot the calibration data along with your predicted temperature conversion curve in the temperature range from 0 to 400°C.

You might be interested in comparing your values for a_1 through a_9 with those established by NIST (see the type K inverse coefficients in the online *NIST ITS-90 Thermocouple Database*). The NIST values are derived from a more complete calibration data set than that used in this problem and so will deviate a bit from your values.

3 **resonant_circuit.py** A resonant circuit composed of an inductor, a resistor, and a capacitor is powered by an AC input voltage whose amplitude V_{in} is constant but whose frequency f can be varied. An experimentalist measures the circuit's output voltage amplitude V_{out} at several frequencies f and then calculates the circuit's gain at each frequency, where gain is defined as the ratio of the output to input voltage amplitudes (i.e., $G \equiv V_{out}/V_{in}$). The results from this experiment are shown in the following data table:

Frequency f (Hz)	Gain G
1500	0.070
1600	0.084
1700	0.102
1800	0.128
1900	0.168
2000	0.235
2100	0.374
2200	0.740
2300	0.861
2400	0.442
2500	0.281
2600	0.206

Frequency f (Hz)	Gain G
2700	0.164
2800	0.136
2900	0.117
3000	0.103

A theoretical analysis of this resonant circuit predicts that the frequency-dependent gain $G(f)$ is given by

$$G = \frac{1}{\sqrt{1 + \left[\frac{1}{\Delta f}\left(f - \frac{f_o^2}{f}\right)\right]^2}}, \tag{10.21}$$

where the constants f_o and Δf are the resonant frequency and full width at half power, respectively (there is one frequency below f_o and one frequency above f_o where $G = 1/\sqrt{2}$; the full width is defined as the difference between these frequencies).

Write a GUI-equipped program that fits the given experimental $G - f$ data to the theoretically predicted relation for $G(f)$ (Eq. (10.21)). Then, on a GUI graph, compare the data and the fitted theoretical curve and also display the optimal values for constants f_o and Δf in indicators. Programming tip: Consider entering the given $G - f$ data into a spreadsheet and then using `np.loadtxt()` to input the data to your program. Alternatively, you can simply code the G and f data as Python lists as part of your program.

4. **discharging_capacitor.py** Beginning at time $t = 0$, a capacitor of unknown value C is discharged through a resistor $R = 47\ \text{k}\Omega$. To determine C, as the capacitor is discharging, the voltage difference V between its plates is measured at 5-ms intervals, resulting in the following data set:

t (s)	0.005	0.010	0.015	0.020	0.025	0.030
V (V)	8.7	6.3	4.5	3.3	2.4	1.7

Theoretically, the relation between V and t is given by

$$V = V_0 e^{-t/RC}, \tag{10.22}$$

where V_0 is the voltage across the capacitor at $t = 0$. Programming tip: In the program you write, enter the two data arrays `t` and `V` directly in your program as Python lists.

(a) **Linearized Model Function** By taking the logarithm of both sides, rewrite Eq. (10.22) in a linearized form, that is, in the form $y = mx + b$. Then, write code that determines C (in μF) and also V_0 from the given V versus t data using this linearized form as your model function. Program `polyfit()` to fit the input arrays to the equation of a straight line, use the resulting optimized parameters m and b to calculate C and V_0, and then display these values in the Shell.

(b) **Nonlinear Model Function** Next, add code that also determines C and V_0 from the given V versus t data, but this time using a nonlinear model function. Use `curve_fit()` to perform a nonlinear fit of the input arrays to the equation $y = ae^{-x/b}$. Use the resulting optimized parameters a and b to calculate C (in μF) and V_0, and then display these values in the Shell.

(c) Run **discharging_capacitor.py**. What values do you obtain for C and V_0 from each model function? Do the two approaches yield the same result?

5. **polynomial_fit_varying_orders.py** Curve fitting of data must be guided by a theory describing the phenomenon under investigation. Consider the following hypothetical situation: An experimenter, investigating the relation between two quantities y and x, uses an experimental system to acquire seven samples of y as a function of x, with the results given in the following table. These data, of course, reflect the true mathematical function $y(x)$ that relates the two quantities, but this relation is somewhat obscured by noise in the experimental system. The experimenter knows that a theory of the phenomenon being studied suggests that the measured quantities should be related by a second-order polynomial $y(x) = a_0 + a_1x + a_2x^2$, where $a_0 = 6.0$, $a_1 = 4.0$, and $a_2 = 2.0$ in appropriate units.

x	1.0	2.0	3.0	4.0	5.0	6.0	7.0
y	10	24	34	54	74	101	133

In the following steps, you will write a GUI-equipped program that plots the acquired data on a graph, along with a fitted polynomial of varying order. Also, the resulting optimized linear expansion coefficients determined by the fitting process can be displayed in the Shell or (better yet) on GUI indicators. Programming tip: In the program you write, enter the acquired data arrays `x` and `y` directly in your program as Python lists.

To compare the experimentalist's results to the theoretically predicted relation, analyze the data in the following ways:

(a) First, use `polyfit()` to fit the acquired data to a second-order polynomial, i.e., the model function is $y(x) = a_0 + a_1x + a_2x^2$, and display the fitted second-order polynomial and the acquired data on the GUI graph. In plotting the fitted equation, use an x-array that ranges from 1 to 7 with a small spacing such as $\Delta x = 0.1$. Also display the optimized values of a_0, a_1, and a_2 in the Shell or GUI indicators.

How do your fitted values of a_0, a_1, and a_2 compare with the theoretically expected values? Why do you suppose that the experimentally determined values for these coefficients don't perfectly match the theoretical values? Characterize qualitatively (in a sentence or two) how well the fitted polynomial describes the measured data points.

(b) A better fit to the acquired data can be obtained using a higher-order polynomial. To demonstrate this fact, reprogram `polyfit()` so that it fits the acquired data to a fourth-order polynomial and run your program. From the GUI graph, does this fitted fourth-order polynomial agree with the experimental data more favorably than the second-order fit carried out in part (a)? What values do you obtain for the optimized coefficients a_0, a_1, a_2, a_3, and a_4? How do the first three of these values compare with the values expected from theory? Explain why, even though the fourth-order polynomial provides a closer match to the experimental data, the second-order fit carried out in part (a) provides the best comparison between the experimental findings and the given theory.

(c) Interestingly, a fitted polynomial of order $N-1$ will perfectly pass through N given data samples. Demonstrate this fact using your program. Here, we are magnifying the effect studied in part (b). By providing too much flexibility in the fitting function, we are fitting experimental noise, at the expense of uncovering the sought-after relation $y(x)$.

CHAPTER 11

Data Analysis: Fast Fourier Transform

11.1 Quick Fast Fourier Transform Example

In this chapter, we will explore the use of the *fast Fourier transform* (FFT), a signal-processing technique used by scientists and engineers in countless applications. In short, the FFT allows one to determine the amplitudes of the component sinusoidal oscillations that add together to produce a given digitized signal. In this section, as a quick introduction to the FFT, let's familiarize ourselves with a few of its properties through a demonstration program.

While NumPy contains a FFT-related module, SciPy does as well, but with greater functionality. Thus, we'll use the SciPy `fft()` function, which (with an unfortunate repetitive name choice) is in the `fft` module. Thus, to take the FFT of a data array `y`, the statement is written as follows:

```
fourier_transform = scipy.fft.fft(y)
```

This function returns an array of complex numbers named `fourier_transform`. Each element in this array encodes the magnitude and phase of oscillation at one of the frequencies in an array of frequencies we will call `freq`.

The array `freq` can be obtained from a function within the `fft` module called `fftfreq()`. The form of this function is the following:

```
freq = scipy.fft.fftfreq(num, delta_t)
```

Here, `num` is the number of data samples in the array `y` and `delta_t` is the time interval between one sample and the next (which equals the inverse of the sampling frequency).

With that minimal background, let's test-drive the FFT in order to gain some experience with its properties. Then, as we work our way through this chapter, we will develop a deeper understanding of its operation and underlying theory.

In the Editor, open the file **waveform_creator.py** in **YourNamePythonWork≫Utilities**. Place the import statement `import scipy.fft as ft` at the top of this file, then add the following definition of the function `fft_mag_only()` to the program.

```
def fft_mag_only(y, num, samp_freq):

    delta_t = 1/samp_freq
    n = num//2
```

The Python-Based Laboratory. John Essick, Oxford University Press. © John Essick (2024). DOI: 10.1093/9780191998478.003.0011

```
    freq = ft.fftfreq(num, delta_t)

    fourier_transform = ft.fft(y)
    amp = (2/num)*np.abs(fourier_transform)

    freq_positive = freq[0:n]
    amp_positive = amp[0:n]

    return freq_positive, amp_positive
```

Here's a brief summary of the operations carried out in `fft_mag_only()` that, for now, is offered with the promise that the underlying reason for each of these operations will be explained later: After determining `delta_t` from the inverse of the sampling frequency, `n` is assigned to be one-half of the number of data samples. The double slash operation `num//2` performs *integer* (or *truncating*) *division* on the dividend `num` by the divisor `2`, with the resulting quotient always being an integer; if `num` is not an even integer, the decimal part of the quotient is truncated, leaving an integer result. Next, the array of frequencies `freq` associated with the FFT is calculated. Then, the FFT of the input data array `y` is found, resulting in an array of complex numbers named `fourier_transform`. The magnitude and phase of each complex number in `fourier_transform` gives the magnitude and phase of oscillation of a corresponding frequency in the `freq` array. We are only interested in the magnitude of oscillation, so the NumPy function `abs()` is used to find the magnitude of each element in `fourier_transform`. Each of these magnitudes is multiplied by a factor of `2/num`, yielding the array of oscillation amplitudes called `amp`. Finally, though slice operations, only the first half of the `freq` and `amp` arrays are returned by `fft_mag_only()` as arrays named `freq_positive` and `amp_positive`, respectively.

Remember to add the import statement `import scipy.fft as ft` at the top of **waveform_creator.py**, then save this program in **YourNamePythonWork≫Utilities** as you close it.

Now, in the Editor, open **waveform_simulator.py** in **YourNamePythonWork**, then use **File≫Save As...** to create a new file named **fft_magnitude.py** in **YourNamePythonWork**. There are only a few changes we need to make to this program. First, modify the code within the `calculate_plot_waveform()` function as shown below. Here, we have added the line with the `wc.fft_mag_only()` function, which takes the FFT of the waveform array `y` created by `wc.create_waveform_data()`, and changed the arrays being plotted by `pc.update_plot()` to `freq_positive` and `amp_positive`. Also, the argument `marker='o'` has selected filled circles as the plot markers.

```
def calculate_plot_waveform():

    num, samp_freq = gc.get_dig_entries(var_num,
                                        var_samp_freq)
    a, f, shape = gc.get_wav_entries(var_a, var_f,
                                     var_shape)

    t = wc.create_time_data(num, samp_freq)
    y = wc.create_waveform_data(t, a, f, shape)
```

```
    freq_positive, amp_positive = wc.fft_mag_only(y, num,
                                                   samp_freq)

    pc.update_plot(ax, canvas, freq_positive, amp_positive,
                   label_x, label_y, marker='o')

    return
```

Second, make the following changes for the root window title and the plot axes labeling:

```
label_x, label_y = 'Frequency (Hz)', 'Amplitude'

root = gc.create_window(title='FFT (Magnitude Only)',
                        size='1000x500')
```

Save your work. The complete program will appear as the following:

```
import Utilities.waveform_creator as wc
import Utilities.gui_creator as gc
import Utilities.plot_creator as pc

def calculate_plot_waveform():

    num, samp_freq = gc.get_dig_entries(var_num,
                                        var_samp_freq)
    a, f, shape = gc.get_wav_entries(var_a, var_f,
                                     var_shape)

    t = wc.create_time_data(num, samp_freq)
    y = wc.create_waveform_data(t, a, f, shape)

    freq_positive, amp_positive = wc.fft_mag_only(y, num,
                                                   samp_freq)

    pc.update_plot(ax, canvas, freq_positive, amp_positive,
                   label_x, label_y, marker='o')

    return


label_x, label_y = 'Frequency (Hz)', 'Amplitude'

root = gc.create_window(title='FFT (Magnitude Only)',
                        size='1000x500')
control_box,right_frame = gc.create_frames(root)

var_num, var_samp_freq = gc.digitizing_control(control_box)
var_a, var_f, var_shape = gc.waveform_control(control_box)

ax,canvas = pc.create_plot(right_frame)
pc.update_plot(ax, canvas, [], [], label_x, label_y)
```

```
gc.action_button(control_box, label='Submit',
                 action=calculate_plot_waveform)

root.mainloop()
```

Your program is now complete and we're ready to test-drive the FFT. Run **fft_magnitude.py**. The program's GUI will appear with the control panel awaiting your entries, along with a blank plot. Enter *1024* and *2000* for **Number of Samples** and **Sampling Frequency (Hz)**, and *4*, *250*, and *Sine* for **Amplitude, Frequency (Hz)**, and **Shape**, respectively. These choices for the parameters simulate sampling a 250-Hz sine wave with amplitude of 4 for a little over one-half second. Click the **Submit** button. If all goes well, you will see the following plot of amplitude versus frequency.

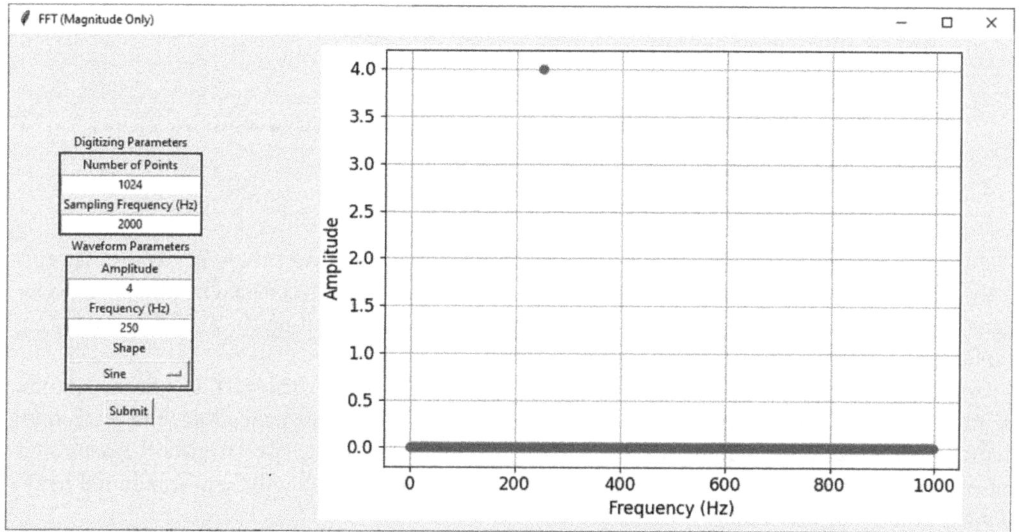

Here, the function `ftt_mag_only()` has given us the *frequency spectrum* of the digitized signal `y` it received from `create_waveform_data()`, where a frequency spectrum is a plot of the amplitudes of the component sinusoidal oscillations present in the given digitized signal. From this spectrum, it appears there is only one sinusoidal oscillation with appreciable amplitude within our signal, and this oscillation has a frequency near 250 Hz and an amplitude of about 4.

To inspect the spectrum more closely and verify there is only a single peak, we can manually rescale the x-axis. In the Editor, open the file **plot_creator.py** in **YourNamePythonWork≫Utilities**. Within the code defining the `update_plot()` function, add the command `ax.set_xlim(200,300)` directly below the statement `ax.grid()` as shown. Save **plot_creator.py** as you close it.

```
ax.tick_params(axis='both', labelsize=size)
ax.set_xlabel(label_x, fontsize=size)
ax.set_ylabel(label_y, fontsize=size)
ax.grid()
ax.set_xlim(200, 300)
```

The scaling of the x-axis will now span the range from 200 to 300 Hz and the individual data will be indicated by circles as before. Rerun **fft_magnitude.py**. The following single-peak plot results:

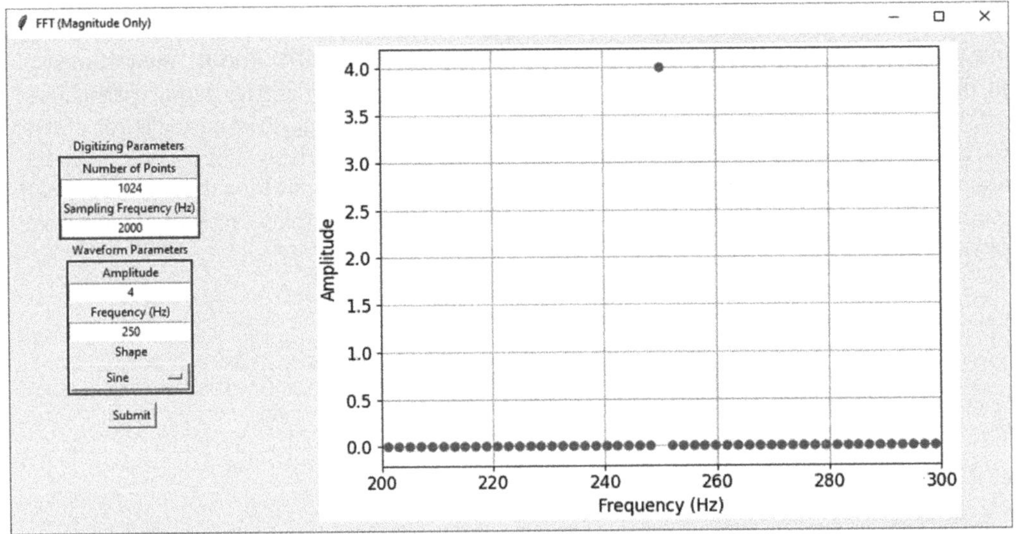

The spectrum looks perfect, doesn't it? The amplitude is equal to zero everywhere, except at the frequency of 250 Hz, where it equals +4.0, just as we expected based on the sine-wave signal we created using `create_waveform_data()`. That's good news! The FFT is easy to implement and its results are easy to understand.

Unfortunately, now I have some bad news for you. Generally, the FFT is not as simple to interpret as has been suggested so far. Try repeating the above procedure, but with one change—set **Frequency** equal to *249*. Surprisingly, the resulting spectrum will no longer appear as a "delta-function" spike with a height of 4.0, but instead will be a broadened peak of maximum height less than 4.0, as shown.

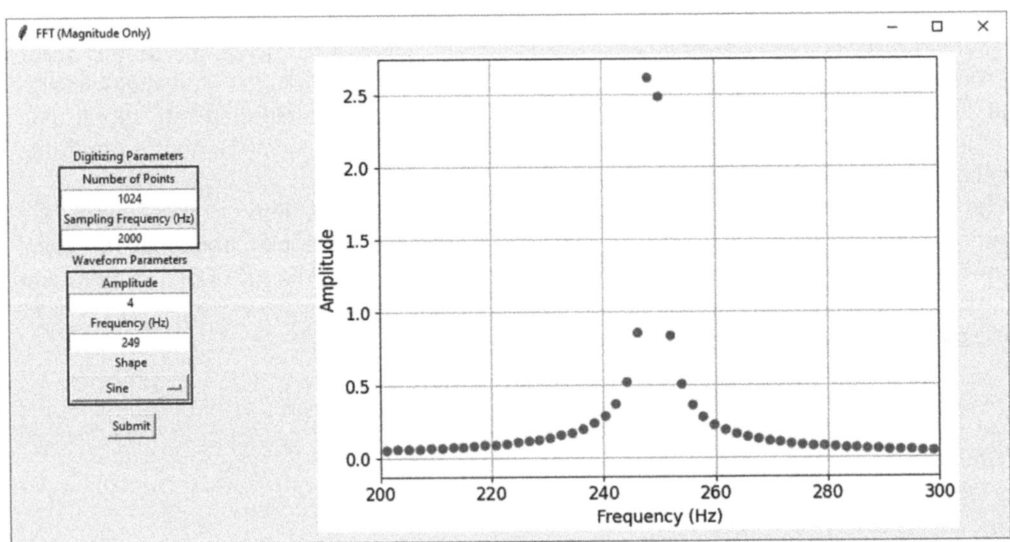

At face value, this spectrum seems to be telling us that the input signal is composed of many sinusoidal oscillations, each with a different frequency in the approximate range of 200 to 300 Hz. However, we know that this is simply untrue. We input a pure sine wave with the exact frequency of 249 Hz.

We are thus faced with a mystery. Why does our FFT-based program work perfectly for an input frequency of 250 Hz, but fails when that frequency is changed to 249 Hz? That's one of the questions we will answer in the course of this chapter.

To close this quick introduction to the FFT, I'll demonstrate a method for mitigating the mysterious problem we just observed. In the Editor, open **waveform_creator.py** in **YourNamePythonWork≫Utilities**. Then, add the following definition for a function named `create_window()`. The first line of `create_window()` executes the `get_window()` function in the `windows` submodule within the `signal` module of SciPy; i.e., the syntax for this function is

```
win = scipy.signal.windows.get_window(win_type, num)
```

`get_window()` creates a mathematical object called a *window* and the particular window is of the type specified by `win_type`. The window is constructed in the form of a `num`-element NumPy array and assigned the name `win`. At the top of **waveform_creator.py** there is already the import statement `import scipy.signal as sig`, so code `create_window()` as follows:

```
def create_window(win_type, num):

    win = sig.windows.get_window(win_type, num)

    coherent_gain = np.mean(win)
    win = win/coherent_gain

    return win
```

The second and third lines of `create_window()` provide a needed normalization of the `win` array by a factor called the `coherent gain`, which is determined using the NumPy `mean()` function. Simply include this two-line normalization procedure now; later in the chapter, we'll explain it fully.

Save you work and close **waveform_creator.py**.

Next, open **fft_magnitude.py** in **YourNamePythonWork**. Modify the code within the `calculate_plot_waveform()` function as shown below, where we have added three window-related lines. The first two of these lines create a window array `win` of the type called *Hann* via the option `win_type='hann'`. The third line then carries out element-wise multiplication of the Hann window array `win` times the data array `y`. In element-wise multiplication, the elements at the equivalent index in each array are multiplied, producing the element at that index in the resultant array. This resultant array is then input to the existing FFT coding.

```
def calculate_plot_waveform():

    num, samp_freq = gc.get_dig_entries(var_num,
                                        var_samp_freq)
    a, f, shape = gc.get_wav_entries(var_a, var_f,
                                     var_shape)
```

```
    t = wc.create_time_data(num, samp_freq)
    y = wc.create_waveform_data(t, a, f, shape)

    win_type = 'hann'
    win = wc.create_window(win_type, num)
    y = y*win

    freq_positive, amp_positive = wc.fft_mag_only(y, num,
                                                  samp_freq)

    pc.update_plot(ax, canvas, freq_positive, amp_positive,
                   label_x, label_y, marker='o')

    return
```

Save your work. Now run **fft_magnitude.py**, again with *1024* and *2000* for **Number of Samples** and **Sampling Frequency (Hz)**, and *4*, *249*, and *Sine* for **Amplitude**, **Frequency (Hz)**, and **Shape**, respectively.

You will find that the spectral peak is still "incorrectly" spread among several frequencies in the neighborhood of 249 Hz, but the spread is now much narrower and the peak amplitude much closer to 4 than when operating without the help of the Hann window.

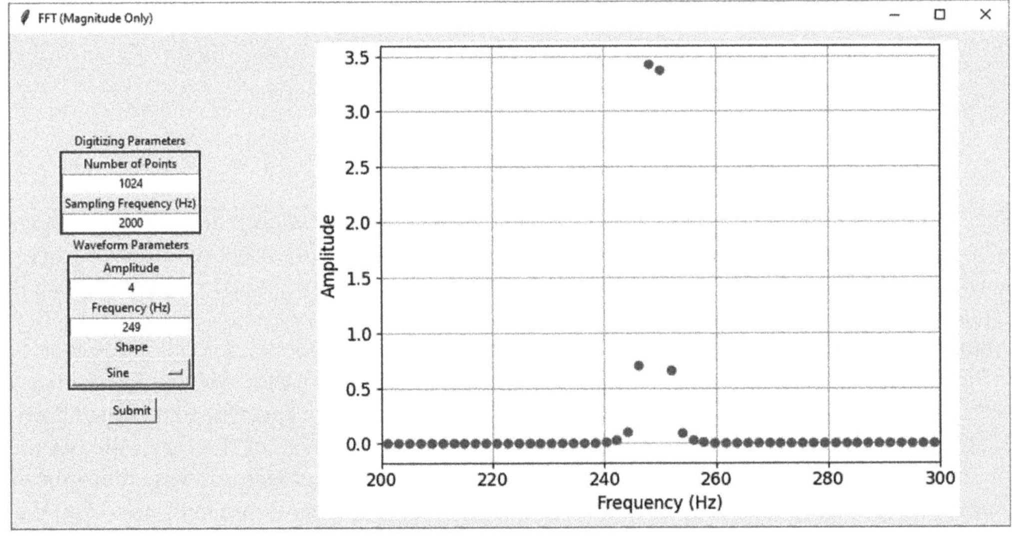

From this brief introduction, hopefully it's apparent that use of the FFT is not as straightforward as one might initially expect. In this chapter, we will develop the basic competence needed to implement and interpret FFT analysis properly. First, we will review the mathematical background of this signal-processing technique, then we will go on to demonstrate how to carry out this method using the relevant SciPy functions. Along the way, we will develop a deeper understanding of the two effects (called *spectral leakage* and *windowing*) that we encountered in this section.

Before moving on, in the `calculate_plot_waveform()` function of **fft_magnitude.py**, comment-out (by placing the hashtag symbol at the start of) the three lines related to windowing, These lines will now be skipped over during program execution:

```
#win_type = 'hann'
#win = wc.create_window(win_type, num)
#y = y*win
```

Save **fft_magnitude.py**, as you close it.

Also, open **plot_creator.py** in **YourNamePythonWork≫Utilities**, comment-out the line we placed within the `update_plot()` function to manually scale the x-axis:

```
#ax.set_xlim(200, 300)
```

Save **plot_creator.py** as you close it.

11.2 Fourier Transform

Imagine you are investigating a physical phenomenon by measuring the analog quantity x that varies continuously with time t, that is, $x = x(t)$. Based on our discussion at the beginning of Chapter 10, we expect that $x(t)$ can be modeled as the linear combination of a set of M basis functions $b(t)$ such that

$$x(t) = \sum_{k=0}^{M-1} a_k \, b_k(t). \tag{11.1}$$

The Fourier transform is an example of just such a model. The pioneering work of nineteenth-century mathematician Baron Jean Baptiste Joseph Fourier, with subsequent elaboration by others, showed that the infinite set of complex exponentials $\exp(i2\pi ft)$, where f includes all of the frequencies from $-\infty$ to $+\infty$ and $i = \sqrt{-1}$, is a complete basis set with which one can form linear combinations to model any arbitrary function (a complex exponential is a form of sinusoidal oscillation, as seen from Euler's formula $e^{ix} = \cos x + i \sin x$). Since an infinite number of exponentials with infinitesimally close frequency spacing is required to model an arbitrary function, the summation in Eq. (11.1) becomes an integral, yielding

$$x(t) = \int_{-\infty}^{+\infty} X(f)\, e^{i2\pi ft}\, df, \tag{11.2}$$

where the *Fourier component* $X(f)$ is the amplitude (I'll be much more precise in my use of the word "amplitude" shortly) of the exponential with frequency f. Generally, $X(f)$ is a complex-valued function, carrying both *magnitude* and *phase* information, a point we will carefully consider in this chapter.

Conversely, the Fourier components $X(f)$ may be obtained by the equation

$$X(f) = \int_{-\infty}^{+\infty} x(t)\, e^{-i2\pi ft}\, dt \tag{11.3}$$

Equations (11.3) and (11.2) are called the *Fourier transform* and *inverse Fourier transform*, respectively.

11.3 Discrete Sampling and the Nyquist Frequency

In a real experiment, one cannot sample the quantity x continuously but must instead settle for discretely sampling it N times. If the samples are taken at equally spaced times, where the time difference between adjacent samples is Δt, then the sampling frequency f_s is given by

$$f_s = \frac{1}{\Delta t} \tag{11.4}$$

and the N consecutive sampling times are

$$t_j = j\,\Delta t \qquad j = 0,\, 1,\, 2,\, \ldots,\, N-1 \tag{11.5}$$

Given such a sampling scheme, there is a maximum-frequency sine wave that we can expect to detect. This limitation is simply due to the fact that to observe a sine signal, we must at the very least, for example, first sample its positive peak, then its negative trough at the next sample, followed by its positive peak at the third sample, and so on. Thus, the minimal sampling of a sine wave is two sample points per cycle. In an experiment, then, where your instruments are acquiring data every Δt seconds (at a sampling rate $f_s = 1/\Delta t$), the maximum-frequency sinusoidal signal you can detect has a period $T = 2\,\Delta t$. This detection limit is called the *Nyquist frequency* $f_{nyquist}$ and is given by

$$f_{nyquist} = \frac{1}{2\,\Delta t} = \frac{f_s}{2} \tag{11.6}$$

Additionally, by sampling a waveform $x(t)$ at N equally spaced times t_j, one can only expect to determine the Fourier components $X(f)$ of N equally spaced frequencies f_k in the Fourier transform. This statement follows from the tacit assumption that $x(t)$ is periodic such that $x(t) = x(t + N\Delta t)$—a criterion that forces the waveform $x(t)$ to be composed only of component frequencies that fit an integer number of cycles into the sequence of N samples. The constant (zero-frequency) function obviously meets this criterion. The next lowest acceptable frequency has a period of $N\Delta t$ and thus a frequency of $f_1 = 1/N\Delta t = f_s/N$. Within the acceptable range of $-f_{nyquist}$ to $+f_{nyquist}$, the rest of the frequencies f_k are the integer multiples of f_1. Thus, defining the spacing between adjacent frequencies to be $\Delta f = f_s/N$, the N equally spaced frequencies f_k determined by discretely sampling data are

$$f_k = k\left(\frac{f_s}{N}\right) = k\,\Delta f \quad k = -\frac{N}{2}, \ldots, 0, \ldots, +\frac{N}{2} - 1 \tag{11.7}$$

Although Eq. (11.7) appears deficient in its neglect of the frequency $+f_{nyquist}$ (given by $k = +N/2$), the two extreme frequencies $\pm f_{nyquist} = \pm f_s/2$ actually describe the same basis function due to the periodic nature of $x(t)$. SciPy arbitrarily chooses to represent this basis function by $-f_{nyquist}$. Thus, the range of k-values in the above relation rightfully runs from $-N/2$ to $+N/2 - 1$.

11.4 Discrete Fourier Transform

With the constraints caused by discrete sampling, as described in Section 11.3, we then approximate the Fourier transform integral (Eq. (11.3)) by the sum

$$X(f_k) = \int_{-\infty}^{+\infty} x(t)\, e^{-i2\pi f_k t} dt \approx \sum_{j=0}^{N-1} x(t_j)\, e^{-i2\pi f_k t_j} \Delta t. \tag{11.8}$$

Note that, using Eqs. (11.4), (11.5), and (11.7),

$$f_k\, t_j = (k\Delta f)\,(j\Delta t) = kj \frac{f_s}{N} \Delta t = \frac{kj}{N}. \tag{11.9}$$

Putting Eq. (11.9) into Eq. (11.8), and realizing that Δt is a constant, we find that

$$X(f_k) \approx \Delta t \sum_{j=0}^{N-1} x(t_j)\, e^{-i2\pi jk/N} \equiv \Delta t X_k \tag{11.10}$$

where we have defined the *discrete Fourier transform* to be

$$X_k \equiv \sum_{j=0}^{N-1} x(t_j)\, e^{-i2\pi jk/N} \quad k = -\frac{N}{2}, \ldots, 0, \ldots, +\frac{N}{2} - 1. \tag{11.11}$$

Finally, we approximate the inverse Fourier transform integral (Eq. (11.2)) by the sum

$$x(t_j) = \int_{-\infty}^{+\infty} X(f)\, e^{i2\pi f t_j} df \approx \sum_{k=0}^{N-1} X(f_k)\, e^{i2\pi f_k t_j} \Delta f. \tag{11.12}$$

Putting Eq. (11.10) into Eq. (11.12), we get

$$x_j \approx \sum_{k=0}^{N-1} X_k\ \Delta t\ e^{i2\pi f_k t_j}\ \Delta f. \tag{11.13}$$

Then, using Eq. (11.9) and also noting that $(\Delta t)(\Delta f) = (1/f_s)(f_s/N) = 1/N$, Eq. (11.13) becomes the *discrete inverse Fourier transform*

$$x_j \approx \sum_{k=0}^{N-1} \frac{X_k}{N} e^{i2\pi f_k t_k} = \sum_{k=0}^{N-1} \frac{X_k}{N} e^{i2\pi kj/N} \equiv \sum_{k=0}^{N-1} A_k e^{i2\pi kj/N.} \tag{11.14}$$

Equations (11.11) and (11.14) provide us with a method of performing spectral analysis on an experimental signal x. Given that x has been sampled discretely at N evenly spaced times t_j, we first calculate the N values of X_k using the definition of the discrete Fourier transform (Eq. (11.11)). Then, in the spirit of Fourier, viewing the signal x as really the composite of oscillations at N different frequencies f_k, Eq. (11.14) tells us that the amplitude of oscillation A_k at each f_k is given by

$$A_k = \frac{X_k}{N}. \tag{11.15}$$

Since A_k is derived from the complex-valued X_k, in general, A_k is a complex-valued function. Let's then name A_k the *complex-amplitude*. The plot of complex-amplitude A_k versus frequency f_k is called the *frequency spectrum* of the sampled signal x_j.

Up to now we have taken k to be all of the integers within the range of $-N/2$ to$+N/2 - 1$. However, it is easy to show that Eq. (11.11) is periodic in k, with a period of N:

$$X_{k+N} = \sum_{j=0}^{N-1} x\left(t_j\right) e^{-i2\pi j(k+N)/N} = \sum_{j=0}^{N-1} x\left(t_j\right) e^{-i2\pi j\,k/N} e^{-i2\pi j} = X_k, \tag{11.16}$$

where we have used the fact that $\exp\left(-i2\pi j\right) = 1$ because j is an integer. Based on Eq. (11.16), the negative k-values in the range of $-N/2$ to -1 are equivalent to the k-value range of $+N/2$ to $+N - 1$. Because the array indices in computer programs generally are positive integers, algorithms to evaluate Eq. (11.11) typically let k vary from 0 to $N - 1$ (which covers one complete period).

The SciPy FFT-related functions `fftfreq()` and `fft()` observe the positive-integer-only convention described above. The discrete Fourier transform values X_k are output from `fft()` as an N-element array, whose elements are indexed 0 to $N - 1$, which are associated with N evenly spaced frequencies (where $\Delta f = f_s/N$) in the range of $-f_{nyquist} \leq f < +f_{nyquist}$. The array elements with indices from 0 to $N/2 - 1$ contain the sequence of X_0 to $X_{N/2-1}$ associated with the positive frequencies from 0 to $+f_{nyquist} - \Delta f$, and the elements with indices from $N/2$ to $N-1$ contain the sequence of $X_{-N/2}$ to X_{-1} associated with the negative frequencies from $-f_{nyquist}$ to $-\Delta f$. The ordering of frequencies produced by `fftfreq()` is consistent with the ordering of X_k produced by `fft()`. That is, `fftfreq()` outputs an N-element array of frequencies, where the elements with indices from 0 to $N/2 - 1$ contain the sequence of positive frequencies from 0 to$+f_{nyquist} - \Delta f$, and the elements with indices from $N/2$ to $N - 1$ contain the sequence of negative frequencies from $-f_{nyquist}$ to $-\Delta f$.

As we found in writing the program **fft_magnitude.py**, the function `fftfreq()` will automatically create the `num`-element array of frequencies (call it `freq`) associated with the FFT with the statement

```
freq = scipy.fft.fftfreq(num, delta_t)
```

Then, if you need all `num` of the frequencies (both positive and negative) in an array, the array `freq` produced by `fftfreq()` is exactly that. However, if you need an array called `freq_positive` consisting of only the positive frequencies, then use this slice operation: `freq_positive=freq[0:num//2]`. The output array produced by `fft()` can be dealt with in an analogous manner.

11.5 Fast Fourier Transform

Since the mid-1960s, a clever algorithm to calculate the discrete Fourier transform (Eq. (11.11)) in an efficient manner has been widely used. This algorithm, known as the *fast Fourier transform* (*FFT*), exploits inherent symmetries in the calculation of Eq. (11.11) that arise because of the periodic way in which the original data were taken (N evenly spaced samples taken with time spacing Δt). Where the "brute-force" evaluation of Eq. (11.11) would require on the order of N^2 complex multiplication operations, the FFT manages this task with only $N\log_2 N$ such operations. The savings in time are immense, especially when N is large. As an example, consider a digital oscilloscope configured to produce a data set of 1024 points (a typical mode of operation). To analyze the frequency spectrum of this set, the FFT algorithm will be about 100 times faster than the straightforward evaluation of Eq. (11.11). For a very large data set of 10^6 points, the FFT is about 50,000 times faster!

In addition to the assumption of evenly spaced data points, there is one extra restriction to the use of FFTs. In deriving a FFT algorithm, one must assume that the size of the data set is a power of 2 (i.e., $N = 2^m$, where $m = 1, 2, 3, \ldots$) to obtain maximum calculational efficiency. If N is not equal to a power of 2, algorithms (called *discrete Fourier transforms*, or *DFTs*) exist to evaluate the discrete Fourier transform, but with a speed much slower than that of the FFT.

11.6 FFT of Sinusoids

To get a handle on how FFT signal processing works, let's synthesize some data consisting of sinusoidal waves with known complex-amplitudes A_k and frequencies f_k, feed these data to the FFT algorithm (i.e., SciPy's `fft()` function), divide the resultant X_k values by N to produce the spectrum of A_k (see Eqs. (11.14) and (11.15)), and determine whether this spectrum matches the known input.

To outfit ourselves for this plan, in the Editor, open the file **waveform_creator.py** in **YourNamePythonWork≫Utilities**, then add this definition of the function `fft_complex()` to the program. Note that `ft.fft(y)` is producing the array X_k of complex numbers, which is then divided by the number of data samples as instructed by Eq. (11.15) in order to yield the array named `complex_amp` of complex-amplitudes A_k.

```
def fft_complex(y, num, samp_freq):

    delta_t = 1/samp_freq

    freq = ft.fftfreq(num, delta_t)
    complex_amp = (1/num)*ft.fft(y)

    return freq, complex_amp
```

Save **waveform_creator.py** in **YourNamePythonWork≫Utilities** as you close the file.

Next, in the Editor, open **fft_magnitude.py** in **YourNamePythonWork**, then use **File>>Save As...** to create a new file named **fft_of sinusoids.py** in **YourNamePythonWork**. Most of the needed changes are in the `calculate_plot_waveform()` function and are shown below. First, the `wc.fft_mag_only()` function is replaced by `wc.fft_complex()` and the three commented-out window-related lines are deleted. The array `complex_amp` produced by `wc.fft_complex()` is the array of complex-amplitudes. Like all Python objects, a NumPy array has *attributes* (e.g., size, number of dimensions), which reflect information intrinsic to the array itself. Pertinent to our present case, the array `complex_amp` has a *real attribute* and an *imaginary attribute*, which can be acquired via the dot notation `complex_amp.real` and `complex_amp.imag`. These operations produce the `amp_real` and `amp_imag` arrays consisting of the real and imaginary portions of each element in `complex_amp`, respectively. Finally, `pc.update_plot()` is called twice, the first time to plot the `amp_real` using filled circle markers and the second time to plot to `amp_imag` using a default solid line. The argument `clear_flag=False` signals that the first plot should not be erased as the second plot is being displayed.

```
def calculate_plot_waveform():

    num, samp_freq = gc.get_dig_entries(var_num,
                                        var_samp_freq)
    a, f, shape = gc.get_wav_entries(var_a, var_f,
                                     var_shape)

    t = wc.create_time_data(num, samp_freq)
    y = wc.create_waveform_data(t, a, f, shape)

    freq, complex_amp = wc.fft_complex(y, num, samp_freq)

    amp_real = complex_amp.real
    amp_imag = complex_amp.imag

    pc.update_plot(ax, canvas, freq, amp_real,
                   label_x, label_y, marker='o')

    pc.update_plot(ax, canvas, freq, amp_imag,
                   label_x, label_y, marker='-',
                   clear_flag=False)

    return
```

Second, make the following changes for the root window title and the plot axes labeling:

```
label_x, label_y = 'Frequency (Hz)', 'Complex-Amplitude'

root = gc.create_window(title='FFT of Sinusoids',
                        size='1000x500')
```

Save your work. As a reminder, note that the real and imaginary parts of the complex-amplitude will be plotted as filled circles and a solid line, respectively.

11.7 Applying FFT to Various Sinusoidal Inputs

Run **fft_of sinusoids.py**. In the **Waveform Parameters** control, program a *Cosine* function with an amplitude of *4.0* and frequency of *250* Hz. *Note*: I will use the word "amplitude" to mean the peak height (relative to the zero level) of a sinusoidal function. For the **Digitizing Parameters**, choose *1024* and *2000* for **Number of Samples** and **Sampling Frequency (Hz)**, respectively. Remember that the FFT algorithm requires the number of samples in your input data to be a power of 2, so we have chosen $N = 2^{10} = 1024$.

Press the **Submit** button. The program will find the real and imaginary parts of the complex-amplitude A_k for positive and negative frequencies f_k up to the Nyquist frequency $\pm f_{nyquist} = \pm f_s/2$. For our choice of $f_s = 2000$ Hz, the Nyquist frequency is 1000 Hz, so the 250-Hz cosine oscillation frequency that you have input will fall within the FFT's range of detection.

After the **Submit** button press, the plot of the Fourier transform of your cosine input will appear in following form: The real part (filled circles) of the complex-amplitude A is equal to zero at all frequencies f, except $A = +2$ at $f = \pm 250$ Hz. The imaginary part (solid line) of A is zero at all frequencies f.

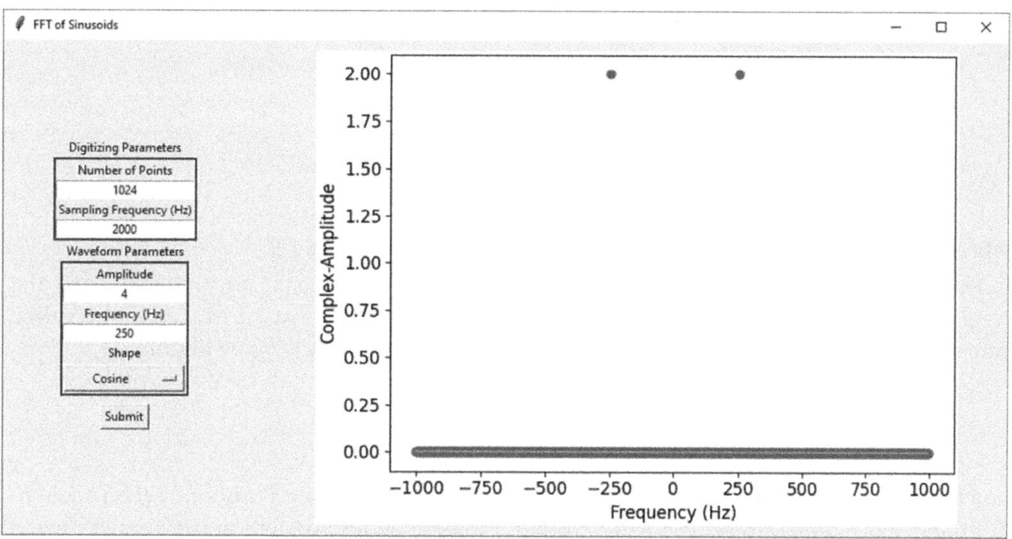

Why is this output the correct representation for the cosine input? Shouldn't the amplitude A be 4, not 2? To understand your result, plug the plot's output back into Eq. (11.14). Then

$$
\begin{aligned}
x_j \approx \sum_{k=0}^{N-1} A_k e^{i2\pi f_k t_j} &= (+2)\, e^{i2\pi(250)t_j} + (+2)\, e^{i2\pi(-250)t_j} \\
&= 4\left[\frac{e^{i2\pi(250)t_j} + e^{i2\pi(-250)t_j}}{2}\right] \\
&= 4\cos\left[2\pi(250)\, t_j\right],
\end{aligned}
$$

where we have used the identity $\cos x = \left(e^{ix} + e^{-ix}\right)/2$ for the last step. So, we see that the output complex-amplitudes are indeed the correct representation for an input cosine function of amplitude 4.0.

Try a sine-wave input and see what you get. Keeping all the other entries the same, change **Shape** to *Sine*, then press the **Submit** button. You should find that the imaginary part (solid line) of the complex-amplitude is $A = -2$ at $f = +250$ Hz, $A = +2$ at $f = -250$ Hz, and zero elsewhere. The real part (filled circles) of A is zero at all frequencies.

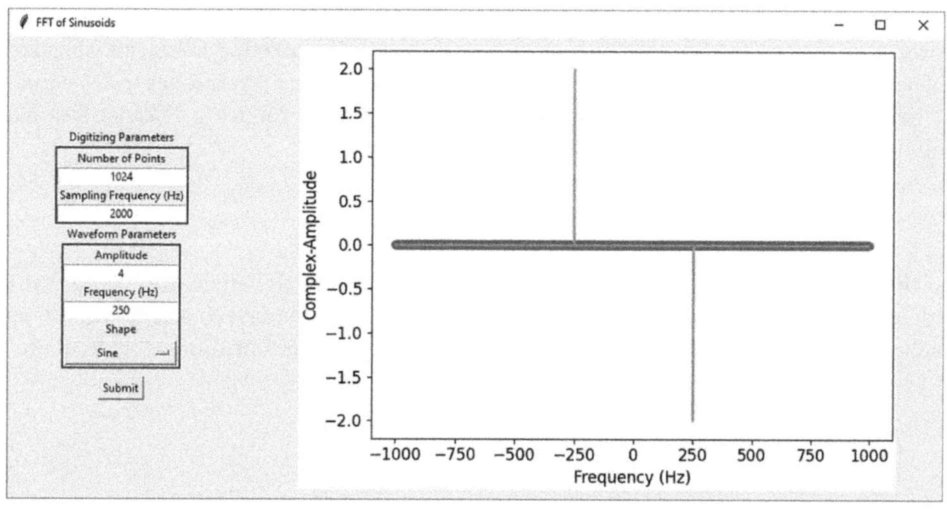

To see that this is the correct representation of your input, we again start with Eq. (11.14) and plug the plot's output into it:

$$
\begin{aligned}
x_j \approx \sum_{k=0}^{N-1} A_k e^{i2\pi f_k t_j} &= (-2i)\, e^{i2\pi(250)\ t_j} + (+2i)\, e^{i2\pi(-250)\ t_j} \\
&= 4\left[\frac{e^{i2\pi(250)\ t_j} - e^{i2\pi(-250)\ t_j}}{2i}\right] \\
&= 4\sin\left[2\pi\,(250)\ t_j\right],
\end{aligned}
$$

where we have used the facts that $i = -1/i$ and $\sin x = \left(e^{ix} - e^{-ix}\right)/2i$.

There are two frequencies that behave differently from the rest: zero frequency and the Nyquist frequency. First, program **fft_of_sinusoids.py** with the constant *DC Level* (also called *zero-frequency*) function (i.e., **Amplitude** equal 4, **Frequency (Hz)** can be left blank),

$$x_j = 4.0,$$

and then run the program with **Number of Samples** and **Sampling Frequency (Hz)** equal to *1024* and *2000*, respectively. You will find that, rather than finding half of the amplitude at a positive and negative frequency as in the previously studied nonzero-frequency cases, the full amplitude (called the *DC component*) appears at the single frequency $f = 0$.

Next, program **fft_of_sinusoids.py** with the following function that oscillates at the Nyquist frequency:

$$x_j = 4.0\cos\left[2\pi\,(1000)\ \ t_j\right],$$

and then run the program again with **Number of Samples** and **Sampling Frequency (Hz)** equal to *1024* and *2000*, respectively. Similar to the zero-frequency case, you'll discover that the full oscillatory amplitude appears at a single frequency, this time $-f_{nyquist}$.

Here is a synopsis of our last observation: The cosine function, which is used to generate the data set, takes on its peak value when the 2000-Hz sampling process begins at time $t = 0$. The 1000-Hz Nyquist-frequency waveform then is sampled twice each cycle, producing a data set that has the following sequence—peak value, trough value, peak value, trough value, and so on. We find that by applying the FFT algorithm of this data set, the correct value for the amplitude of oscillation (i.e., 4.0) is determined at $-f_{nyquist}$.

Unfortunately, by exploring the Nyquist-frequency situation a little further, a troublesome problem emerges. Keeping **Number of Samples** and **Sampling Frequency (Hz)** equal to *1024* and *2000*, respectively, program **fft_of_sinusoids.py** to produce a Nyquist-frequency oscillation, but this time one that follows the sine function

$$x_j = 4.0\sin\left[2\pi\,(1000)\ t_j\right].$$

Press the **Submit** button. Note the order of magnitude of the y-axis values. You'll find that the FFT detects virtually no amplitude of oscillation at 1000 Hz in this case. If a mysterious diagonal solid line appears in the resulting FFT plot, it is due to the discontinuous way frequencies are distributed in the `freq` array; i.e., the data move from the highest frequency to the lowest frequency at the midpoint of the frequency array. Can you explain why the 1000-Hz sinusoidal oscillation is "invisible" in this data set?

If interested, you can try including a phase constant δ (in radians) in the sine function's argument to produce a data set that takes on neither its peak value nor its zero-value (like the cosine and sine function, respectively) at time $t0$. To create the desired function, open **waveform_creator.py** in **YourNamePythonWork≫Utilities** and program the **User-Defined** option within the `create_waveform_data()` function to be

$$x_j = 4.0 \sin\left[2\pi (1000)\ t_j + \delta\right].$$

Save **waveform_creator.py**, then run **fft_of_sinusoids.py** with **Shape** selected as *User-Defined*, with **Number of Samples** and **Sampling Frequency (Hz)** equal to *1024* and *2000*, respectively (remember **Amplitude** and **Frequency (Hz)** can be left blank for the **User-Defined** selection). Does the FFT yield the correct amplitude of oscillation (i.e., 4.0) for any nonzero value of δ?

To summarize our Nyquist-frequency findings, in taking the FFT of a discretely sampled data set, the resultant value for the amplitude at $-f_{nyquist}$ is only accurate when the Nyquist-frequency oscillation is "in phase" with the sampling process. Since one cannot guarantee that this will be the case in an actual experimental situation, the complex-amplitude at $-f_{nyquist}$ should always be viewed with suspicion. Thus, in our work below, we will ignore the Fourier component determined at the Nyquist frequency.

Next, program the **User-Defined** option in **waveform_creator.py** to be a function that includes two simultaneous sinusoidal oscillations, along with a DC level, such as

$$x_j = 8.0 + 4.0 \sin\left[2\pi (250)\ t_j\right] + 6.0 \cos\left[2\pi (500)\ t_j\right].$$

Save **waveform_creator.py**. Run **fft_of_sinusoids.py**, selecting **Shape** as *User-Defined*, with **Number of Samples** and **Sampling Frequency (Hz)** equal to *1024* and *2000*, respectively. By selecting **Shape** as *User-Defined*, the **Amplitude** and **Frequency (Hz)** entry boxes will not be read and so can be left blank. Press the **Submit** button. Do you understand the output?

Finally, what happens when the input contains a sine and cosine oscillation, both at the same frequency? Run **fft_of_sinusoids.py** with the **User-Defined** function

$$x_j = 4.0 \sin\left[2\pi (250)\ t_j\right] + 6.0 \cos\left[2\pi (250)\ t_j\right].$$

Can you decipher the real and imaginary plots?

11.8 Magnitude of Complex-Amplitude

With Section 11.7 as background, we are now positioned to understand the operation of **fft_magnitude.py**, the program we wrote at the beginning of this chapter. To this point, we have been representing a complex number as the sum of a real and an imaginary part, that is,

$z = x + iy$. However, as you know, the complex number z may also be represented as a *magnitude* r times a *phase factor* $e^{i\,\phi}$, where $r = \sqrt{x^2 + y^2}$ and $\tan\phi = y/x$. Let's say that the oscillation of a system at frequency f is, as in the final example of Section 11.7, the composite of a sine and cosine function with (real) amplitudes B and C, respectively:

$$x = B\sin(2\pi ft) + C\cos(2\pi ft) \tag{11.17}$$

This expression is the most general form of sinusoidal oscillation at frequency f. Then, defining $\theta = 2\pi ft$ and using the complex exponential representation of the sine and cosine functions, we have

$$\begin{aligned} x &= B\frac{e^{i\theta} - e^{-i\theta}}{2\,i} + C\frac{e^{i\theta} + e^{-i\theta}}{2} \\ &= \frac{1}{2}\,(C - i\,B)\;e^{i\theta} + \frac{1}{2}\,(C + i\,B)\;e^{-i\theta}. \end{aligned}$$

From Fig. 11.1, we see that $(C + iB) = re^{+i\phi}$ and $(C - iB) = re^{-i\phi}$, where $r = \sqrt{B^2 + C^2}$ and $\phi = \tan^{-1}(B/C)$. Thus,

$$\begin{aligned} x &= \frac{1}{2}\left(re^{-i\phi}\right)\,e^{i\,\theta} + \frac{1}{2}\left(re^{+i\phi}\right)\,e^{-i\theta} \\ &= r\frac{e^{i(\theta-\phi)} + e^{-i(\theta-\phi)}}{2} \\ &= r\;\cos(\theta - \phi)\,. \end{aligned} \tag{11.18}$$

Then, equating Eq. (11.17) with Eq. (11.18) and remembering $\theta = 2\pi ft$, $r = \sqrt{B^2 + C^2}$, and $\phi = \tan^{-1}(B/C)$, we find that

$$x = B\sin(2\pi f t) + C\cos(2\pi f t) = \sqrt{B^2 + C^2}\cos(2\pi ft - \phi) \tag{11.19}$$

Equation (11.19) tells us that a sine and cosine oscillation at the frequency f combine to produce a single phase-shifted cosine oscillation at frequency f with a net amplitude equal to the vectorial sum of the sine and cosine amplitudes.

The above derivation provides the theoretical underpinning of the beginning-of-the-chapter program **fft_magnitude.py**. That program, given an input array representing the displacements of a waveform, finds the net amplitude of oscillation at an array of associated (positive) frequencies, but ignores the phase information (i.e., neglects whether this oscillation is in the form of a pure sine wave or a pure cosine wave or a composite of sine plus cosine).

To test this theory, open **waveform_creator.py** in **YourNamePythonWork≫Utilities** and program the **User-Defined** option within the `create_waveform_data()` function to calculate the sum of a sine and cosine function, both having frequency $f = 250$ Hz, with amplitudes 3.0 and 4.0, respectively:

$$x = 3.0\sin\left[2\pi\,(250)t\right] + 4.0\cos\left[2\pi\,(250)t\right].$$

From Eq. (11.19), the above waveform should be equivalent to

$$x = \sqrt{3^2 + 4^2}\cos\left[2\pi\,(250)\,t - \tan^{-1}(3/4)\right].$$

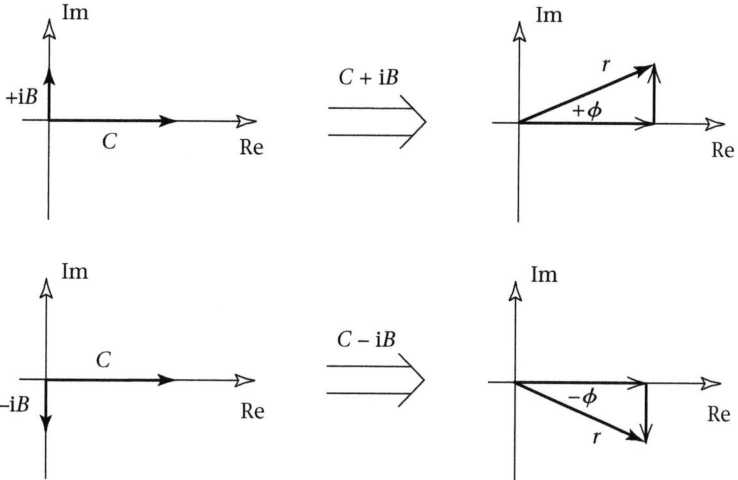

Fig. 11.1 Graphical representation of $C + iB$ and $C - iB$.

Save **waveform_creator.py**, then run **fft_magnitude.py**. Enter **Number of Samples** and **Sampling Frequency (Hz)** equal to *1024* and *2000*, respectively. Select **Shape** as *User-Defined*; both **Amplitude** and **Frequency (Hz)** can be left blank. Press the **Submit** button. Is the output representative of a single 250-Hz cosine curve with net amplitude $\sqrt{3^2 + 4^2} = 5$?

The Fourier transform method represents a function as the linear combinations of the complex exponential basis set. Because $\cos(2\pi ft) = \left(e^{i2\pi ft} + e^{-i2\pi ft}\right)/2$, a cosine function's amplitude is equally divided between the complex-amplitudes of the positive- and negative-frequency basis functions $e^{\pm i2\pi ft}$. Two exceptions to this "equal division of amplitude" occur, of course, for the basis functions at zero frequency (i.e., $e^{i2\pi(0)\,t} = 1$) and at the Nyquist frequency. When displaying the spectrum of a given input data set, this symmetry of complex-amplitudes in frequency space is commonly exploited by simply doubling the magnitude of the positive frequency's complex-amplitude and only plotting the positive-frequency axis. Such a plot displays the net amplitude of sinusoidal oscillation at each frequency $|f_k|$ directly, while phase information is completely neglected.

As you hopefully recognize, the previous paragraph outlines the approach we took in writing the following statements in **waveform_creator.py's** function `fft_mag_only()`:

```
n = num//2
freq = ft.fftfreq(num, delta_t)

fourier_transform = ft.fft(y)
amp = (2/num)*np.abs(fourier_transform)

freq_positive = freq[0:n]
amp_positive = amp[0:n]
```

Here, the complex-amplitude $A_k = \frac{X_k}{N}$ at frequency f_k is found by `(1/num)*ft.fft (fourier_transform)`. Then, at both positive and negative frequencies, the magnitude of the complex magnitude A_k (i.e., the amplitude of oscillation `amp`) is determined by `(1/num)*np.abs(fourier_transform)`. Since we will plot amplitude only in the positive-frequency range, a factor of 2 is included to account for the contribution of negative frequencies. The array of associated (positive and negative) frequencies is calculated by `freq = ft.fftfreq(num,delta_t)`, but this array is sliced as follows: `freq_positive = freq[0:n]`, where `n=num//2`, in order to send only the positive frequency to the GUI plot. The `amp` array is sliced similarly.

In regard to the two "exception" frequencies, we decided the amplitude at the Nyquist frequency is problematic and so should be ignored, while the amplitude at zero frequency should not be doubled. In the above procedure, the SciPy function `fftfreq()` chooses to represent the Nyquist frequency as the negative frequency $-f_{nyquist}$ and, since we only plot amplitudes associated with positive frequencies, the Nyquist frequency is already being ignored. However, the amplitude at zero frequency is being doubled, along with the amplitudes at all of the other positive frequencies. Since $f=0$ is the first element (index-0) in the `freq_positive` array, a simple way to account for this zero-frequency exception is to include the statement `amp[0]=amp[0]/2`, after the entire array `amp` is determined.

Open **waveform_creator.py** and add this zero-frequency-related statement within the `fft_mag_only()` function. The code for the completed function will appear as shown here:

```
def fft_mag_only(y, num, samp_freq):

    delta_t = 1/samp_freq
    n = num//2

    freq = ft.fftfreq(num, delta_t)

    fourier_transform = ft.fft(y)
    amp = (2/num)*np.abs(fourier_transform)
    amp[0] = amp[0]/2

    freq_positive = freq[0:n]
    amp_positive = amp[0:n]

    return freq_positive, amp_positive
```

Save your work, then run **fft_magnitude.py**. Taking **Number of Samples** and **Sampling Frequency (Hz)** equal to *1024* and *2000*, respectively, execute the program with several choices of waveforms with frequencies less than 1000 Hz, including *DC Level*, to verify that it performs correctly.

11.9 Observing Spectral Leakage

Let's revisit the good news/bad news mystery we observed at the beginning of the chapter. We'll be inspecting a peak in the frequency spectrum near 250 Hz, so zoom in on this peak by manually scaling the *x*-axis to this region. To accomplish this feat, simply open **plot_creator.py** in **YourNamePythonWork≫Utilities** and remove the prepending hashtag symbol we used to comment out the line within the `update_plot()` that controlled x-axis scaling. That line will now appear as shown here:

```
ax.set_xlim(200, 300)
```

Save **plot_creator.py** as you close it. For now, keep the three lines related to windowing within the `fftcreate_plot_waveform()` function of **fft_magnitude.py** commented out:

```
    #win_type = 'hann'
    #win = wc.create_window(win_type, num)
    #y = y*win
```

Run **fft_magnitude.py**. Enter *1024* and *2000* for **Number of Samples** and **Sampling Frequency (Hz)**, respectively, as well as *4*, *250*, and *Sine* for **Amplitude**, **Frequency (Hz)**, and **Shape**, respectively. Press the **Submit** button and observe the resulting frequency spectrum. It's perfect, with the amplitude equal to zero everywhere, except at the frequency of 250 Hz, where it equals +4.0, just as we expected. That's the good news.

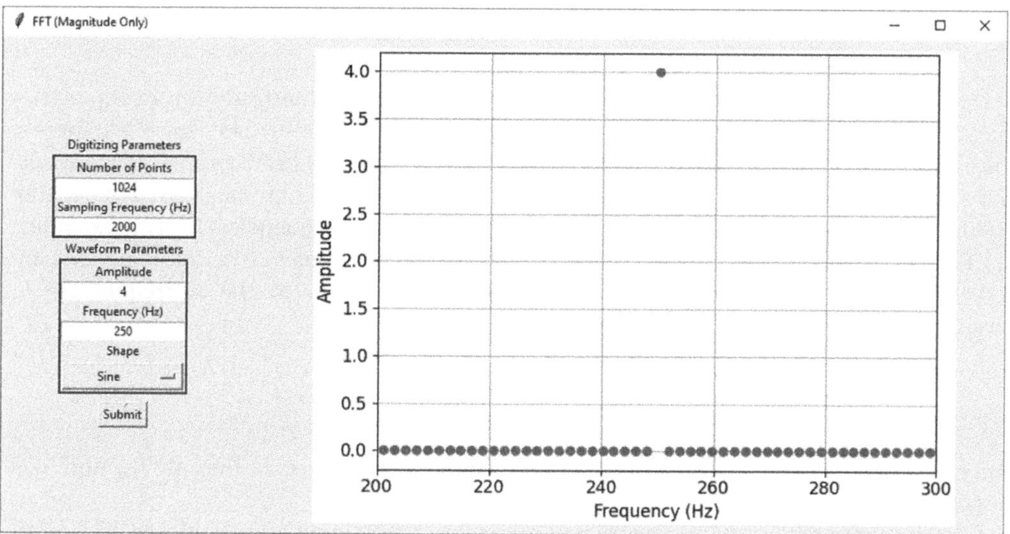

Now, the bad news. Repeat the above procedure after programming **fft_magnitude.py** with the function

$$x = 4.0 \cos [2\pi (249) t] .$$

That is, simply change **Frequency (Hz)** to 249 Hz and press **Submit**.

As we saw at the chapter opening, the resulting spectrum no longer appears as a "delta-function" spike with a height of 4.0, but instead is a broadened peak of maximum height less than 4.0. This spectrum seems to be telling us that the input data are oscillating at several different frequencies in the approximate range of 200 to 300 Hz. However, we know that this is incorrect. We input a pure sine wave with the exact frequency of 249 Hz.

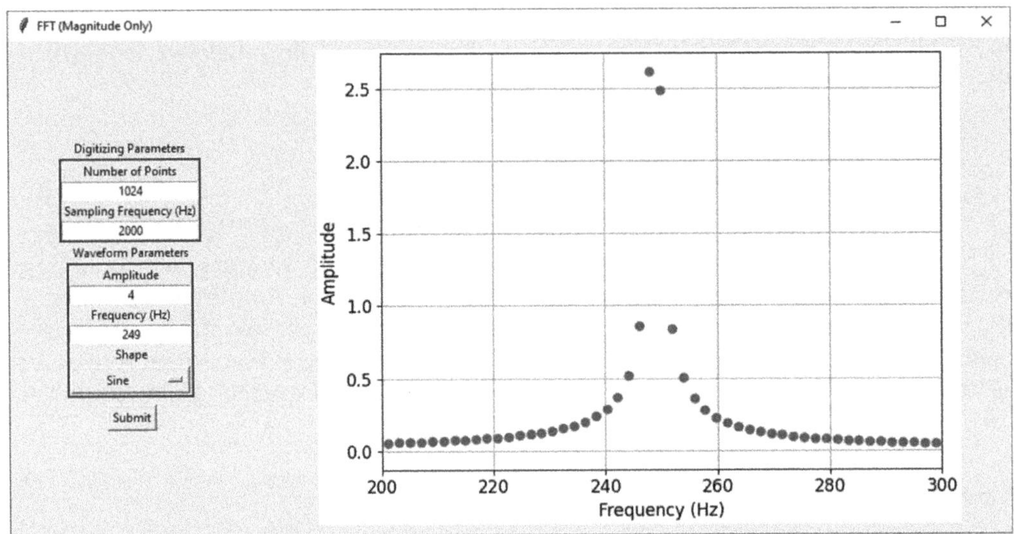

That is the mystery. Why does our FFT-based program work perfectly for an input frequency of 250 Hz, but fails when that frequency is changed to 249 Hz? Well, here's the resolution to this paradox: For most of this chapter, I've been sneaky and always, with the exception of f = 249 Hz, asked you to generate input data using sine and cosine functions that oscillate at one of the N discrete frequencies f_k. For example, via Eq. (11.7), when acquiring 1024 equally spaced data samples at a sampling rate of 2000 Hz, the FFT algorithm produces complex-amplitudes A_k at the N frequencies f_k given by

$$f_k = k\left(\frac{2000\ \text{Hz}}{1024}\right) \qquad k = -512, \ldots, 0, \ldots, +511.$$

You can verify that 250 and 500 Hz are both one of the f_k. In particular, they are f_{128} and f_{256}, respectively.

A way of summarizing our above observations is this: If you input a sinusoidal oscillation at exactly one of the frequencies f_k—for example, f_{128} = 250.00 Hz or f_{256} = 500.00 Hz—the FFT algorithm will perfectly produce the frequency spectrum of that data (i.e., a "delta-function" spike of the correct height at the correct frequency). However, if instead one inputs a sinusoidal oscillation at a frequency f not equal to one of the f_k—such as 249 Hz, which falls between f_{127} = 248.05 Hz and f_{128} = 250.00 Hz—the resulting spectrum is imprecise. In fact, it is as if its spectral amplitude, which rightfully should be a spike at frequency f, has diffused from this central point and distributed itself among the neighboring f_k. This smearing of spectral information—termed *spectral leakage*—is an artifact of the finite number N of data samples contained in our discretely sampled data set. You can prove this for yourself by rerunning **fft_magnitude.py** programmed to calculate a 249-Hz cosine wave, first making **Number of**

Samples equal to *512*, then *1024*, then *2048*, followed by *4096*, and so on. You will find that as N increases, the spectrum becomes much more delta-function-like.

11.10 Windowing

The finite sample size of a data set (which causes spectral leakage) can be viewed as a windowing effect. That is, when we acquire a finite number N of discretely sampled points for FFT spectral evaluation, we are in effect observing an infinite set of data d_j (where $j = -\infty, \ldots, -1, 0, +1, \ldots, +\infty$) through a rectangular viewing window in time. Defining the rectangle window function $w(t_j)$ to be zero at all times $t_j = j\Delta t$, except during the "data-viewing" time interval from $j = 0$ to $j = N - 1$, when it is equal to 1, our finite set of N sampled data points x_j is then given by the product $x_j = d_j w_j$. This idea is illustrated in Fig. 11.2.

A detailed mathematical analysis of the situation illustrated in Fig. 11.2 leads to the following conclusion: The discontinuous jumps (i.e., the sharp turn-on and turn-off) at the edges of the rectangle window function are the chief cause of the broad spectral leakage we observed in Section 11.9.

But, as we saw at the beginning of the chapter, there is a way to mitigate this problem. Thanks to the power of your high-speed computer, you have the option of windowing your finite-length data set with some function other than a rectangle. Once the rectangular-windowed data set $(x_{rectangle})_j$ has been collected into your computer, all you have to do is construct a more desirable window w_j in software and then form the product $x_j = w_j\,(x_{rectangle})_j$. Then spectral leakage can simply be suppressed by choosing a software window with a much less abrupt turn-on and turn-off than the rectangle window. For example, the *Hann* window, which is defined to be

$$(w_{hann})_j = 0.5\left[1 - \cos\left(\frac{2\pi j}{N}\right)\right] \qquad j = 0, 1, \ldots, N-1, \tag{11.20}$$

is a popular choice for such a software window. The "smooth-edged" Hann and "abrupt-edged" Rectangle windows are plotted in Fig.11.3 for comparison.

Let's reimplement windowing of our data set in **fft_magnitude.py** and see how this improves things. At the beginning of the chapter, we used `create_window()` in **waveform_creator.py** to create a Hann window and found that, by multiplying this window by our waveform data set, frequency spread of spectral leakage can be decreased.

The function `create_window()` is reproduced here; the import statement `import scipy.signal as sig` is at the top of **waveform_creator.py**. We see that `create_`

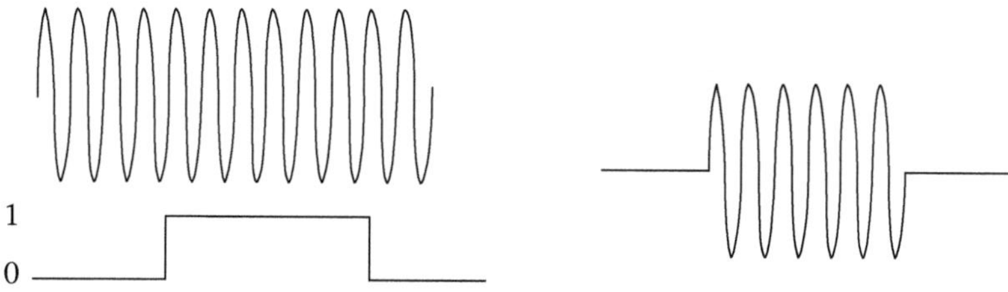

Fig. 11.2 Finite-length sine wave produced by rectangular viewing window.

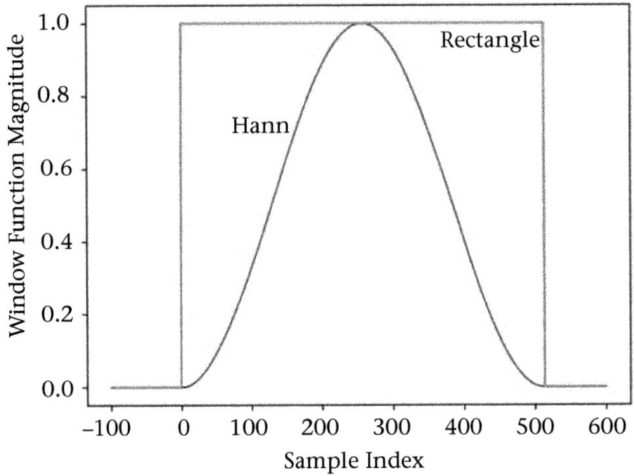

Fig. 11.3 Comparison of Rectangle and Hann windows with $N = 512$.

`window()` creates a requested window by calling the `get_window()` function in the `windows` submodule within the `signal` module of SciPy. To be proved later in this section, in order to produce correct waveform amplitudes, the window array must be normalized (through division) by a constant called the *coherent gain*. The coherent gain is simply the mean value of the window array elements, which is determined using NumPy's `mean()` function.

```
def create_window(win_type, num):

    win = sig.windows.get_window(win_type, num)

    coherent_gain = np.mean(win)
    win = win/coherent_gain

    return win
```

The `get_window()` function provides access to over twenty different types of windows. Each window has its own properties such as central-lobe width and turn-on/turn-off rate, which are designed to optimize its use in particular applications. The Hann window is a satisfactory choice in most situations. This window suppresses spectral leakage, allowing frequencies within the input signal to be well resolved. Table 11.1 lists some of the commonly used windows, along with the applications for which they are most suitable.

Table 11.1 Commonly Used Windows

Window	Optimal application
Hann	General-purpose applications
Hamming	Resolution of closely spaced sine waves
Flat Top	Accurate amplitude measurement of isolated frequency
Blackman	Resolution of two closely spaced frequencies with widely differing amplitudes; used in audio applications
Rectangle (also called Boxcar)	Resolution of two closely spaced frequencies with almost equal amplitudes

By adding a option menu widget called **Window Menu** to our GUI, we can give a user the ability to choose from among several possible windows. In the Editor, open **gui_creator.py** in **YourNamePythonWork≫Utilities**. Then, add the following definition for a function named `window_control()`. See the function `waveform_control()` for similar code.

```
def window_control(container):

    lbl_1 = tk.Label(container, text='Window Menu')
    lbl_1.pack()

    frame = tk.Frame(container, bg='blue', bd=2)
    frame.pack()

    var_window = tk.StringVar()
    options = ['Rectangle', 'Hann', 'Hamming', 'Flattop',
               'Blackman']
    opm = tk.OptionMenu(frame, var_window, *options)
    opm.pack(fill=tk.X)

    return var_window
```

Then, add the following definition for a function named `get_window_option()`:

```
def get_window_option(var_window):

    win_type = var_window.get()

    return win_type
```

Save you work and close **gui_creator.py**.

Next, in the Editor, open **waveform_creator.py** in **YourNamePythonWork≫Utilities**. Then, within the function `create_window()`, add the `if-else` statement shown. The `if` portion changes the name of the Rectangle window to the name used by SciPy (`'boxcar'`). The `else` portion switches the capitalized version of the window names used in the **Window Menu** control to the lowercase versions used by the `get_window_option()` function. The built-in string method `lower()` converts a given string to all lowercase. `get_window_option()` creates the `num`-element window array `win` of the requested `win_type`, which is then properly normalized.

```
def create_window(win_type, num):

    if win_type == 'Rectangle':
        win_type = 'boxcar'
    else:
        win_type = win_type.lower()

    win = sig.windows.get_window(win_type, num)
```

```
    coherent_gain = np.mean(win)
    win = win/coherent_gain

    return win
```

Save **waveform_creator.py** as you close it.

Finally, in the Editor, open **fft_magnitude.py** in **YourNamePythonWork**. Below the two lines that create the **Digital Parameters** and **Waveform Parameters** controls, add a third line that creates the **Window Menu** control as shown here:

```
var_num, var_samp_freq = gc.digitizing_control(control_box)
var_a, var_f, var_shape = gc.waveform_control(control_box)
var_window = gc.window_control(control_box)
```

Then, within `create_plot_waveform()`, as shown below, add the statement that gets the **Window Menu** entry, uncomment-out the window-related commands, and delete the line `win_type = 'hann'`:

```
def calculate_plot_waveform():

    num, samp_freq = gc.get_dig_entries(var_num,
                                        var_samp_freq)
    a, f, shape = gc.get_wav_entries(var_a, var_f,
                                     var_shape)
    win_type = gc.get_window_option(var_window)

    t = wc.create_time_data(num, samp_freq)
    y = wc.create_waveform_data(t, a, f, shape)

    win = wc.create_window(win_type, num)
    y = y*win

    freq_positive, amp_positive = wc.fft_mag_only(y, num,
                                                  samp_freq)

    pc.update_plot(ax, canvas, freq_positive, amp_positive,
                   label_x, label_y, marker='o')

    return
```

Save **fft_magnitude.py**.

Let's explore the positive effects of windowing. Run **fft_magnitude.py** and assign **Number of Samples** and **Sampling Frequency (Hz)** as *1024* and *2000*, respectively, and select **Waveform Parameters** that will produce the waveform $4.0 \sin[2\pi(249\text{ Hz})\,t]$. Finally, opt for the program to run without the benefit of windowing by selecting **Window Menu≫Rectangle**. Press the **Submit** button.

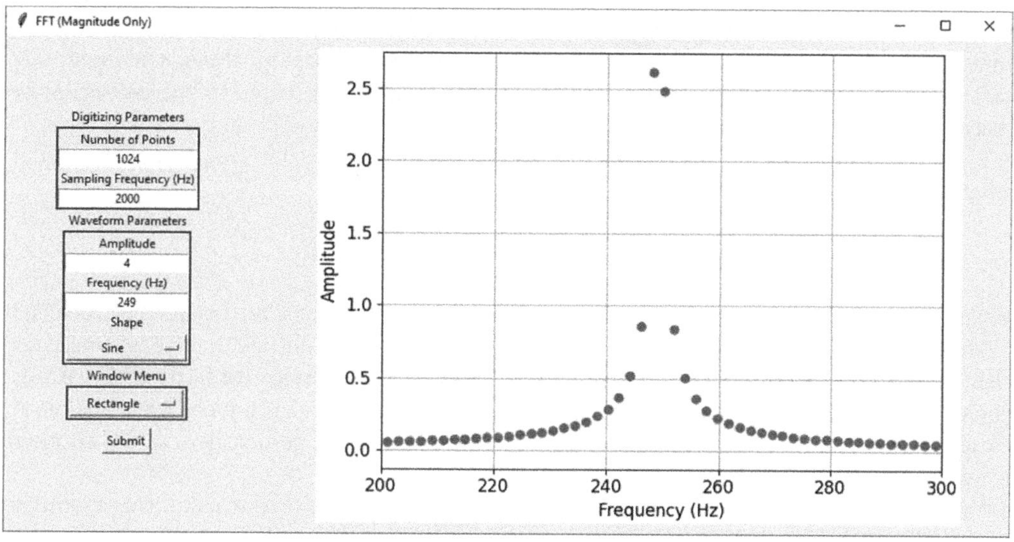

Now select the *Hann* window and rerun the program. Note the dramatic decrease in spectral leakage through the use of a Hann window.

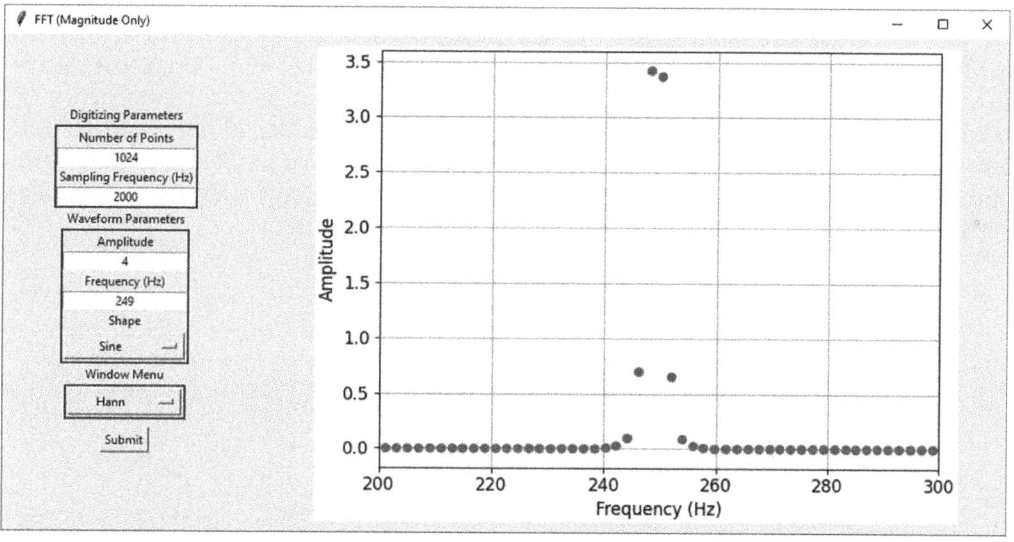

To demonstrate that certain windows perform better in a particular application, consider the challenge of resolving two sine waves of nearly the same frequency with widely differing amplitudes, a problem that might be encountered in, say, speech recognition research. Open **waveform_creator.py** and program the **User-Defined** option within `create_waveform_data()` to input the following waveform to **fft_magnitude.py**:

$$x = 4.0\sin\left[2\pi(249)t\right] + 0.5\sin\left[2\pi(255)t\right].$$

Here, it may be helpful to change the plot marker to a solid line, i.e., choose `marker'-'` in `calculate_plot_waveform()`. Run **fft_magnitude.py** several times, each time with a different type of window. Which windows perform best for this particular application? For which windows does the smaller 255-Hz peak get obscured by the 249-Hz peak (because of large spectral leakage)?

Change the input waveform to two sine waves of nearly the same frequency with equal amplitudes:

$$x = 4.0\sin\left[2\pi\,(249)\,t\right] + 3.0\sin\left[2\pi\,(253)\,t\right].$$

Which window best resolves these peaks now (e.g., shows the deepest dip between the peaks)?

The *Flat Top* window is optimized to produce a peak with an amplitude that closely matches that of the input waveform. Input $4.0\sin\left[2\pi\,(249\text{ Hz})\,t\right]$ to **fft_magnitude.py** with various choices for the type of window. Is the amplitude of the spectral peak produced from this input closest to the value of 4.0 when using the Flat Top window (in comparison to when other windows are used)?

To conclude, here is the theory behind the quantity called coherent gain. This quantity accounts for the fact that, by windowing a given waveform data array, the average amplitude of the resultant "windowed" array is smaller than the given data array, as shown in Fig. 11.4. Since the FFT is taken on the windowed array, the amplitudes determined by the FFT are smaller than those of the given waveform data array.

To account quantitatively for this effect, assume that the given waveform data array $x\left(t_j\right)$ represents a single sinusoidal wave of frequency f_k and amplitude A that has been digitized at the sequence of times t_j, where $j = 0, 1, 2, \ldots, N-1$. The goal of our FFT analysis then is to determine A. In the following derivation, we will only consider the positive frequency $+f_k$ needed to represent a sinusoidal wave, knowing that the contribution of the negative $-f_k$ is accounted for by a factor of 2 at the end of the amplitude determination. Then, only the term at frequency $+f_k$ has a nonzero amplitude (of A) in the sum of complex exponentials in Eq. (11.14) and so, the jth term in the waveform data array x is given by

$$x_j = Ae^{i2\pi f_k t_j} = Ae^{i2\pi kj/N}. \tag{11.21}$$

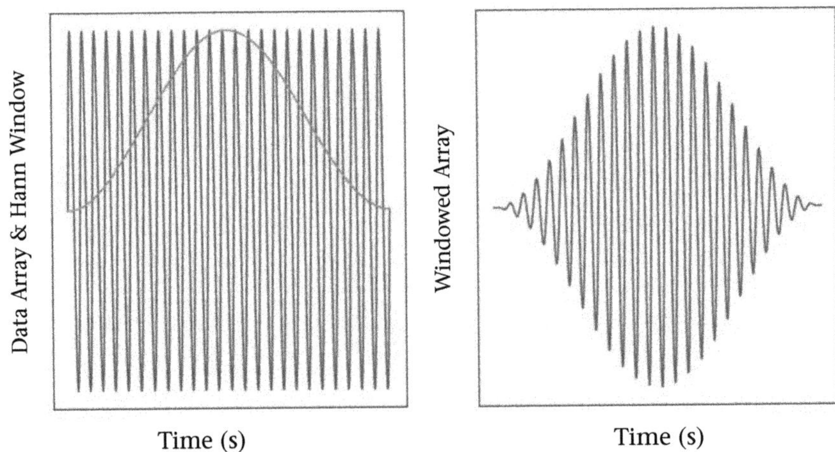

Fig. 11.4 (left) Sine-wave data array and Hann window array; (right) windowed array from product of data array and Hann window array.

Next, prior to application of the FFT, the data array is element-wise multiplied by a window array $w(t_j)$. Calling the jth element of the window array w_j, the discrete Fourier transform [Eq. (11.11)] gives the Fourier component associated with the frequency f_k to be

$$X_k = \sum_{j=0}^{N-1} \left[\left(Ae^{i2\pi kj/N} \right) w_j \right] e^{-i2\pi kj/N} = A \sum_{j=0}^{N-1} w_j \tag{11.22}$$

In our FFT program, the quantity $A_k = X_k/N$ is determined and assigned as the value for the amplitude at frequency f_k. However, from Eq. (11.22), for a windowed array,

$$A_k = \frac{X_k}{N} = \frac{A \sum_{j=0}^{N-1} w_j}{N} \equiv A \times CG, \tag{11.23}$$

where the coherent gain CG is defined as

$$CG \equiv \frac{\sum_{j=0}^{N-1} w_j}{N}. \tag{11.24}$$

Note, the right-hand side of this equation simply calculates the mean value of the window array elements.

Thus, from Eq. (11.23), the amplitude A of the sinusoidal data is found from $A = A_k/CG$. In our work, we chose to apply the "amplitude correction" factor CG in the window array creation function `create_window()` within **waveform_creator.py**.

11.11 Estimating Frequency and Amplitude

In general, an input sinusoidal signal of frequency f and amplitude A will produce a spectral peak that is broadened over a range of f_k-values, where a_k is the amplitude at each frequency f_k (after including the coherent-gain correction). For this situation, here is a final nifty trick for obtaining an accurate estimate for f and A. First, within the spectral peak, define the *peak power* at frequency f_k to be $P_k \equiv a_k^2$. Then, the input sinusoidal frequency f that generated this finite-width spectral peak can be estimated by a weighted sum, with each observed f_k weighted by the peak power at that frequency. That is,

$$f \approx \frac{\sum_k f_k a_k^2}{\sum_k a_k^2}, \tag{11.25}$$

where the sum is over the k values that span the peak.

Also, from signal processing theory, the peak power $P \equiv A^2$ of the sinusoid with amplitude A can be estimated by

$$P \approx \frac{\sum_k P_k}{\text{ENBW}} = \frac{\sum_k a_k^2}{\text{ENBW}}, \tag{11.26}$$

where $ENBW$ is the *equivalent noise bandwidth* of the N-element window array defined by the relation

$$ENBW = N \frac{\sum_{j=0}^{N-1} w_j^2}{\left(\sum_{j=0}^{N-1} w_j \right)^2}, \tag{11.27}$$

where w_j is the jth element of the N-element window. Once P is determined, then the sinusoidal amplitude is $A = \sqrt{P}$.

In the Editor, open the file **waveform_creator.py** in **YourNamePythonWork≫Utilities**. Add the function called `estimated_freq_amp()` shown below, which implements Eqs. (11.25), (11.26), and (11.27). Because it is easy to do, we will perform the sums in these equations over all k, but it is only necessary to include k values (by slicing the frequency and amplitude arrays) for which the a_k are significantly nonzero, that is, k in the neighborhood of the peak.

```
def estimated_freq_amp(freq_positive, amp_positive, num,
                       win):

    f = freq_positive
    a = amp_positive

    freq_est = np.sum(f*a**2)/np.sum(a**2)

    enbw = num*np.sum(win**2)/np.sum(win)**2
    amp_est = np.sqrt(np.sum(a**2)/enbw)

    return freq_est, amp_est
```

Here, the NumPy function `sum(array)` performs a sum of all of the elements in the array `array`. Save **waveform_creator.py** as you close it.

In the Editor, open the file **fft_magnitude.py** in **YourNamePythonWork**. In the statements that create GUI controls, add two `create_indicator()` functions to display the estimated frequency and amplitude as shown here:

```
var_num, var_samp_freq = gc.digitizing_control(control_box)
var_a, var_f, var_shape = gc.waveform_control(control_box)
var_window = gc.window_control(control_box)

lbl_freq = gc.create_indicator(control_box,
                    label_1='Estimated Frequency (Hz)',
                    char_width=15)
lbl_amp = gc.create_indicator(control_box,
                    label_1='Estimated Amplitude',
                    char_width=15)
```

Then, just above the return line in `calculate_plot_waveform()`, add these three lines to calculate and display the estimated frequency and amplitude:

```
freq_est, amp_est = wc.estimated_freq_amp(freq_positive,
                                          amp_positive,
                                          num, win)
lbl_freq.config(text=f'{freq_est:.2f}')
lbl_amp.config(text=f'{amp_est:.2f}')
```

Save your work. The completed **fft_magnitude.py** (with `marker='o'`) is as follows:

```
import Utilities.waveform_creator as wc
import Utilities.gui_creator as gc
import Utilities.plot_creator as pc

def calculate_plot_waveform():

    num, samp_freq = gc.get_dig_entries(var_num,
                                        var_samp_freq)
    a, f, shape = gc.get_wav_entries(var_a, var_f,
                                     var_shape)
    win_type = gc.get_window_option(var_window)

    t = wc.create_time_data(num, samp_freq)
    y = wc.create_waveform_data(t, a, f, shape)

    win = wc.create_window(win_type, num)
    y = y*win

    freq_positive, amp_positive = wc.fft_mag_only(y, num,
                                                  samp_freq)

    pc.update_plot(ax, canvas, freq_positive, amp_positive,
                   label_x, label_y, marker='o')

    freq_est,amp_est = wc.estimated_freq_amp(freq_positive,
                                             amp_positive,
                                             num, win)

    lbl_freq.config(text=f'{freq_est:.2f}')
    lbl_amp.config(text=f'{amp_est:.2f}')

    return


label_x, label_y = 'Frequency (Hz)', 'Amplitude'

root = gc.create_window(title='FFT (Magnitude Only)',
                        size='1000x500')
control_box,right_frame = gc.create_frames(root)

var_num, var_samp_freq = gc.digitizing_control(control_box)
var_a, var_f, var_shape = gc.waveform_control(control_box)
var_window = gc.window_control(control_box)

lbl_freq = gc.create_indicator(control_box,
                      label_1='Estimated Frequency (Hz)',
                      char_width=15)
lbl_amp = gc.create_indicator(control_box,
                      label_1='Estimated Amplitude',
                      char_width=15)
```

```
ax,canvas = pc.create_plot(right_frame)
pc.update_plot(ax, canvas, [], [], label_x, label_y)

gc.action_button(control_box,label='Submit',
                 action=calculate_plot_waveform)

root.mainloop()
```

With *1024* and *2000* for **Number of Samples** and **Sampling Frequency (Hz)**, and *4*, *249*, and *Sine* for **Amplitude**, **Frequency (Hz)**, and **Shape**, respectively, run **fft_magnitude.py** using various windows. I think you'll be very pleased with the estimated values of **Calculated Frequency (Hz)** and **Calculated Amplitude**!

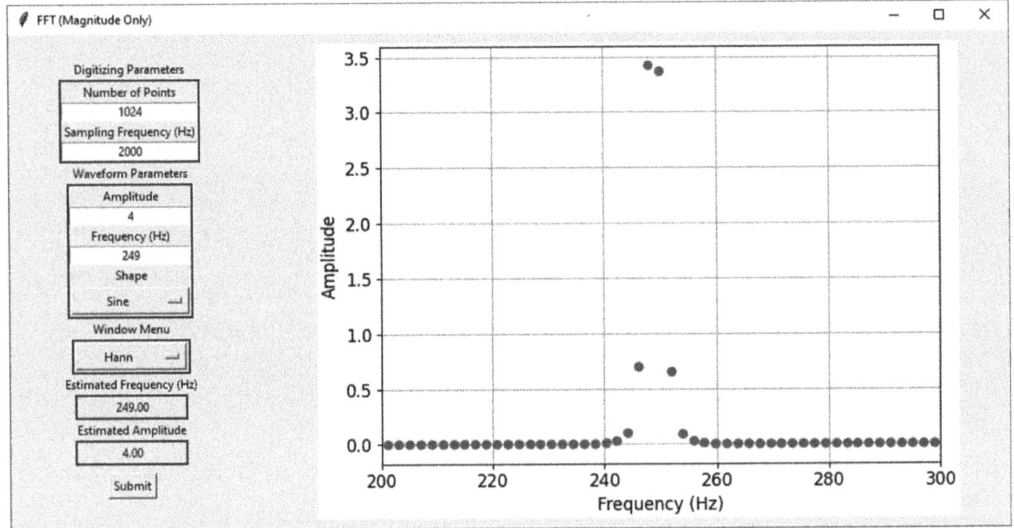

11.12 Aliasing

In preparation for this section, open **plot_creator.py** in **YourNamePythonWork≫Utilities**, then comment-out the line we placed in within the `update_plot()` function to manually scale the *x*-axis:

```
#ax.set_xlim(200, 300)
```

Save **plot_creator.py** as you close it.

Let's use **fft_magnitude.py** to demonstrate aliasing, the phenomenon in which an incoming signal is digitally sampled too infrequently, resulting in a digitized waveform whose frequency is much lower than that of the actual signal. In Chapter 6, we found that when sampling a signal of frequency f at sampling frequency f_s, the digitized waveform will also have the frequency f only if $f \leq f_{nyquist}$, where $f_{nyquist} = f_s/2$ is the Nyquist frequency. If, instead, the frequency of the incoming signal is $f > f_{nyquist}$, then the digitized waveform will have frequency f_{alias} $(\neq f)$ given by

$$f_{alias} = |f - nf_s| \qquad n = 1,\ 2,\ 3,\ \ldots, \tag{11.28}$$

where f_{alias} is in the range $0 \leq f_{alias} \leq f_{nyquist}$.

Program **fft_magnitude.py** so that the input waveform is a sine wave of amplitude 4.0. Choose your favorite window, say, the general-purpose *Hann*. In **Digitizing Parameters**, set **Number of Samples** and **Sampling Frequency (Hz)** to be *1024* and *2000*, respectively. Then we expect aliasing to occur when the frequency of our input sine wave exceeds the Nyquist frequency $f_{nyquist} = f_s/2 = 2000\,\text{Hz}/2 = 1000\,\text{Hz}$. Run **fft_magnitude.py** several times with the following sequence of input sine-wave frequencies: 1200, 1500, 1800, 2000, 2800, 3300, and 7700 Hz. Does Eq. (11.28) correctly predict the aliased frequency you observe for each input? For one of these frequencies, recall the result of sampling a sine wave at the Nyquist frequency. Try switching your input to a cosine wave at that frequency and see if you obtain the expected result.

In this exercise, we have the benefit of knowing the input signal's true frequency and so can use Eq. (11.28) to figure out why the digitizing process is producing an incorrect lower-frequency output. In a real experiment, however, where we would have no prior knowledge of the input signal's true frequency, we would have no other recourse than to assume that the digitized waveform has the same frequency as the incoming signal. That is, we would assume the signal is not being aliased in the digitizing process. To assure that this "no-aliasing" assumption is valid, it is imperative that an experimentalist first determine the digitizing system's Nyquist frequency (based on the known sampling rate) and then ensure (e.g., through the use of a lowpass filter as in the Use It! example in Chapters 7 and 8) that no signals with frequencies greater than the Nyquist frequency are input into the digitizer.

Do It Yourself

spectrum_analyzer.py Build a computer-based spectrum analyzer that digitizes an incoming voltage waveform, takes the Fourier transform of these acquired data, and displays the resulting frequency spectrum. Complete the following steps:

1. Write the **spectrum_analyzer.py** program: With **digital_oscilloscope_ni.py** (or **digital_oscilloscope_mcc.py**) open in the Editor, use **File≫Save As...** to create and save the new program in **YourNamePythonWork**. This code will be configured to perform a digitally triggered analog input voltage operation on a particular AI channel of your DAQ device (e.g., *ai0*, CH0) with a GUI that includes a **Digitizing Parameters** control, where data-acquisition parameters **Number of Samples** and **Sampling Frequency (Hz)** can be selected, along with a Matplotlib graph for use in displaying an output frequency spectrum.

   Using **fft_magnitude.py** as your guide, modify the program's code to place a **Window Menu** control and **Estimated Frequency (Hz)** and **Estimated Amplitude** indicators on the GUI as well as to carry out the following procedure: After the waveform array `y` is acquired, it is multiplied by a selected window, a FFT is taken of the windowed waveform, and the resulting frequency spectrum is displayed on the GUI graph (with proper labeling of the axes). Include code to calculate and display the estimated frequency and amplitude in the GUI indicators. Repeat these actions, after being initiated by the press of a toggling **Start** button, until the **Stop** button is pressed.
2. Operate the computer-based spectrum analyzer: Using a function generator, input a sine-wave voltage with a frequency of about 100-Hz input into the analog input channel of

your DAQ device that was specified in your program. For digital triggering, input your function generator's sync output to the DAQ device's appropriate pins. With **Number of Samples** and **Sampling Frequency (Hz)** set to *1024* and *2000*, respectively, and your choice in the **Window Menu** control, run **spectrum_analyzer.py** to see the spectral makeup of your voltage signal. How do the values of **Estimated Frequency (Hz)** and **Estimated Amplitude** compare with the known value (from the function generator's control panel) of your input signal? Try changing the input sine-wave frequency f to values within the range $0 \leq f \leq 1000$ Hz. Does the instrument perform well within this frequency range?

3. Explore interesting inputs: Increase the input sine wave's frequency beyond the Nyquist frequency to see the aliasing effect. Explain your observations using Eq. (11.28).

Try inputting a square wave with a frequency of about 180 Hz and amplitude of 1 V (i.e., oscillates between −1 and +1 V). Fourier analysis of a square wave of frequency f and unity amplitude predicts that this waveform is equivalent to the sum of sine waves

$$\frac{4}{\pi}\sum_{n \text{ odd}} \frac{1}{n} \sin\left[2\pi n f t\right] = \frac{4}{\pi}\left\{\sin\left[2\pi f t\right] + \frac{1}{3}\sin\left[2\pi\left(3f\right) t\right] + \frac{1}{5}\sin\left[2\pi\left(5f\right) t\right] + \ldots..\right\}.$$

Note that this sum of sine waves only includes the odd harmonic frequencies f, $3f$, $5f$, and so on. With a square-wave input, is the output of your program consistent with this Fourier sum? Do you note any aliasing of the square wave's higher harmonics?

Use It!

digital_filter.py In addition to the fast Fourier transform, the SciPy module offers another frequency-related signal-processing capability—digital filtering. Filtering is used to alter the frequency content of a signal. Based on the frequency range they transmit and block (called the *passband* and *stopband*, respectively), filters are classified as the following types: *lowpass, highpass, bandpass, and bandstop*. For example, a lowpass filter transmits low-frequency signals and blocks high-frequency signals, and the shift between these two regimes takes place over a range of frequencies called the *transition band* (see the Use It! example in Chapters 7 and 8). The transition band begins near the filter's cutoff frequency f_c and its width is determined by the shape of the particular filter's response curve. For each filter type, there are various possible shapes of the response curve, where each shape is optimized for a particular signal-processing purpose. For example, the popular lowpass filter shapes ("topologies") are named Butterworth, Chebyshev, Elliptic, and Bessel.

The following program allows you to explore the response curve for the Butterworth filter using the `butter()` function found in the `signals` module of SciPy. This function constructs a Butterworth filter based on the values input for its arguments. The syntax for this function is as follows:

```
sos = scipy.signal.butter(filter_order, cutoff_freq,
                          filter_type, samp_freq,
                          output='sos')
```

Here, `filter_order` is an integer that controls the width of the transition band, `cutoff_freq` is the cutoff frequency (defined to be where the output amplitude has fallen to $1/\sqrt{2} \approx 0.7$), `filter_type` can be `'lowpass'` or `'highpass'`, `samp_freq` is the sampling frequency, and `output` determines the method used to construct the filter. By choosing the `'sos'` construction method, the function returns the object `sos`, which consists of items that mathematically represent the filter. Then, the function `filtered_y = scipy.signal.sosfilt(sos,y)` applies the filter to a waveform array `y` and returns the result as the array `filtered_y`.

The program shown below gives the Butterworth filter's response curve defined to be the output signal root-mean-square (rms) amplitude as a function of frequency, assuming an input signal with a rms amplitude of 1. Using a `for` loop, a sequence of unity-amplitude digitized sine-wave signals is input to the filter, with frequencies ranging from 100 Hz up to the Nyquist frequency, and the amplitudes of the digitized sine-wave output by the filter are recorded in a Python list. The incoming sine wave is sampled for one second at a sampling frequency f_s, so the number of samples equals the numerical value of the sampling frequency (= sample per second) and the Nyquist frequency is one-half of f_s. The result is plotted on a Matplotlib graph.

```
import matplotlib.pyplot as plt
import numpy as np
import scipy.signal as sig

cutoff_freq = 5000
filter_order = 5
filter_type = 'lowpass'

start, stop = 0, 1
samp_freq = 40000
num = samp_freq
nyquist_freq = samp_freq//2

t = np.linspace(start, stop, num, endpoint=False)

amp = []
freq = list(range(100, nyquist_freq, 100))

for f in freq:

    y = np.sqrt(2)*np.cos(2*np.pi*f*t)

    sos = sig.butter(filter_order, cutoff_freq,
                     filter_type, fs=samp_freq,
                     output='sos')
    filtered_y = sig.sosfilt(sos, y)
    rms = np.sqrt(np.mean(filtered_y**2))
    amp.append(rms)

plt.xlabel('Frequency (Hz)')
```

```
plt.ylabel('Filter Output RMS Amplitude')

plt.plot(freq, amp, marker='o')
plt.show()
```

Save **digital_filter.py** as written above, where a Butterworth lowpass filter of order 5 with f_c = 5000 Hz is selected. When the program is run, the resulting plot will appear as shown here:

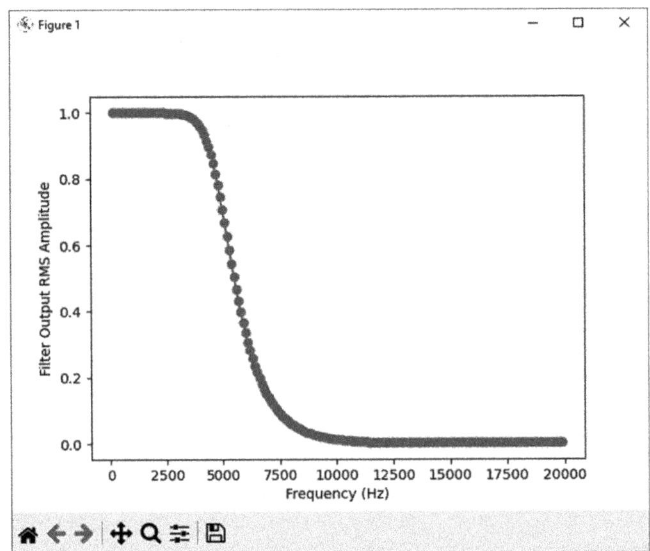

As you'll find, the Butterworth lowpass filter is designed to have a "flat" plateau in its passband (i.e., almost constant unity-amplitude output at low frequencies). Try increasing the order of the filter and to demonstrate that the transition region becomes narrower as the order is increased. By selecting `filter_type='highpass'`, a high-pass filter is created and the effect of changing the value of `cutoff_freq` can be explored. Both a `'bandpass'` and `'bandstop'` filter can be created. In these cases, `cutoff_freq` is a tuple consisting of a low and high cutoff frequency, e.g., `cutoff_freq=(5000,8000)`.

SciPy's signals module also provides functions for creating other filters that are, for example, optimized to have a narrow transition region, but as a consequence have "ripples" in their passband and/or stopband. The Elliptic filter has the narrowest transition region of all of the available filters—the selling point of that filter.

Problems

1. **Real Versus Aliased Peaks** Use **fft_magnitude.py** to investigate aliasing. In **waveform_creator.py**, program the **User-Defined** option of `create_waveform_data()` to produce the digitized signal `y` composed of four incoming sine waves of frequency f_1 = 25 Hz, f_2 = 70 Hz, f_3 = 160 Hz, and f_4 = 510 Hz, each with an amplitude of 1. That is, code the **User-Defined** option as follows:

```
y = (np.sin(2*np.pi*25*t)
     + np.sin(2*np.pi*70*t)
     + np.sin(2*np.pi*160*t)
     + np.sin(2*np.pi*510*t))
```

Note the outside parentheses are necessary to alert Python that this is a multiline equation. Save your work, then run **fft_magnitude.py** with `marker='-'`. Set **Number of Samples, Sampling Frequency, Shape,** and **Window Menu** equal to *4096, 100, User-Defined,* and *Rectangle*, respectively, then press **Submit**.

Determine the frequencies of the four peaks that you observe in the resulting spectrum. Identify each peak; that is, tell whether the peak is associated with a true frequency in the signal `y` or with an aliased frequency produced by the digitizing process. For each aliased frequency, use Eq. (11.28) to identify the true frequency (f_1, f_2, f_3, or f_4) that was aliased.

2. **Frequency Resolution** Use **fft_magnitude.py** (with `marker='-'`) to explore the improved frequency resolution that results from long-time sampling of an input signal.

   In **waveform_creator.py**, program the **User-Defined** option of `create_waveform_data()` under the assumption that the input signal `y` from an experiment consists of two closely spaced sine waves of frequency 399 and 401 Hz, each with an amplitude of 4, that is, $y = 4\sin[2\pi(399)t] + 4\sin[2\pi(401)t]$. Also, in **plot_creator.py**, zoom in x-axis scaling to the relevant region by activating the command `ax.set_xlim(380,420)` within the `update_plot()`. Then, run **fft_magnitude.py**, setting **Sampling Frequency (Hz), Shape,** and **Window Menu** equal to *2000, User-Defined,* and *Rectangle*, respectively.

   (a) Simulate sampling the input signal for longer and longer times by running **fft_magnitude.py** with the following succession of values for **Number of Samples**—128, 256, 512, 1024, 2048, 4096, 8192—each time noting whether the 399- and 401-Hz peaks are resolved in the resulting spectrum.

   (b) At what threshold value for **Number of Samples** do the two peaks begin to be resolved? From your knowledge of FFT theory presented in Section 11.3 (in particular, Eq. (11.7)), explain how this threshold value is determined by the various parameters values used in this investigation.

3. **FFT of Square Wave** Observe the Fourier transform of a square wave. Run **fft_magnitude.py**, with **Number of Samples** and **Sampling Frequency (Hz)** equal to *4096* and *3000*, and **Frequency (Hz), Amplitude,** and **Shape** equal to *180, 1,* and *Square*, respectively, to simulate a square wave with a frequency f = 180 Hz, which oscillates between −1 and +1 V. Fourier analysis predicts that this waveform is a sum of sine waves as follows:

$$\frac{4}{\pi}\sum_{n\,\mathrm{odd}}\frac{1}{n}\sin[2\pi nft] = \frac{4}{\pi}\left\{\sin[2\pi ft] + \frac{1}{3}\sin[2\pi(3f)t] + \frac{1}{5}\sin[2\pi(5f)t] + \ldots\right\}.$$

Note that the sum of sine waves only includes the odd harmonic frequencies $f, 3f, 5f, \ldots$, with amplitudes of $4/\pi$, $4/3\pi$, $4/5\pi$, ... Finally, set **Window Menu** to *Flattop* so that the Fourier peaks will have accurate amplitudes.

Determine the frequency and amplitude of every peak that you observe in the resulting Fourier spectrum. Then identify the harmonic to which each peak corresponds, including those peaks present as a result of aliasing. Comment on whether each observed peak has the amplitude expected from Fourier analysis. Finally, change **Sampling Frequency (Hz)** to *5000* to see if some of the previously aliased peak now appear at their correct frequencies.

4. **Frequency Doubling** Use **fft_magnitude.py** to explore the *frequency doubling* (and *tripling*) phenomenon that occurs when a pure sine-wave signal is input to a nonlinear detector/amplifier (e.g., radio wave on diode detector, monochromatic light on nonlinear crystal). To investigate this effect, in **waveform_creator.py** use the **User-Defined** option of `create_waveform_data()` to simulate the following process: A pure sinusoid $y(t) = A\sin(2\pi ft)$ is passed through a nonlinear amplifier producing an output signal $a(y)$ given by

$$a(y) = \alpha y + \beta y^2,$$

where α and β are constants and, in practice, α is often significantly larger than β.
Run **fft_magnitude.py** with some choice of f, α, and β (e.g., $f = 250$ Hz, $\alpha = 10$, and $\beta = 1$ with **Sampling Frequency (Hz)** equal to *2000*) to find the spectrum of $a(y)$.
   (a) In terms of f, what frequencies do you observe in the spectrum?
   (b) The y^2 term in $a(y)$ produces a product of sine waves that can be rewritten using the trigonometric identity

$$\sin\theta\sin\phi = \frac{1}{2}\left[\cos(\theta - \phi) - \cos(\theta + \phi)\right].$$

   With the help of this relation, predict the frequencies that should appear in the spectrum of $a(y)$. Are these the frequencies that you observed?
   (c) Explore what happens when the amplifier is even more nonlinear so that $a(y) = \alpha y + \beta y^2 + \gamma y^3$, with $\gamma < \beta < \alpha$. In terms of f, what frequencies do you observe now in the spectrum? In experimental optics, laser light of frequency f is commonly input on a nonlinear crystal to produce output light of frequency $2f$ and $3f$.

5. **Sidebands of AM Wave** In this problem, we will explore the *sidebands* that exist in the frequency spectrum of an *amplitude modulated* (AM) wave. The displacement y of an amplitude-modulated (AM) wave obeys the relation

$$y(t) = A\left\{1 + \sin\left[2\pi f_{\text{mod}} t\right]\right\} \sin\left[2\pi f_{\text{sig}} t\right],$$

where A is the amplitude, and f_{mod} and f_{sig} are the modulation and signal frequencies, respectively. Typically $f_{\text{sig}} >> f_{\text{mod}}$. For example, in AM radio systems, f_{sig} is a radio station's broadcasting frequency on the order of megahertz, while f_{mod} is the audio information being broadcast and so is on the order of kilohertz.
   (a) In **waveform_creator.py**, use the **User-Defined** option of `create_waveform_data()` to simulate the following AM wave with $f_{\text{sig}} = 10 f_{\text{mod}}$:

$$y(t) = \left\{1 + \sin\left[2\pi(40)t\right]\right\} \sin\left[2\pi(400)t\right].$$

   Then, to view a plot of this function, run the **waveform_simulator.py** with **Number of Samples, Sampling Frequency (Hz)**, and **Shape** equal to *1000*, *10000*, and *User-Defined*, respectively.

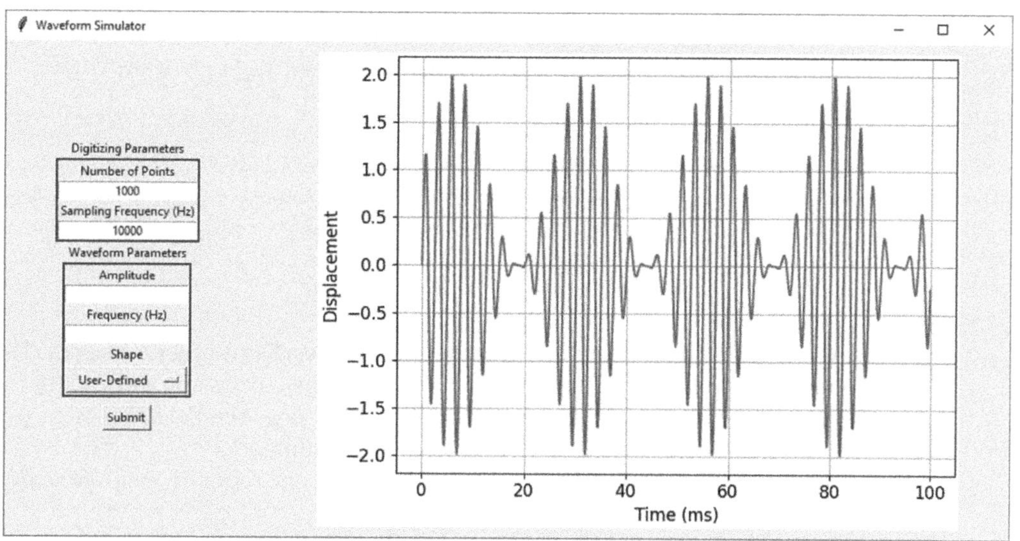

(b) Next, run **fft_magnitude.py** with the following choices: **Number of Samples** and **Sampling Frequency (Hz)** equal to *1024* and *2000*; select **Shape** and **Window Menu** equal to *User-Defined* and *Hann*, respectively. The spectrum that you observe consists of a central peak and "sidebands." In terms of f_{sig} and f_{mod}, what is the central peak's frequency and the spacing of the sidebands from the central peak? To determine the frequencies, you may wish to activate `ax.set_xlim(350,450)` in **plot_creator.py**. Comment-out that statement when you are finished.

(c) In an AM radio receiver, the AM wave received from an antenna is "demodulated" by passing it through a nonlinear diode detector. By making an appropriate modification of the **User-Defined** option in **waveform_creator.py**, simulate this demodulation method by applying a nonlinear amplification $a(y) = y^2$ to your AM wave $y(t)$. Run **fft_magnitude.py**, and identify all of the frequencies in the resultant spectrum in terms of f_{sig} and f_{mod}. You may wish to, for example, activate `ax.set_xlim(0,100)` in order to accurately determine the frequencies. Note that one of these frequencies is the modulation frequency f_{mod} and that, by proper (lowpass) filtering, it can be selected out for distribution to the radio's speaker. In optics, this technique is implemented to produce two optical laser frequencies ($\sim 10^{14}$ Hz) separated by a small "radio-frequency" ($\sim 10^6$ Hz) difference.

6. **Lock-In Detection** Given an input consisting of the sum of many AC voltages with various frequencies, a *lock-in amplifier* can selectively measure the oscillatory amplitude of a single one of those frequencies. The theory of operation of this instrument can be simulated as follows. First, assume that the "input signal" y_{sig} to the lock-in consists of the sum of three sine waves with frequencies 100, 200, and 300 Hz and amplitudes 4, 6, and 8, respectively, that is, $y_{\text{sig}} = 4\sin[2\pi(100)t] + 6\sin[2\pi(200)t] + 8\sin[2\pi(300)t]$. Then, if a user programs the lock-in to measure the amplitude of the 200-Hz sine wave, the instrument internally produces a "reference signal" $y_{\text{ref}} = 2\sin[2\pi(200)t]$ and produces the product $y = y_{\text{sig}}y_{\text{ref}}$. Finally, the instrument filters y to determine the magnitude of the DC component present in y and outputs this DC value.

(a) Use **fft_magnitude.py** to gain insight into the above-described theory. First, in **waveform_creator.py**, program the **User-Defined** option of `create_waveform_data()` to produce the given y_{sig} and y_{ref} and then the product $y = y_{sig}y_{ref}$ as shown here:

```
elif shape == 'User-Defined':
    y_sig = (4*np.sin(2*np.pi*100*t)
            + 6*np.sin(2*np.pi*200*t)
            + 8*np.sin(2*np.pi*300*t))
    y_ref = 2*np.sin(2*np.pi*200*t)
    y = y_sig*y_ref
```

Next, run **fft_magnitude.py** with **Number of Samples, Sampling Frequency (Hz),** and **Shape** equal to *1024* and *2000*, and *User-Defined*, respectively. You will find that several frequencies, including $f = 0$, are present in y. How does the DC component compare with the amplitude of the 200-Hz sine wave present in y_{sig}?

(b) Change the reference signal to $y_{ref} = 2 \sin [2\pi (300) t]$ and rerun **fft_magnitude.py**. What is the resulting DC component equal to now?

(c) From the trigonometric identity $\sin\theta \sin\phi = \frac{1}{2} [\cos(\theta - \phi) - \cos(\theta + \phi)]$, can you explain the frequencies you observe in y?

7. **phase_plot.py** Our function `wc.complex_fft()` determines the complex-amplitude associated with an input signal `y`, where the complex-amplitude contains the magnitude and phase of oscillation at frequencies up to the Nyquist frequency. As we saw in the chapter text, the magnitude of the complex-amplitude can be determined by the `abs()` function. In this problem, we will show that the phase of the complex-amplitude can be found using NumPy's `angle()` function as shown in the code:

```
import numpy as np
import Utilities.waveform_creator as wc
import matplotlib.pyplot as plt

num = 1024
samp_freq = 2000

t = wc.create_time_data(num, samp_freq)
y = np.cos(2*np.pi*250*t)

freq, complex_amp = wc.fft_complex(y, num, samp_freq)

phase = np.angle(complex_amp)

for i in range(num):
    if np.abs(complex_amp[i]) < 0.05:
        phase[i] = 0

plt.plot(freq,phase)
plt.show()
```

Here, the input signal `y` is a 250-Hz unity-amplitude cosine waveform. The `for` loop is necessary to set the phase equal to zero at frequencies where the magnitude is zero. If you run this code without the `for` loop, you will find that the phase randomly fluctuates at such frequencies.

(a) Program **phase_plot.py** using the code given above, then run it. Determine the resulting values of phase at $f = 250$ Hz for this input cosine signal (carefully note the y-axis scaling). Next, change the input signal to `y = np.sin(2*np.pi*250*t)`. Rerun the program to determine the phase at $f = 250$ Hz for this input sine signal.

(b) In the chapter text, you found that the complex-amplitude of a 250-Hz cosine and sine function at $f = 250$ Hz is $A = +2$ and $A = -2i$, respectively. Rewrite these two complex numbers in the form $A = |A|\, e^{i\theta}$ and thereby predict the phase θ expected for the FFT of a cosine and a sine function. How do these predicted values compare with the values you observed in part (a)?

(c) Finally, change the input signal to `y=np.cos(2*np.pi*250*t+np.pi/6)` and then run your program. Is the observed value for the phase at $f = 250$ Hz in agreement with the value you expect? Programming tip: Due to NumPy's element-wise manner of evaluation, the entire `for` loop can be replaced by the single command `phase[np.abs(complex_amp) < 0.05]=0`. Give this shorthand statement a try, if interested.

8. **First-Order Anti-Aliasing Filter** (a) Using a function generator, input a square-wave voltage with a frequency of about 180 Hz input into the first-order anti-aliasing filter (see the Use It! example of Chapter 7 and 8) shown in Fig. 11.5 with $R = 15$ kΩ and $C = 0.015$ μF, whose cutoff frequency is given by $f_c = 1/2\pi RC$. Attach the output of this filter to the analog input channel of your DAQ device that was programmed into **spectrum_analyzer.py** (i.e., the program you wrote in this chapter's Do It Yourself project). With **Number of Samples** and **Sampling Frequency (Hz)** set to *1024* and *2000*, respectively, run **spectrum_analyzer.py** to see the spectral makeup of your filtered voltage signal.

(b) Calculate f_c. Given this frequency, what harmonics of the square wave do you expect to observe and what should be the ratio of their amplitudes compared with the first harmonic (see the discussion in this chapter's Do It Yourself project)? Which harmonics do you actually observe and what are their amplitudes (in comparison with the amplitude of the first harmonic)? Comment on any deviation between your expectations and your observations.

(c) Are any of the observed peaks caused by the aliasing of harmonics with frequencies higher than $f_{nyquist}$? If so, should you increase or decrease C in your filter to eliminate these peaks? (You might enjoy repeating this exercise with the eighth-order anti-aliasing filter given in the Use It! example of Chapter 7 and 8; make $f_c \approx 650$ Hz by choosing $C = 0.5$ nF.)

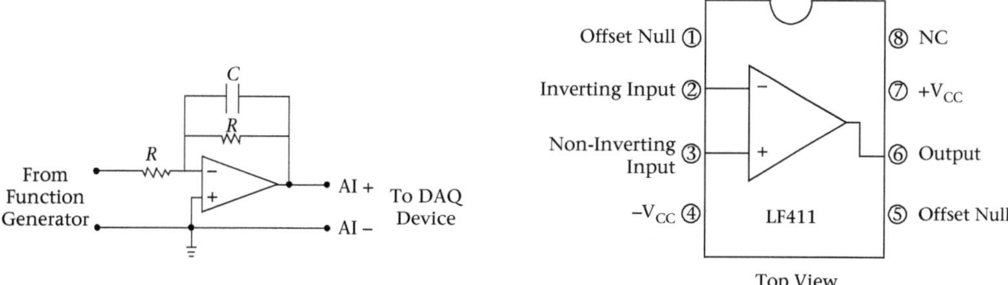

Fig. 11.5 Lowpass filter for Problem 8. Suggested op-amp is LF411.

CHAPTER 12

Control of Stand-Alone Instruments Using VISA

To perform the exercises in this chapter, you must have a stand-alone instrument equipped with a Universal Serial Bus (USB) interface and/or General Purpose Interface Bus (GPIB), ideally a digital multimeter (DMM) that is software-compatible with the Keysight/Agilent 34410A (or 34401A) DMM. For USB communication, only an appropriate USB cable is needed to connect your instrument and computer's USB port. For GPIB communication, you must also have a NI GPIB device connected to your computer. This device might be a PCIe-GPIB board plugged into a PCI Express expansion slot and linked to your instrument via a GPIB cable or a GPIB-USB device attached between your instrument and a USB port on your computer. Also, make sure that the device driver **NI-VISA** is installed.

If, while performing the chapter exercises, the communication between the interface bus and instrument breaks down (e.g., caused by an accidental error in your programming), communication can often be restored by turning the instrument off and then on again, unplugging and then replugging the cabling, or restarting your computer.

In addition, a few commands in this chapter (which do not possess parentheses) are too long to fit on one line of the book. For those commands, we will use Python's method of *explicit line continuation* in which a backslash character (\) allows the command to be split over two lines. As an example of this method, the command

```
resource_name = 'USB0::0x0957::0x0607::MY47022122::INSTR'
```

can be equivalently written via explicit line continuation as

```
resource_name = \
          'USB0::0x0957::0x0607::MY47022122::INSTR'
```

We previously learned that, for a command that possesses parentheses (e.g., a function), that command can be continued onto multiple lines through the use of the parentheses, a method called *implicit line continuation*.

The Python-Based Laboratory. John Essick, Oxford University Press. © John Essick (2024). DOI: 10.1093/9780191998478.003.0012

12.1 VISA-Based Instrument Control Using PyVISA

In previous chapters, you used Python software to transform a personal computer (connected to an appropriate DAQ device) into several handy laboratory instruments. In particular, you programmed this system to become a DC voltmeter, digital oscilloscope, waveform generator, spectrum analyzer, digital thermometer, and frequency meter. Pause to consider the following tantalizing prospect: Perhaps the only instrument required in a modern-day laboratory is a DAQ device-equipped computer controlled by Python software. That is, by simply writing a collection of appropriate programs, it might be possible for you—the cutting-edge experimentalist—to satisfy all of your laboratory instrumentation needs with this single Python-based data-acquisition and generation system. This system's tremendous flexibility would then obviate the need to purchase an expensive collection of stand-alone electronic equipment such as power supplies, function generators, picoammeters, spectrum analyzers, and oscilloscopes.

The functioning programs that you have written in previous chapters demonstrate that the above "tantalizing prospect" can, at least in certain situations, be realized. But don't discard your stand-alone instruments just yet. The timing, speed, sensitivity, and simultaneous data-taking requirements of many contemporary research experiments are beyond the capabilities of your DAQ device. For instance, although the Python-based digital oscilloscope we constructed worked well for observing audio-range frequencies (less than 20 kHz), it would prove miserably inadequate at displaying the several nanosecond-wide voltage pulses emanating from a silicon photomultiplier. In this latter situation, a stand-alone digital oscilloscope with a very fast analog-to-digital converter (on the time scale of several gigasamples per second) will do the job nicely. Thus, stand-alone instruments play a central role in state-of-the-art research, and so it might not surprise you to find that they, too, fall under the scope of Python.

Over the past several decades, a message-based communications standard has evolved by which stand-alone instruments can be software-controlled using a personal computer. In this communications scheme, a particular instrument obeys an array of manufacturer-defined ASCII character commands that represent all the possible ways of manually pressing buttons, turning dials, and viewing output data on its front panel (recall that ASCII characters are the first 128 characters of Unicode). Although the hardware conduit (called an *interface bus*) through which these ASCII messages are passed between the PC and laboratory instrument can take on various guises (including RS-232, GPIB, LAN/Ethernet, and USB), there is a universal library of software commands available to control this communication process. This communication standard is named *VISA* (short for Virtual Instrument Software Architecture) and its commands are accessed in Python programming through a package called *PyVISA*.

To carry out VISA-based communication with a stand-alone instrument, a VISA software driver is required. Various instrument manufacturers provide their own version of this low-level software, the most common of which is *NI-VISA*, but there is also Keysight I/O Layer, Tektronix TekVISA, Agilent I/O Library Suite, and Rohde & Schwarz VISA. In addition, there is the noncommercial Python-based driver *PyVISA-py*. Each of the commercial low-level drivers is designed to operate with the proprietary high-level programming of its manufacturer. For example, NI-VISA (which can be downloaded for free from the NI website) is designed to work directly with NI's LabVIEW programming language. Then, a LabVIEW program that acquires data through the control of a stand-alone instrument follows the path from high-level LabVIEW command to low-level VISA control as shown:

Here, high-level commands within a LabVIEW program are internally converted to proper C-based coding, which is input to the NI-VISA driver software. Hidden from the user, LabVIEW itself provides the needed interfacing between the high-level code and NI-VISA, facilitating user-friendly control of the stand-alone instrument.

With the general-purpose programming language Python, "under-the-hood" interfacing between high-level Python commands and NI-VISA is not built-in. Thus, an additional interfacing layer, termed a *Python binding*, is required. This Python binding is a software package named *PyVISA*, which is available as a free download from the Python Package Index (PyPI). When using Python to control a stand-alone instrument, the chain from high-level Python command to low-level VISA control follows the path shown next:

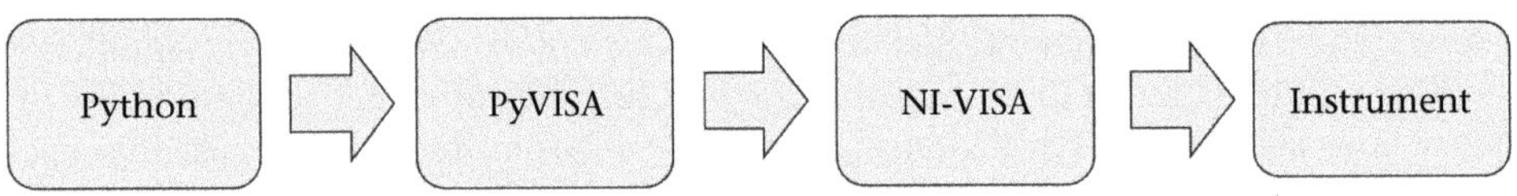

Here, the PyVISA software package provides a library of high-level Python-based commands, which can be called from a Python program. This library of commands covers the full range of available NI-VISA options. When one of these commands is called, PyVISA coverts the Python code to proper C-based coding understood by NI-VISA, and inputs this converted code to the NI-VISA driver software. NI-VISA then communicates with the stand-alone instrument to carry out the desired operation.

For this chapter, I've chosen to use the NI version of VISA. Thus, for our study of the Python control of a stand-alone instrument, two free software packages are required—NI-VISA and PyVISA.

- To download NI-VISA, go to https://www.ni.com/downloads and search for NI-VISA. Download the version of NI-VISA for your computing system's platform—Windows, Mac OS, or Linux. PyVISA adapts to all of these versions.
- To download PyVISA, use `pip` at the Command Prompt as follows:

```
pip install pyvisa
```

Mac users: change `pip` to `pip3`. Full information on these two packages and how to download them can be found by searching "PyVISA: Control your instruments with Python".

In this chapter, you will learn how to use PyVISA to control the message-based communication between a stand-alone instrument and your computer. You will explore generic features of this communication process such as the *Standard Commands for Programmable Instruments (SCPI)* language and various synchronization methods while writing code that controls a particular type of stand-alone instrument—a *Digital Multimeter* (*DMM*)—using two interface buses, the *Universal Serial Bus (USB)* and the *General Purpose Interface Bus (GPIB)*.

12.2 VISA Session

When using the VISA standard to facilitate message-based communication between a computer and a particular stand-alone instrument, the instrument is termed a *VISA resource* and the communication activity is called a *VISA session*. To *query* a VISA resource (i.e., send it a command and then receive back its response), the required VISA session consists of the following four steps: *open the session, write the command message to the resource, read the response from the resource, close the session*. In PyVISA, these four given steps are implemented through the use of two Python classes: `ResourceManager()` and `Resource()`. As with all classes, `ResourceManager()` and `Resource()` can be used to create objects, which possess attributes and can be operated on by associated methods.

To open a VISA session, the following two-step process involving `ResourceManager()` is used:

```
rm = pyvisa.ResourceManager()

instrument = rm.open_resource(resource_name)
```

The first command creates an instance (i.e., an object) of the `ResourceManager()` class named `rm`, while the second line implements a method of the `ResourceManager()` class called `open_resource()`, which begins a VISA session between your computer and the resource identified by the string `resource_name`. The `resource_name` string consists of text that specifies the interface type being used (e.g., USB or GPIB), followed by resource information (such as the instrument's interface address, manufacture ID, and/or serial number), and ending with the *resource type*. For our work, the resource type will be a stand-alone instrument denoted by *INSTR*. The VISA session that has been opened is the object named `instrument`, which is an instance of the `Resource()` class appropriate to the interface type that has been specified in `resource_name`.

Now that the VISA session named `instrument` has been opened, the write and read steps of a query can be carried out using the following `write()` and `read()` methods of the `Resource()` class:

```
    instrument.write(command)

    response = instrument.read()
```

In this two-step write–read operation, the `write()` method performs the transfer of the ASCII message `command` from your computer to the stand-alone instrument. This ASCII string is one of the commands recognized by the instrument and, when received by the instrument, configures it properly for a desired data-taking measurement. After the instrument completes the operation requested by `command`, the `read()` method transfers the ASCII message response from the stand-alone instrument into your computer's memory and the message is assigned to the object named `response`. This ASCII string typically contains the results of a data-taking measurement performed by the instrument.

Alternatively, the two-step write–read operation can be carried out via the single `query()` method:

```
    response = instrument.query(command)
```

However, throughout our work, we will use the two-step write–read operation to perform a query in order to accommodate synchronization methods (to be discussed later in the chapter).

Finally, the resource manager and VISA session named `rm` and `instrument`, respectively, are closed by the `close()` method of the `ResourceManager()` and `Resource()` classes:

```
instrument.close()
rm.close()
```

An example of the four-step VISA session to query an instrument is accomplished by this code:

```
import pyvisa

rm = pyvisa.ResourceManager()
instrument = rm.open_resource('GPIB0::22::INSTR')

instrument.write('*IDN?')
response = instrument.read()

print(response)

instrument.close()
rm.close()
```

In this example, the message **IDN?* is sent over a GPIB interface to an instrument at address 22. **IDN?*, a command recognized by most instruments, instructs the instrument to identify itself. After receiving this command, the instrument's response (which is an ASCII string consisting of identification information) is received by the computer over the GPIB and displayed in the Shell.

Because VISA-based programming is so robust, you can write highly dependable data-taking programs with just the information already presented. However, with a bit more grounding in the message-based communication scheme, you'll be able to create programs in which you can have near-total confidence. The following sections will take you to the next level of sophistication in stand-alone instrument control.

12.3 IEEE 488.2 Standard

When remote control of laboratory instruments first became possible, there was a chaotic period during which, more or less, each instrument manufacturer defined its own communications protocol through a unique blend of parallel and serial modes, positive and negative polarities, and assorted handshaking signals. In 1965, Hewlett Packard (which then became Agilent and now Keysight) ended this cacophony by designing a universal instrument interface called the Hewlett Packard Interface Bus (HP-IB) and offered it as the only option on all of its new computer-programmable instruments. Because of its (for that era) high transfer rates, HP-IB quickly gained popularity with other instrument manufacturers and, in 1975, was accepted as an industry-wide standard known as IEEE 488 or, more commonly, the General Purpose Interface Bus (GPIB). In 1987, an improved version of this standard called *IEEE 488.2* was adopted, which enhanced and strengthened message-based communication by specifically defining an instrument's minimally required communication capabilities, a protocol for message exchange, a generic set of commonly needed commands, and a status-reporting system. Today, most computer-controlled laboratory instruments are IEEE 488.2 compliant, even those that communicate over interface buses other than the GPIB (e.g., USB and LAN/Ethernet).

12.4 Common Commands

One important innovation of the IEEE 488.2 standard was the introduction of a standardized set of *common commands* for the many generic operations that all instruments must perform. The common commands begin with asterisks to delineate them from the other device-specific commands recognized by a particular instrument. All IEEE 488.2 compliant instruments, at the very least, are required to recognize the subset of thirteen common commands given in Table 12.1. Many of these commands are related to the reporting of events using two status registers called the *Status Byte Register* and *Standard Event Status Register*, which will be described in detail starting in Section 12.5.

12.5 Status Reporting

Another IEEE 488.2 innovation is a standardized scheme for *status reporting*. This status report system is available to inform you of significant events that occur within each instrument connected to an interface bus. In this scheme, an instrument is equipped with two status registers called the *Standard Event Status Register* (*SESR*) and the *Status Byte Register* (*SBR*). Each bit in these registers records a particular type of event that may occur while the instrument is in use, such as an execution error or the completion of an operation. When the event of a given type occurs, the instrument sets the associated status register bit to a value of 1, if that bit has previously been enabled (as described in the following paragraphs). Thus, by reading the status registers, you can tell what events have transpired.

Table 12.1 Common Commands for IEEE488.2 Compliant Instruments

Mandatory common commands	Function
*IDN?	Reports instrument identification string.
*RST	Resets instrument to known state.
*TST?	Performs self-test and reports results.
*OPC	Sets operation complete (OPC) bit in SESR upon completion of the command.
*OPC?	Returns "1" to the output buffer upon completion of the command.
*WAI	Waits until all pending operations complete execution.
*CLS	Clears status registers.
*ESE	Enables event-recording bits in SESR.
*ESE?	Reports enabled event-recording bits in SESR.
*ESR?	Reports value of SESR.
*SRE	Enables a SBR bit to assert the SRQ line.
*SRE?	Reports SBR bits that are enabled to assert the SRQ line.
*STB?	Reports the contents of the SBR.

The Standard Event Status Register, which is schematically shown here, records eight types of events that can occur within a data-taking instrument.

Standard Event Status Register (SESR)

7	6	5	4	3	2	1	0
PON	URQ	CME	EXE	DDE	QYE	RQC	OPC

The eight events associated with the eight bits of the SESR are described in Table 12.2. In our work, the OPC bit will be most useful.

The SESR exists as an event-signaling tool for you to use in your programs. However, this status register completely lacks initiative and will not perform any work unless you request it to do so. Thus, when initiating communications with an instrument, one of the messages that you may wish to send is an instruction that activates the subset of event-reporting SESR bits that are of interest to you. For instruments that conform to the IEEE 488.2 standard, this activation process is accomplished via the **ESE* (Event Status Enable) command. For example, suppose you wish the QYE bit to be activated and thus record any execution errors in the SESR's bit 2. Remembering that the bit count starts at 0, bit 2 is switched on by the byte $00000100_2 = 4_{10}$. Thus, the QYE bit can be activated by performing a VISA Write of the ASCII command **ESE 4* to the instrument. In our work to come, we will activate the OPC bit (which is the SESR's bit 0) with the command **ESE 1*, since $00000001_2 = 1_{10}$.

The Status Byte Register, which is schematically shown next, records whether data are available in the instrument's output buffer, whether the instrument requests service, and whether the SESR has recorded any events.

Status Byte Register (SBR)

7	6	5	4	3	2	1	0
—	RQS	ESB	MAV	—	—	—	—

The functions of the eight bits of the SBR are described in Table 12.3. The SBR bits are studious, performing their status-reporting duties without need of a request from you.

Table 12.2 Eight Standard Event Status Register (SESR) Events

Bit	Associated events of SESR
7 (MSB)	**PON** (Power on): Instrument was powered off and on since the last time the event register was read or cleared.
6	**URQ** (User request): Front-panel button was pressed.
5	**CME** (Command error): Instrument received a command with improper syntax.
4	**EXE** (Execution error): Error occurred while the instrument was executing a command.
3	**DDE** (Device error): Instrument is malfunctioning.
2	**QYE** (Query error): Attempt was made to read the instrument's output buffer when no data were present, or a new command was received before previously requested data had been read from the output buffer.
1	**RQC** (Request control): Instrument requests to be controller.
0 (LSB)	**OPC** (Operation complete): All commands prior to and including an *OPC command have been executed.

Table 12.3 Functions of Eight Standard Byte Register (SBR) Bits

Bit	Function of SBR Bit
7 (MSB)	May be defined for use by instrument manufacturer.
6	**RQS** (Request service): The instrument has asserted a SRQ because it requires service from the interface bus controller.
5	**ESB** (Event status bit): An event associated with an enabled SESR bit has occurred.
4	**MAV** (Message available): Data are available in the instrument's output buffer.
3–0	May be defined for use by the instrument manufacturer.

An instrument can be configured to assert a *Service Request (SRQ)* in response to either of two events—an event detected by the Standard Event Status Register or the presence of previously requested data in the instrument's output buffer (i.e., the assertion of the Event Status Bit (ESB) or Message Available (MAV) bit, respectively). This configuration process is accomplished on IEEE 488.2 instruments using the **SRE* (Service Request Enable) command. For example, if you wish an event detection by the SESR to trigger a request for service by the instrument, initialize the instrument by writing the ASCII command **SRE 32* to the instrument. Remembering that the bit count starts at 0, since $00100000_2 = 32_{10}$, the setting of the SBR's bit 5 (ESB) will then be the criterion for the instrument asserting a SRQ. If, instead, you wish the presence of data in the output buffer (signaled by the MAV bit being set) to trigger a SRQ, then write **SRE 16*, since $00010000_2 = 16_{10}$. Finally, **SRE 0* will disable the instrument's ability to assert a SRQ.

The relationship between the Standard Event Status Register, the instrument's output buffer, and the Status Byte Register (along with the common commands that configure and query each) is illustrated in Fig. 12.1.

As an alternative to the use of the SRQ, *serial polling* is a common method for determining the status of an instrument. In a serial poll process, the interface bus queries an instrument, and the instrument responds by returning the value of the bits in its Status Byte Register. After opening a VISA session called `instrument`, the method `read_stb()` allows one to easily carry out a serial poll and acquire the value of the status byte formatted as a one-byte integer. The form of the command is shown here:

```
status_byte = instrument.read_stb()
```

Then, for example, by checking the value of STB's bit 5, you can determine when an event has occurred (i.e., when the value switches from 0 to 1).

12.6 Device-Specific Commands

Finally, each stand-alone instrument is designed for a specialized purpose and has its own idiosyncratic methods for accomplishing its objectives. Thus, every programmable instrument comes with a set of *device-specific commands* that allow the user to control its functions remotely and to transfer the information it produces into a computer's memory. The array of device-specific commands for an instrument is listed in its user manual. This set of commands is defined by the instrument's maker . . . and therein lies a problem. When surveying the user manuals for programmable instruments of varying models and manufacturers, you

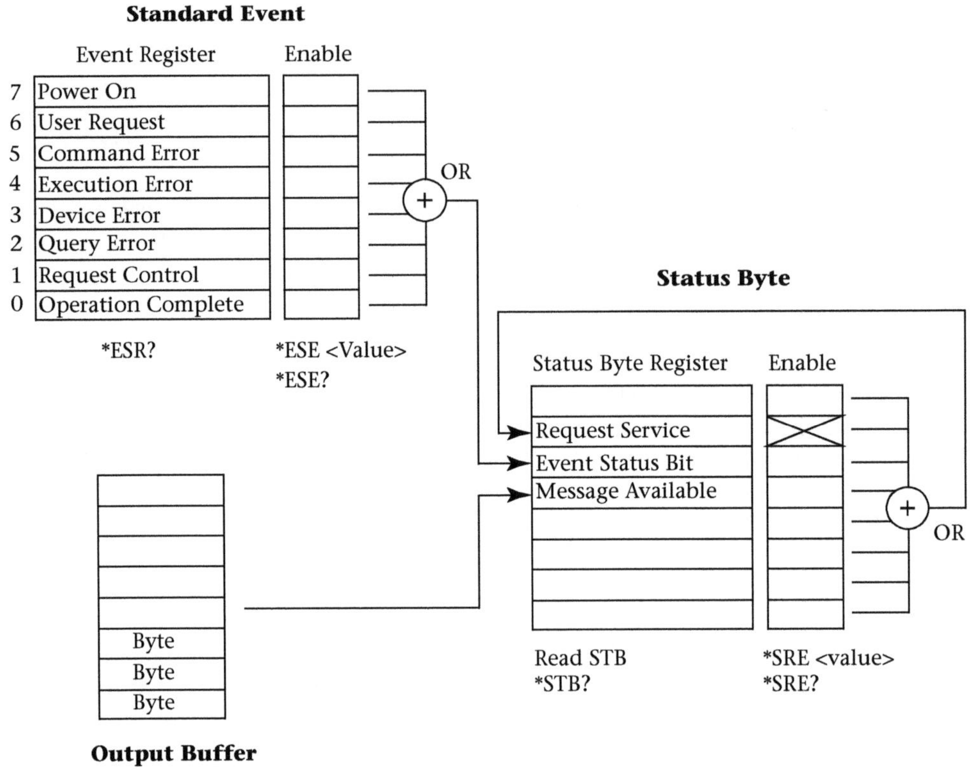

Standard Event

Event Register Enable

7 Power On
6 User Request
5 Command Error
4 Execution Error
3 Device Error
2 Query Error
1 Request Control
0 Operation Complete

OR

*ESR?

*ESE <Value>
*ESE?

Status Byte

Status Byte Register Enable

Request Service
Event Status Bit
Message Available

OR

Read STB
*STB?

*SRE <value>
*SRE?

Byte
Byte
Byte

Output Buffer

Fig. 12.1 Relationship between Standard Event Status Register (SESR) and Status Byte Register (SBR).

will find a great diversity in the style of the various command sets. Some (especially those associated with older model instruments) are an alphabetized collection of cryptic one- or two-character strings (the designers' thinking was obviously "shorter commands yield quicker and, therefore, better computer-instrument communication"). At the other extreme are the user-friendly sets, with similar commands logically grouped, each represented by an easy-to-read-and-remember mnemonic.

As programmable instruments have come into wider use, it has become apparent that development costs and unscheduled delays can be diminished markedly by simplifying the instrument programmers' task whenever possible. Thus, user-friendly device-specific command sets are the rule, rather than the exception, for instruments currently being manufactured. Commonly, these command sets are organized in a hierarchical *tree structure*, similar to the file system used in computers. Each of an instrument's major functions—such as **TRIG**ger, **SENS**e (alternatively, **MEAS**ure), **CALC**ulate, and **DISPlay**—defines a *root*, and all commands associated with that root form its *subsystem*. So, for example, to configure the Keysight/Agilent 34410A Digital Multimeter to measure a DC voltage whose value is expected to fall within the range of ± 10 V (an action within its **SENS**e subsystem), the appropriate command is

```
SENSe:VOLTage:DC:RANGe 10
```

Here, *SENSe* is the root keyword and colons (:) represent the descent to the lower-level *VOLTage*, then *DC*, then the lowest-level *RANGe* keywords. Finally, *10* is a parameter associated with *RANGe*. Although the full command given here can be sent to the instrument, it is only absolutely necessary to send the capitalized characters.

In 1990, a consortium of equipment manufacturers defined the *Standard Commands for Programmable Instruments* (abbreviated *SCPI* and pronounced "skippy") in an effort to standardize the device-specific command sets of computer-controlled instrumentation. Although this standard has not been universally adopted, it is not uncommon to discover that your post-1990 instrument is SCPI compliant. As a means of categorizing generally applicable command groups, the SCPI standard posits the following model for a generic programmable instrument. An instrument that performs measurements on an input signal is assumed to have the root functions shown in Fig. 12.2. Here, for example, **SENSe** includes any action involved in the actual conversion of an incoming signal to internal data, such as setting the range, resolution, and integration time, while **INPut** consists of actions that condition the signal prior to its conversion, such as filtering, biasing, and attenuation.

Alternatively, an instrument that generates signals is modeled by Fig. 12.3.

The SCPI command set is organized in a hierarchical tree structure using the syntax illustrated above by the *SENSe:VOLTage:DC:RANGe 10* command. In this command format, the sequence of colon-separated keywords is called the *header*, which is separated by a *<Space>* from the *parameter* portion of the command. For commands with more than one parameter, the parameters are separated by commas. You'll learn more about the SCPI command syntax as you work your way through this chapter. But maybe now is a good time to dive in and actually control a stand-alone instrument.

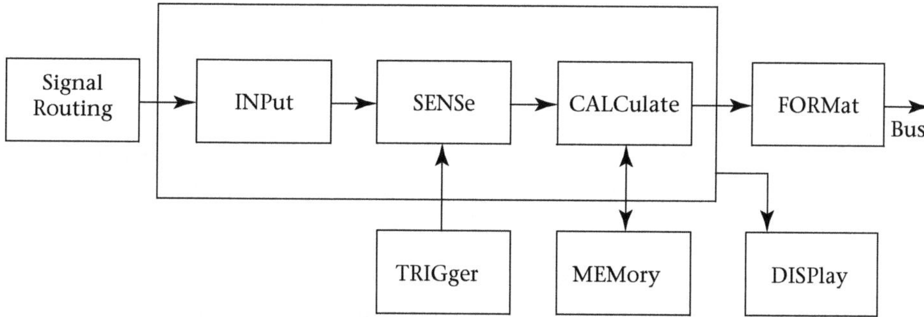

Fig. 12.2 Signal measurement instrument.

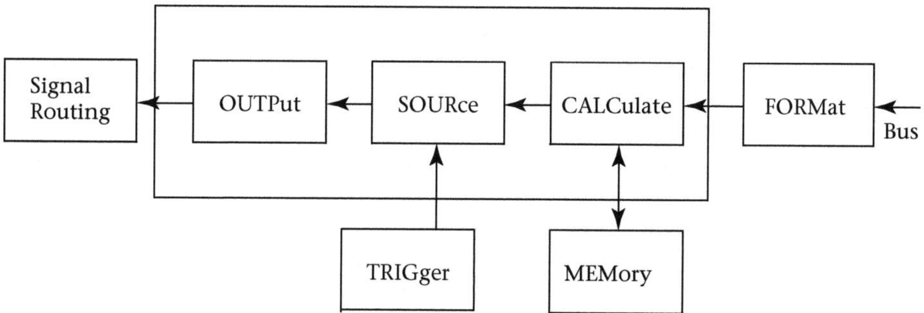

Fig. 12.3 Signal generation instrument.

12.7 Specific Hardware Used in this Chapter

In designing this chapter, I faced the following problem. There are thousands of computer-controllable stand-alone instruments available for purchase from the myriad of worldwide scientific instrument makers. Each of these instruments is capable of communicating using one (or many times, several) of the available interface buses. I have a small subset of these instruments in my laboratory and I can use them there to practice the art of message-based communication. You also, hopefully, have a small subset of such instruments available to practice with in your own laboratory. What's the problem? Well, because of the high cost and specialized nature of such equipment, the probability that my subset and your subset have some common instrument is most likely very small. The unfortunate thing about this situation is that each stand-alone instrument is designed to take specialized measurements and understands its own unique set of ASCII commands (which are defined by its maker and are listed in its user manual). Thus, before attempting to control a particular instrument using an interface bus, the programmer must have a detailed understanding of the measurement that the instrument is designed to take, the procedure that it implements in doing its work, and the command set that it recognizes. All of these considerations greatly constrain the writing of a set of generic laboratory exercises that everyone can perform.

That said, I still was faced with the fact that I had to choose a particular instrument and interface bus to work with in this chapter's exercises. With regard to the interface bus, the USB is a must. It is the interface you will almost certainly encounter in your computer-based laboratory work and is currently a widely used interface bus for new laboratory equipment. Although GPIB rarely comes standard anymore, there are an estimated 10 million GPIB-equipped instruments in use in research and industry worldwide, and so the GPIB warrants some attention (we'll see that VISA makes that easy). Configuring a RS-232 and LAN/Ethernet interfaces involves some distracting details and so I'd like to not get sidetracked with those details in this presentation.

For the following reasons, a Keysight Digital Multimeter (DMM) has many features that make it ideal for our work and so I have chosen it as our "device to practice with" in this chapter's exercises. The particular model I've chosen has gone through several iterations, starting in 1991 as the Hewlett Packard 34401A, then evolving to the Agilent/Keysight 34410A, and now the Keysight 34461A. My reasons for making this choice are as follows: First, this instrument measures voltage, current, and resistance—vanilla-flavored quantities that require no specialized knowledge to understand (unlike, for instance, the control of grating angle and slit size in a spectrometer). Second, the various iterations of this instrument are equipped to communicate via the USB and/or GPIB interface, so no matter which vintage, either USB or GPIB control will be possible. Third, for such a high-quality instrument standardly equipped with several interfaces, its price tag of approximately $1500 makes it affordable (at least by scientific instrument standards). Every lab should have one and many do! Fourth, this instrument is both IEEE 488.2 and SCPI compliant and so we will be able to use it to gain experience with generic features of widely adopted communication schemes such as SCPI command syntax and IEEE 488.2 status reporting. Better still, the original Hewlett Packard 34401A was the longtime industry-standard DMM. Thanks to the manufacturer's interest in maintaining backward compatibility between latter versions of this instrument, the commands I have chosen to use in this chapter work with all of the Hewlett Packard/Agilent/Keysight model numbers mentioned above. And even better still, because the HP 34401A was the long-standing industry standard digital multimeter, DMMs from many other manufacturers have been designed with the option to be controlled using the 34401A command set.

Examples of such instruments are the Rigol DM3058/3058E, Keithley 2100/DMM6500, Fluke 45/8845A/8846A, and Tektronix DMM 4040/4050. Thus, a wide variety of DMM owners will be able to carry out this chapter's exercises on their instrument using either the USB or GPIB interface.

In the best circumstance, a DMM that can be controlled with the HP 34401A command set is already available for your use, or, with a modest investment, you can purchase this worthwhile instrument. Then, without need for modification, you can straightforwardly work your way through the given exercises to learn the basics of message-based communication. If, instead, you have some other interface-equipped instrument available, try reading the following pages to understand the generic issues being investigated. Then, by consulting your instrument's user manual, it may be fairly easy, for instance, to use the interface-appropriate VISA resource name and substitute an ASCII command string here and there to adapt the exercises to your particular instrument and interface bus. If neither of the above describes your situation, simply read through the following pages. I believe you will learn some valuable features of instrument control that will serve you well in future work.

12.8 Instrument Resource Name and Identification String

To carry out the exercises in this chapter, you must have a stand-alone instrument equipped with a USB and/or GPIB interface. To communicate over the GPIB, you must also have a NI GPIB controller (e.g., PCI-GPIB or PCIe-GPIB) connected properly to your computer, which is in turn connected to the stand-alone instrument by a GPIB cable. To use the USB, simply connect an appropriate USB cable between the stand-alone instrument's USB connector and a USB port on your computer. Additionally, the NI-VISA software driver must be correctly installed, along with PyVISA.

To verify that these conditions are met, in the Shell, enter the following sequence of PyVISA commands, where the `list_resources()` method of the `ResourceManager()` class returns a tuple listing all of the instruments connected to your computer system's interfaces:

```
>>> import pyvisa
>>> rm = pyvisa.ResourceManager()
>>> rm.list_resources()
('ASRL1::INSTR', 'USB0::0x0957::0x0607::MY47022122::INSTR')
```

Above, the tuple on the final line results when an Agilent 34410A DMM is connected to my computer via its USB interface. The tuple's first entry is the COM 1 port with the VISA resource name `ASRL1`. The second entry is the VISA resource name for the DMM, which follows the syntax for a USB-interfaced instrument:

```
USB0::manufacturer ID::model code::serial number::INSTR
```

`USB0` identifies *USB Board 0*, while the `manufacturer ID` and `model code` are hexadecimal code numbers corresponding to *Agilent* and *34410A*, respectively. `INSTR` is the resource type, i.e., an instrument. If using a USB interface, the VISA resource name for your particular instrument will conform to this syntax.

The Agilent 34410A DMM also has a GPIB interface. When that interface is used instead, the same sequence of commands gives the following result:

```
>>> import pyvisa
>>> rm = pyvisa.ResourceManager()
>>> rm.list_resources()
('ASRL1::INSTR', 'GPIB0::22:INSTR')
```

This time, the VISA resource name for the DMM follows the syntax for a GPIB-interfaced instrument:

```
GPIB0::<GPIB address>::INSTR
```

`GPIB0` identifies the *GPIB 0* controller, `<GPIB address>` gives the instrument's assigned GPIB address. Again, `INSTR` is the resource type. Because up to fifteen instruments can be connected to a single GPIB controller, each instrument has an identification number called its *GPIB address*. A GPIB address can be any integer between 0 and 30 and is typically defined via a hardware DIP switch setting within the instrument or a sequence of button-pressing and/or knob-turning on its front panel. The instrument's user manual will describe the method for setting its address. Above, we see that my DMM's GPIB address is *22*. If using a GPIB interface, the GPIB address for your particular instrument will possibly differ.

Let's write a simple program to query the DMM. In the Editor, use **File≫New File** to create a new program and save it under the name **query_instrument_id.py** in **YourNamePythonWork**. Enter the following code, using the resource name that you found a few moments ago for your particular instrument on the particular communication interface (USB or GPIB) you're using. Note that explicit line continuation has been used to render the second command because it was too long for the book's page width and it contains no parentheses. As you write your program, however, you may find that your Editor allows you to fit that command on one line (i.e., explicit line continuation may not be needed).

```
import pyvisa

resource_name = \
             'USB0::0x0957::0x0607::MY47022122::INSTR'

rm = pyvisa.ResourceManager()
dmm = rm.open_resource(resource_name)

dmm.write('*IDN?')
response = dmm.read()

print(response)

dmm.close()
rm.close()
```

This program first opens a VISA session named `dmm` for the instrument identified by `resource_name` (i.e., the Agilent 34410A DMM). Then, the write-then-read action of a query operation is carried out, where in this case, the ASCII string `*IDN?` is written over the USB interface to the DMM, and then the instrument's ASCII response is read back over the USB interface to the computer and assigned to the object named `response`. The IEEE 488.2 common command **IDN?* instructs an instrument to identify itself via an identification string.

This received identification string is printed and then the VISA session and resource manager are closed.

Save your work and then run **query_instrument_id.py**. If your interface communication is configured properly, the identification string received from the instrument will appear in the Shell; my result is shown here:

```
Agilent Technologies,34410A,MY47028136,2.35-2.35-0.09-46-09
```

For the Agilent 34410A DMM, this identification string identifies the instrument's manufacturer, model, and serial number, followed by some integers that denote the version numbers of installed firmware that controls the multimeter's internal microprocessors. Note that this identification string is fifty-nine characters long.

Before leaving this section, let's move all PyVISA-related operations to a module in **YourNamePythonWork≫Utilities**. In the Editor, select **File≫New File** and save it under the name **visa_operations.py** in **YourNamePythonWork≫Utilities**. Then add the following code, which defines functions to open a VISA session for a generic instrument named `instr`, write a command to the instrument, read the response from the instrument, and close the VISA session.

```
import pyvisa

def open_session(resource_name):

    rm = pyvisa.ResourceManager()
    instr = rm.open_resource(resource_name)

    return rm, instr


def write_instrument(instr, command):

    instr.write(command)

    return


def read_instrument(instr):

    response = instr.read()

    return response


def close_session(rm, instr):

    instr.close()
    rm.close()

    return
```

Save your work and close **visa_operations.py**.

Open **query_instrument_id.py** in **YourNamePythonWork** and modify it as shown below. Note that `import pyvisa` is changed to `import Utilities.visa_operations as vo`.

```
import Utilities.visa_operations as vo

resource_name = \
            'USB0::0x0957::0x0607::MY47022122::INSTR'

rm, dmm = vo.open_session(resource_name)

command = '*IDN?'

vo.write_instrument(dmm, command)
response = vo.read_instrument(dmm)

print(response)

vo.close_session(rm, dmm)
```

Save **query_instrument_id.py**, then run it to verify that the instrument identification string is still output correctly to the Shell.

12.9 Simple VISA-Based Query Program with GUI

Let's begin by writing a VISA-based program that carries out the query (i.e., write-then-read) action that you just coded, but this time equipping the program with a GUI that accepts any desired command.

We can reuse some of the code in **query_instrument_id.py**, so open that program in the Editor, then use **File≫Save As...** to create a new program called **simple_visa_query.py** and save it in **YourNamePythonWork**. We'll use some functions in **gui_creator.py** to produce widgets on our GUI window, so import that module from **YourNamePythonWork≫Utilities** at the top of the file and use the `create_window()` function to create the GUI window named `root`.

```
import Utilities.visa_operations as vo
import Utilities.gui_creator as gc

root = gc.create_window(title='Simple Visa Query',
                        size='500x150')

resource_name = \
            'USB0::0x0957::0x0607::MY47022122::INSTR'

rm, dmm = vo.open_session(resource_name)
```

Then, at the bottom of the program, add the following three functions from **gui_creator.py**, which will place widgets on the `root` window: A control and an indicator, each with sixty-character widths, named `Command` and `Response`, respectively, as well as a button labeled `Submit`. The control is chosen wide enough to accept long ASCII commands, while the indicator has sufficient width to display the instrument identification string we saw a few moments ago. The function activated by the button is the yet-to-be-coded `perform_query()`. Also, add `root.mainloop()`.

```
var_ent = gc.create_control(root, label='Command',
                            char_width=60)

lbl_resp = gc.create_indicator(root, label_1='Response',
                               char_width=60)

btn = gc.action_button(root, label='Submit',
                       action=perform_query)

root.mainloop()

vo.close_session(rm, dmm)
```

Finally, right above the **gui_creator.py** functions you just coded, define the function `perform_query()` as shown here. When the button is pressed, this function retrieves the `command` a user has placed in the control, writes this `command` to the DMM, reads the instrument's `response`, then displays this `response` in the indicator:

```
def perform_query():

    command = gc.get_control_string(var_ent)
    vo.write_instrument(dmm, command)
    response = vo.read_instrument(dmm)
    lbl_resp.config(text=response)

    return
```

Save your work. The completed **simple_visa_query.py** appears as follows:

```
import Utilities.visa_operations as vo
import Utilities.gui_creator as gc

root = gc.create_window(title='Simple Visa Query',
                        size='700x150')

resource_name = \
        'USB0::0x0957::0x0607::MY47022122::INSTR'

rm, dmm = vo.open_session(resource_name)

def perform_query():

    command = gc.get_control_string(var_ent)
    vo.write_instrument(dmm, command)
    response = vo.read_instrument(dmm)
    lbl_resp.config(text=response)

    return
```

```
var_ent = gc.create_control(root, label='Command',
                             char_width=60)

lbl_resp = gc.create_indicator(root, label_1='Response',
                                char_width=60)

btn = gc.action_button(root, label='Submit',
                        action=perform_query)

root.mainloop()

vo.close_session(rm, dmm)
```

Pause for a moment to appreciate the almost-too-good-to-be-true fact: The VISA functions in our program are interface-independent. That is, once the appropriate VISA `resource name` is input to `open_resource()` in our `open_session()` function, the rest of our VISA-based code will be the same, regardless of whether we are communicating over the USB, GPIB, or several other possible interface buses (including LAN/Ethernet, RS-232, PXI, VXI). All of the low-level details related to the specific interface bus being used are taken care of "under the hood" by the VISA functions. We are left to simply direct the communication between the instrument and computer at a generic high level, for example, orchestrating what, and at what time, commands and data are written and read. For the programs in the rest of this chapter, my selected VISA resource name will be appropriate for USB communication. As just described, however, it is a simple matter to change the VISA resource name choice to switch to communication over another interface, if desired. For example, in the above program, if instead I wanted to query the DMM over its GPIB interface, I'd simply have to make the substitution `resource_name='GPIB0::22:INSTR'`. With this single substitution, the VISA commands will automatically switch to using the GPIB interface; i.e., no other changes to the program are needed.

Test drive **simple_visa_query.py**. Run the program and enter **IDN?* in the **Command** control. Press the **Submit** button. If all works properly, the instrument's identification string will appear in the **Response** indicator.

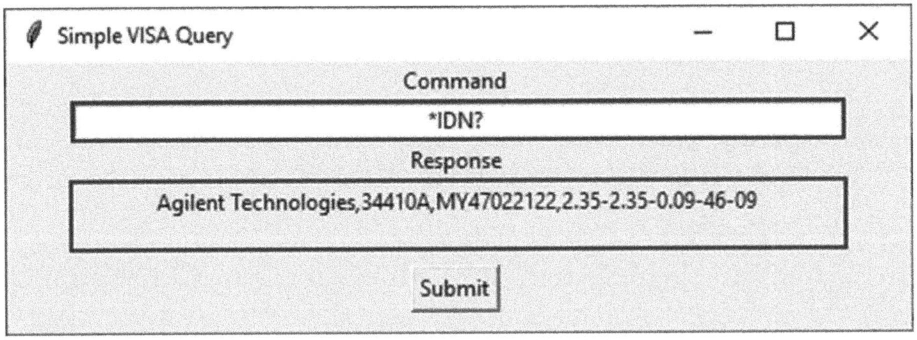

Although the instrument identification string is correctly displayed in the **Response** indicator, you may notice that the string is shifted upward within the indicator. We will explore why this is so in Section 12.10.

For some DMM models, **simple_visa_query.py** will leave the multimeter in *remote mode* with its triggering circuitry "idled." You can return to *local* mode, which continuously "triggers" measurements, by pressing the instrument's front-panel Shift/Local key.

12.10 Message Termination

At the conclusion of a message-transfer process, some method must be used to signal that the complete message has been passed. The IEEE 488.2 standard appoints the ASCII LF (line feed, also called new line) as its special *end-of-string* (EOS) character. That is, when receiving a message string, the LF escape sequence `\n` is always interpreted by the receiver as the last byte of a message. Thus, appending `\n` to a command string is the method of signaling message termination in IEEE 488.2 communication. When using VISA functions to control an IEEE 488.2-compliant instrument, message termination is taken care of automatically, allowing you to remain blissfully ignorant of this lower-level activity.

However, in Section 12.9 we did observe an effect of the appended EOS character. This line feed character (`\n`) instructed a new line to be inserted at the end of the display within the **Response** indicator, causing the instrument identification string to shift upward. This (usually invisible) message termination character can be viewed as follows: In the `perform_query()` function of **simple_visa_query.py**, replace `lbl_resp.config(text=response)` with `lbl_resp.config(text=repr(response))`. The function `repr()` creates a string representation of its argument, including explicit rendering of escape sequences, a quantity termed a *raw string literal*. Running **simple_visa_query.py** with this modification yields the following.

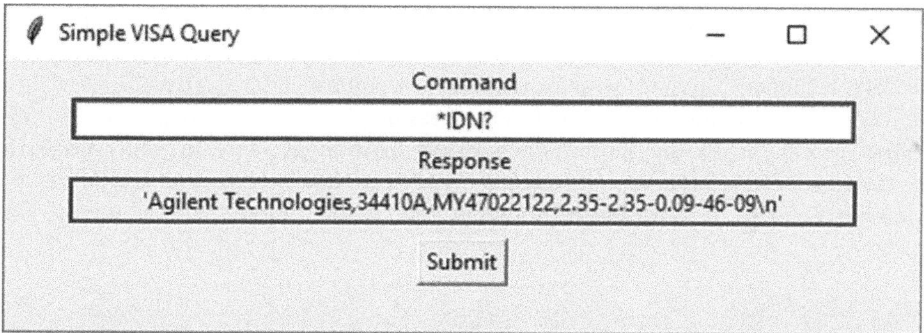

The EOS character can be easily removed from the `response` string. Thinking of `response` as a sequence of characters, the slice operation `response[:-1]` removes its last character. Thus, modify the `perform_query()` function as shown here, noting the `repr()` has been removed:

```
def perform_query():

    ent_string = gc.get_control_string(var_ent)
    vo.write_instrument(dmm, ent_string)
    response = vo.read_instrument(dmm)
    response = response[:-1]
    lbl_resp.config(text=response)

    return
```

Run **simple_visa_query.py**, verifying that the upward shift of text in **Response** no longer occurs.

12.11 Getting and Setting Communication Properties Using VISA Attributes

In addition to message termination, there are other low-level functions connected with message-based communication. These low-level functions can be implemented using methods of the `Resource()` class and each method may have one (or more) associated *resource attribute* corresponding to a communication parameter setting. The VISA standard (or your instrument) assigns default values for these resource attributes, and as long as the VISA-based programs that you write fall within the scope of these default settings, the VISA functions will automatically take care of these low-level functions without any programming effort needed by you (as demonstrated by the message termination example shown in Section 12.10). At times, however, you will most likely write programs that fall outside the range of the default resource-attribute settings and so will need to assign nondefault values to these quantities. Reading ("getting") and writing ("setting") resource-attribute values can be done within your programs as we will now demonstrate.

As a concrete example of a resource attribute, consider *timeout*, which is a fail-safe feature of that prevents a program from running endlessly, if an error occurs. If, for example, an instrument being queried doesn't seem to be responsive (perhaps there's some temporary malfunction in the instrument or interface bus), the query operation will only wait for the instrument's response for a certain number of milliseconds (given by the value of the resource attribute `timeout`) before aborting the read operation and issuing an error message.

The default timeout value for your system can be determined by `instrument.timeout`, where `instrument` is the name of your VISA session. To write a short program to implement this command, in the Editor, open **query_instrument_id.py** in **YourNamePythonWork**, then use **File≫Save As...** to create a new program and save it under the name **get_timeout_value.py** in **YourNamePythonWork**. Modify the code so that it appears as shown here:

```
import pyvisa

resource_name = \
          'USB0::0x0957::0x0607::MY47022122::INSTR'

rm = pyvisa.ResourceManager()
dmm = rm.open_resource(resource_name)

to = dmm.timeout

print(to)

dmm.close()
rm.close()
```

Save your work, then run **get_timeout_value.py**. The default timeout value (in milliseconds) will be printed in the Shell. For my system, the value is 2000 ms = 2 s.

You can set the value of timeout to something other than its default value. The command `instrument.timeout=XXXX` accomplishes that feat, where XXXX is the number of milliseconds for the desired timeout value. In the Editor, open **get_timeout_value.py**, then use **File≫Save As...** to create a new program named **set_timeout_value.py** in **YourNamePythonWork**. In **set_timeout_value.py**, add the command as shown, which sets the timeout value to 1000 ms = 1 s:

```
import pyvisa

resource_name = \
        'USB0::0x0957::0x0607::MY47022122::INSTR'

rm = pyvisa.ResourceManager()
dmm = rm.open_resource(resource_name)

dmm.timeout = 1000

to = dmm.timeout

print(to)

dmm.close()
rm.close()
```

Save your work, then run **set_timeout_value.py**. The timeout value (in milliseconds) will be printed in the Shell. Is it now equal to 1000 ms = 1 s? Try setting `timeout` to a few other values. When finished, run **set_timeout_value.py** one last time, setting `timeout` equal to its original default value.

12.12 Performing Measurement Over Interface Bus

If you are using a DMM that is software-compliant with the HP 34401A command set, configure your DMM as follows: For the Rigol, write the command *CMDSET AGILENT* or select Agilent command set under **Utility≫Cmd** on the front panel; for the Keithley, write the command **lang SCPI34401* or select 34401 command set under **Menu≫System** on front panel; for Fluke and Tektronix, write the command *L1* or select 34401 command set under **Setup≫Commands** on front panel.

Now that **simple_visa_query.py** has given us a template for the VISA query process, let's try controlling a real measurement. Hook up some known DC voltage difference, say, 5 V, between the HI and LO Voltage Inputs of your DMM. For instruments compliant with the 34401A command set, delivering the following sequence of ASCII commands will result in one DC voltage sample being acquired and then loaded into the instrument's output buffer (which is part of its interface circuitry):

```
CONF:VOLT:DC<Space>10,0.00001
INIT
FETC?
```

Here is the meaning of this secret code. First, the digital multimeter can be programmed to perform many different types of measurement functions, including DC voltage, AC voltage, DC current, AC current, resistance, frequency, capacitance, and temperature. Given these options, the first command instructs the instrument that we wish to take a DC voltage measurement. The full command is *CONFigure:VOLTage:DC <Space><Range>,<Resolution>* (this command actually executes a collection of commands drawn from the 34410A DMM's **INP**ut, **SENS**e, **TRIG**ger, and **CALC**ulate root subsystems). The command *CONFigure:VOLTage:DC* is constructed in the hierarchical tree structure, typical of SCPI-compliant instruments, and is the command's *header*. Within the header, *CONFigure* is the root-level keyword and colons (:) represent the descent to the lower-level *VOLTage* and then the lowest-level *DC* keywords. Although the full command can be sent to the instrument, it is only absolutely necessary to send the capitalized characters. Separated from the header *CONF:VOLT:DC* by a *<Space>* (i.e., one tap on the keyboard's spacebar), the numeric values for two measurement parameters—*<Range>* and *<Resolution>*—are specified. *<Range>* selects among the instrument's five available voltage measurement scales. Each scale offers a different sensitivity, with *<Range>* giving the maximum measurable value on a particular scale. The five available ranges are 100 mV, 1 V, 10 V, 100 V, and 1000 V (check the manual for your DMM for available ranges; for example, for the Rigol DM3058/3058E, the available ranges are 200 mV, 2 V, 20 V, and 200 V). In our situation of measuring a signal of approximately 5 V, the 10-V scale is appropriate. *<Resolution>* specifies the precision of the measurement, with 10 levels of accuracy available on the 34410A DMM. For our programs to be compatible with both the 34410A and the older-model 34401A DMM, we will consider only two resolution levels for now. These resolutions have 5½ and 6½ digits of precision (where the ½ digit means that the most significant decimal place can only take on a value of *1* or *0*), and so, on the 10-V scale, voltages can be resolved at the level of either 0.0001 or 0.00001 V, respectively (again, check the manual for your DMM for available resolutions; for example, for the Rigol DM3058/3058E, only 5½ digits of precision is available). The trade-off in requesting higher accuracy is that the measurement takes a longer time. In the command sequence above, the highest resolution of 6½ digits is selected by setting *<Resolution>* equal to *0.00001* when *<Range>* equals *10*. Note that the syntax of the *CONF* command obeys the conventions of the SCPI language: a *comma* (,) separates the parameters from each other and a *<Space>* separates the command header from the parameters.

Once the multimeter has been configured for the desired measurement function as described in the previous paragraph, the data-taking process is begun by sending the *INITiate* command (from the **TRIG**ger root subsystem). Upon receipt of *INIT*, the multimeter will acquire the requested voltage sample and then store this value in its internal memory. Finally, the *FETCh?* command (from the **MEM**ory root subsystem) instructs the instrument to transfer the reading in its internal memory to its interface-related output buffer.

We would now like to place this command sequence into **simple_visa_query.py**. Since there are three commands to be sent, followed by receiving back the instrument's response, it appears that we must modify the program to include a sequence of three successive implementations of PyVISA's `write()` command.

Although you are free to do so, a much easier solution is available. The SCPI language allows the programmer to concatenate several commands together into one long multicommand string that can be sent in a single statement. The syntax for this concatenation process is as follows:

- Use a semicolon (;) to separate two commands within the string.
- Begin a command with a colon (:) if it has a different root level than the command preceding it. The first command in the concatenated string and IEEE 488.2 common commands (which begin with an asterisk) do not require a leading colon.

Since each of our three commands has a different root level, applying the above rules results in the concatenated string

```
CONF:VOLT:DC<Space>10,0.00001;:INIT;:FETC?
```

With a DC voltage difference being input to the DMM, run **simple_visa_query.py**, Then, type the above command into its **Command** control as shown next; remember *<Space>* means a space character created by the keyboard's spacebar and modify the *<Range>* and *<Resolution>* values in the command to comply with your particular DMM, if needed. Press the **Submit** button. Your computer will instruct the multimeter to acquire a 6½ digit voltage reading on the 10-V range, retrieve this value, and then display it in the **Response** indicator. Cool, eh?

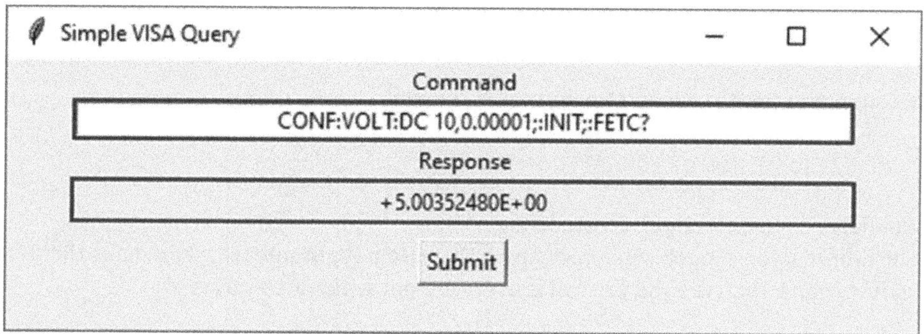

Note that, although extra digits are displayed, the value within **Response** is only accurate to the fifth decimal place.

As shown above, the 34410A DMM reports its data samples in the form of an ASCII character string using the exponential format SD.DDDDDDDDESDD, where S is a positive or negative sign, D is a numeric digit, and E is an exponent. For future reference, note that the string that represents a data sample is 15 bytes long. If you want to use this reading as input to a mathematical calculation (a common situation), you will need to convert the string representation (call it `string`) into a numeric format such as float. Such conversion operations can be easily accomplished in Python, e.g., using the command `float_number = float(string)`.

In the following sections, we will acquire multiple readings from the DMM and so need a much larger **Response** display. The scrolled text widget is a scalable rectangular text display and so serves such a purpose. This widget is created with the command `ScrolledText()`, which is found in the `scrolledtext` module of `tkinter`; the syntax for this command is shown here:

```
tkinter.scrolledtext.ScrolledText(container, width,
                                  height)
```

The `width` and `height` arguments give the number of characters displayed horizontally and the number of lines vertically, respectively. If there is more text than can be displayed within those constraints, a vertical scrollbar automatically appears, which allows viewing of the entire text.

In the Editor, open **gui_creator.py** in **YourNamePythonWork≫Utilities**, then add the following definition for `create_text_box()`. This code creates a scrolled text widget named `stx` inside a blue-bordered-frame with default `width` and `height` of *60* and *20*, respectively. A `label` is placed above the frame.

```
def create_text_box(container, label,
                    char_width=60, lines=20):

    lbl = tk.Label(container, text=label)
    lbl.pack()

    frame = tk.Frame(container, bg='blue', bd=2)
    frame.pack()

    stx = tkst.ScrolledText(frame, width=char_width,
                            height=lines)
    stx.pack(fill=tk.X)

    return stx
```

At the top of **gui_creator.py**, add the import statement

```
import tkinter.scrolledtext as tkst
```

Save you work as you close **gui_creator.py**.

In the Editor, open **simple_visa_query.py** in **YourNamePythonWork**, then make the following modifications. Increase the vertical size of the root window to, say, *450*:

```
root = gc.create_window(title='Simple VISA Query',
                        size='700x450')
```

Replace the `create_indicator()` command with the following command to create a scrolled text widget of given width and height:

```
stx = gc.create_text_box(root, label='Response',
                         char_width=55, lines=20)
```

Finally, in the `perform_query()` function, replace `lbl_resp.config(text=response)`, which wrote `response` to the single-line indicator, with the two `stx`-related commands shown below. The first command deletes text in the scrolled text widget from position *1.0* to the *end*, i.e., all of the existing text, and the second command inserts `response` into the widget, starting at position *1.0*, i.e., the beginning position.

```
def perform_query():

    ent_string = gc.get_control_string(var_ent)
    vo.write_instrument(dmm, ent_string)
    response = vo.read_instrument(dmm)
    response = response[:-1]
    stx.delete(1.0, 'end')
    stx.insert(1.0, response)

    return
```

Save your work. The completed **simple_visa_query.py** program is as follows:

```
import Utilities.visa_operations as vo
import Utilities.gui_creator as gc

root = gc.create_window(title='Simple VISA Query',
                        size='700x450')

resource_name = \
        'USB0::0x0957::0x0607::MY47022122::INSTR'

rm, dmm = vo.open_session(resource_name)

def perform_query():

    ent_string = gc.get_control_string(var_ent)
    vo.write_instrument(dmm, ent_string)
    response = vo.read_instrument(dmm)
    response = response[:-1]
    stx.delete(1.0, 'end')
    stx.insert(1.0, response)

    return


var_ent = gc.create_control(root, label='Command',
                            char_width=60)

stx = gc.create_text_box(root, label='Response',
                         char_width=55, lines=20)

btn = gc.action_button(root, label='Submit',
                       action=perform_query)

root.mainloop()

vo.close_session(rm, dmm)
```

To verify your coding is correct, run **simple_visa_query.py** with *CONF:VOLT:DC<Space>10,0.00001;:INIT;:FETC?* entered in the **Command** control:

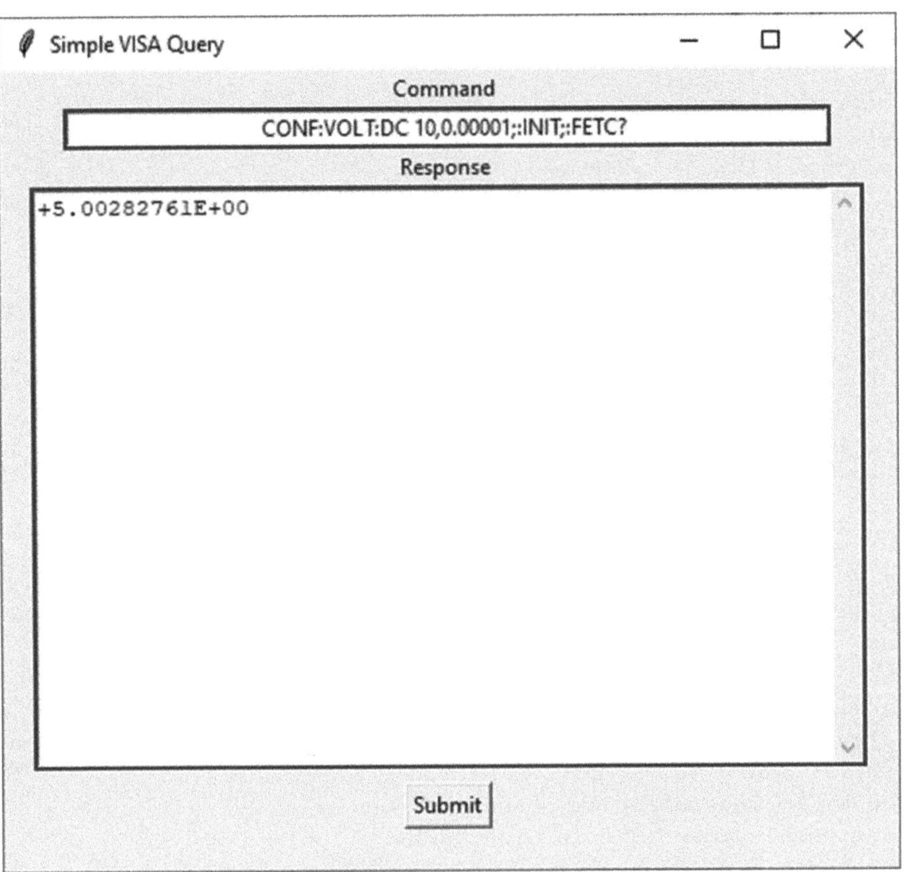

Going forward, I will put an actual space in SCPI commands, rather than writing *<Space>*. Please remember that for a command with parameters, by SCPI convention, a space is placed between the header and the first parameter.

12.13 Synchronization Methods

Although most ASCII commands are completed quickly after being received by a programmable instrument, some commands start a process that requires a significant amount of time (such as acquiring a large amount of data or moving an object from point *A* to point *B*). The time required for such processes must be taken into account when writing a data-acquisition program, or else, upon execution, the program may request data before they are available, induce undesirable motion, or cause some other chaotic outcome.

As an example, in its default configuration, the 34410A DMM acquires one data sample after receipt of the *INITiate* command and then stores this measured value in its internal memory. However, through use of the *SAMPle:COUNt* *<Space><Value>* command, the multimeter can be instructed to take and store multiple data samples upon receiving *INITiate*, where *<Value>* is the number of requested samples. The 34410A DMM is configured to acquire 1500 DC voltage samples with 6½-digit resolution via the following concatenated string of commands:

```
CONF:VOLT:DC 10,0.00001;:SAMP:COUN 1500;:INIT;:FETC?
```

The *FETCh?* command will load the 1500 acquired samples from the multimeter's internal memory (which, by the way, can hold up to a maximum number of 50000 measured values) into the instrument's interface-related output buffer. (For 34401A DMM owners: Your instrument can store only 512 values and performs more slowly, so configure it to acquire 15, rather than 1500, i.e., use *SAMP:COUN 15* in your command string.)

Before we send the command string instructing the DMM to gather a sequence of 1500 voltage samples, let's consider the following issue: Simply stated, voltage sampling takes time. According to its user manual, when the 34410A multimeter is configured for 6½-digit resolution, it takes 0.2 power line cycles (PLC) to acquire each voltage sample. Additionally, the 34410A DMM has an optional autozero feature, which operates as follows: After each voltage measurement, the multimeter internally disconnects the input signal and takes a zero reading. The instrument then subtracts the zero reading from the preceding measured value to prevent offset voltages in the multimeter's internal circuitry from affecting measurement accuracy. Since the zero reading also takes the same number of PLC as a regular voltage measurement, in comparison to when autozeroing is not activated, each complete voltage sample by the multimeter takes twice as long. Again, consulting the DMM user manual, we find that the autozeroing feature is turned off by default for the resolution we have chosen, so each voltage sample takes 0.2 (not 0.4) PLC. Assuming this instrument is plugged into a 60-Hz power source (i.e., 60 PLC per second), 1500 voltage samples will take about

$$1500 \times \left(\frac{0.2 \text{ PLC}}{60 \text{ PLC/s}}\right) = 5 \text{ s}.$$

That's a problem! A few moments ago, we found that the default timeout value for a read operation is 2 s, but the measurement initiated by our command takes about 5 s. Thus, before all of the requested data are available, the read operation will time out. (For 34401A DMM owners: Each data sample by your DMM takes 10 PLC and autozeroing is turned on by default. Thus, 15 samples takes $15 \times (20\,\text{PLC}/60\,\text{PLC/s}) = 5\,\text{s}$.)

There are a couple of crude solutions to this dilemma. First, within the program, a delay greater than 5 s can be inserted between the issuance of the data-taking command and the order to read the gathered data samples. That coding would appear as follows for a 10-s delay.

```
vo.write_instrument(dmm, command)
time.sleep(10.)
response = vo.read_instrument(dmm)
```

Second, for a slightly more elegant fix, you can use **set_timeout_value.py** in **YourNamePythonWork** to change the timeout value from its default value (on my system) of 2000 ms to something larger than 5 s.

In both of these fixes, knowledge of the time span is required for the particular measurement process being implemented and this time span will vary from measurement to measurement (e.g., depending on the number of samples being acquired). Thus, there is no universal delay or timeout value that will account for every possible measurement. Please note that lack of communication (in particular, the interface bus not correctly knowing in every case when the instrument's data will be available) is the root problem being crudely solved by these fixes.

Fortunately, powerful tools exist that allow one to monitor the status of tasks being performed by a programmable instrument. For IEEE 488.2-compliant instruments, these tools are the Standard Event Status Register (SESR) and Status Byte Register (SBR) that were discussed at the beginning of this chapter. With proper use of the SESR and SBR, many potential data-taking glitches, such as the timing one just discussed, can be avoided.

The status-reporting capabilities of the SESR and SBR can be employed in several ways. We will explore two commonly used techniques: the Serial Poll and Service Request Methods. The core operation for both of these methods is the same—the completion of an assigned task triggers the Operation Complete (OPC) bit in the SESR to be set, which in turn sets the Event Status Bit (ESB) of the SBR.

In the Serial Poll Method, the setting of ESB is detected by directly checking the SBR, whose state is obtained by serial polling the instrument. The complete step-by-step process of this method is shown in Fig. 12.4.

In the Service Request Method, the SBR is configured such that, when its ESB is set, the Request Service bit is induced to be set also. This action then causes the instrument to assert a SRQ, which alerts the interface bus that the assigned operation is complete. This method is pictured in Fig. 12.5.

In the following sections, we will write programs that implement both of these approaches to status reporting.

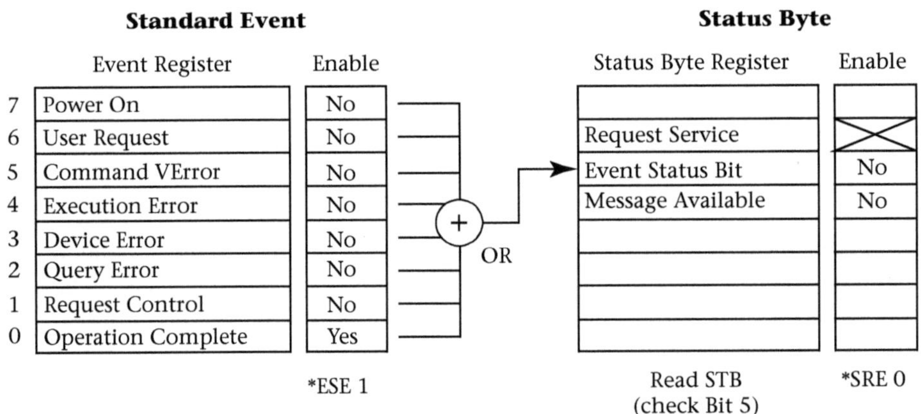

Fig. 12.4 Serial poll method.

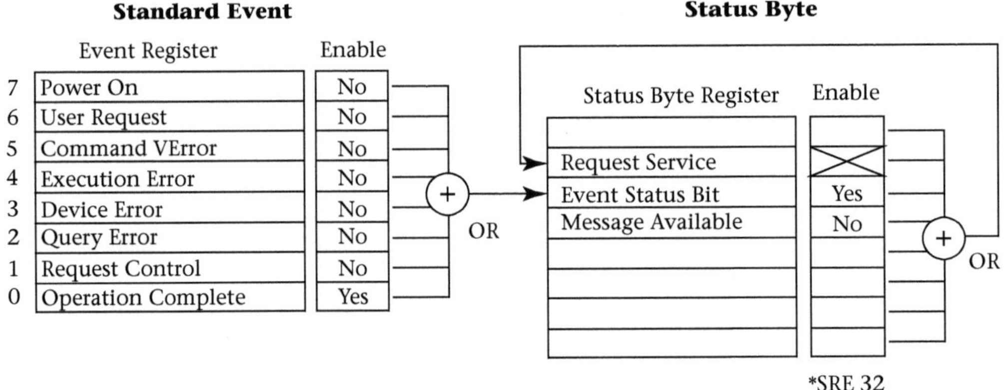

Fig. 12.5 Service request method.

12.14 Measurement Operation Based on Serial Poll Method

Let's try the Serial Poll Method first by writing a program named **visa_query_serial_poll.py**. For this program, we'll need to add a few new functions to **visa_operations.py**. So, in the Editor, open **visa_operations.py** in **YourNamePythonWork≫Utilities**. Then, to configure our instrument for status reporting using the Serial Poll Method, add the definition of the following function called `status_config_serial_poll()`:

```
def status_config_serial_poll(instr):

    instr.clear()

    instr.write('*CLS;*ESE 1;*SRE 0;*OPC?')
    response = instr.read()

    return response
```

Save your work. Here's how this function works, assuming that the instrument referenced by VISA session named `instr` is IEEE 488.2 compliant. First, the method `instr.clear()` executes. Although not an absolute necessity for inclusion in the serial poll configuration process, this method performs the precautionary actions of "clearing" the instrument. That is, `instr.clear()` instructs the instrument to abort all measurements in progress, disable its triggering circuitry, clear its interface-related output buffer, and prepare to accept a new command string.

Next, `instr.write()` sends the concatenated command string **CLS;*ESE 1;*SRE 0;*OPC?* to configure the instrument for status reporting using the Serial Poll Method. Note that since the component strings are all IEEE 488.2 common commands, leading colons are not required in the concatenation. In this sequence of common commands, **CLS* clears the contents of the SESR and SBR. As described in the beginning of this chapter, **ESE 1* enables the SESR's OPC bit to set the ESB in the SBR and **SRE 0* disables the instrument from asserting a SRQ. Then **OPC?* requests the instrument to return a "1" to the instrument's output buffer after this command is completed. This last command is included simply as a method of checking that the entire sequence of commands has been executed. If this function completes its operations successfully, `instr.write()` will return `response` with a value equal to a single ASCII character "1".

At the bottom of **visa_operations.py**, insert the following test-block programming:

```
if __name__ == '__main__':

    resource_name = \
                'USB0::0x0957::0x0607::MY47022122::INSTR'

    rm, dmm = open_session(resource_name)
    response = status_config_serial_poll(dmm)
    print(response)
    close_session(rm, dmm)
```

Save your work, then run **visa_operations.py**. If your coding is correct, the Shell will display an ASCII character "1".

To continuously read the SBR of an instrument until a given bit is set, in **visa_operations.py** add the following definition of a function called `serial_poll()`. Here, we make the SBR's fifth bit the default bit to check:

```
def serial_poll(instr, check_bit=5):

    bit_test = 0
    status_bit = 2**check_bit

    while bit_test != status_bit:

        status_byte = instr.read_stb()
        bit_test = status_byte & status_bit

    return
```

Save your work as you close the file. The method `instr.read_stb()` is the workhorse of this function. With each iteration of the `while` loop, it assigns the current values of the SBR's eight bits in the form of an unsigned integer to the object `status_byte`. For example, if the SBR's fifth bit (ESB) is set, then `status_byte` equals the binary number $XX1XXXXX_2$, where X is the value of bits other than the fifth bit and can be 0 or 1.

As a trick for checking whether the *n*th bit of the SBR (called `check_bit`) has been set, we first create an object called `status_bit`, which is an integer equivalent to an eight-bit binary number with just the *n*th bit set to *1* and all other bits equal to *0*. For example, if we are interested in the $n = 5$ bit, then `status_bit` equals $00100000_2. = 32_{10}$. Then, to check whether the SBR's *n*th bit is set in `status_byte`, we apply the *bitwise logical AND* operation to `status_byte` and `status_bit`. The bitwise AND is a binary operator that produces a ("resultant") binary number by evaluating an AND operation at each corresponding bit of two binary numbers (called the *operands*). At a particular bit location, if both of the operands' bits are *1*, that particular bit of the resultant is *1*; otherwise, it is *0*. Because all bits of `status_bit` are *0*, except the *n*th bit, for the bitwise AND operation involving `status_byte` and `status_bit`, the only possible nonzero bit in the resultant is the *n*th bit. Thus, if the *n*th bit of `status_byte` is *1*, the corresponding *1* in the `status_bit` will produce a resultant with the *n*th bit equal to *1* and all others equal *0*, i.e., the decimal integer 2^n. Otherwise, if the *n*th bit of `status_bit` is *0*, then the AND operation at the *n*th bit produces *0* and the resultant becomes zero. In Python, the bitwise AND operator is denoted by the ampersand symbol (`&`) and the *not equal* operator is `!=`. Above, in `serial_poll()`, we see that `bit_test` is the resultant of the bitwise AND operation between `status_byte` and `status_bit`, and that the `while` loop executes until `bit_test` equals `status_bit`, that is, until `check_bit` is set.

Finally, here is the reason we have chosen to use the two-step "`write()`, then `read()`" operation to query an instrument rather than simply `query()`, the single function that carries out the complete write-then-read process. To query the instrument using the Serial Poll Method requires sandwiching the serial poll between the write and read operations, which is not possible with `query()`. That is, the Serial Poll Method consists of the following three-step operation: (1) Use `write()` to send a SCPI-based message to an instrument, instructing it to carry out an operation, which is possibly time-consuming, (2) perform a serial poll via `serial_poll()` to determine when the operation has completed, then (3) use `read()` to receive back from the instrument the result of the operation.

We're now ready to write **visa_query_serial_poll.py**. This program will implement the Serial Poll Method to synchronize the interface-bus activities necessary in acquiring 1500 voltage samples using a 34410A multimeter.

In the Editor, open **simple_visa_query.py**, and then use **Save As...** to create **visa_query_serial_poll.py** in **YourNamePythonWork**. The only required changes are as follows: In the `perform_query()` function, the inclusion of `vo.status_config_serial_poll()` to configure the instrument for the Serial Poll Method (note, by not assigning its return value to an object named `response`, the ASCII "1" generated by this function is ignored and it will be "garbage-collected"); and the placing of `serial_poll()` between the `write_instrument()` and `read_instrument()` functions. Note, in line with the Serial Poll Method, `serial_poll()` is programmed to detect the setting of the SBR's fifth bit, i.e., the Event Status Bit:

```
def perform_query():

    vo.status_config_serial_poll(dmm)
    command = gc.get_control_string(var_ent)
    vo.write_instrument(dmm, command)
    vo.serial_poll(dmm, check_bit=5)
    response = vo.read_instrument(dmm)
    response = response[:-1]
    stx.delete(1.0, 'end')
    stx.insert(1.0, response)

    return
```

Also, change the argument of `create_window()` to `title='VISA Query (Serial Poll)'`.

Save your work. The complete program **visa_query_serial_poll.py** should appear as follows:

```
import Utilities.visa_operations as vo
import Utilities.gui_creator as gc

root = gc.create_window(title='VISA Query (Serial Poll)',
                        size='500x450')

resource_name = \
          'USB0::0x0957::0x0607::MY47022122::INSTR'

rm, dmm = vo.open_session(resource_name)

def perform_query():

    vo.status_config_serial_poll(dmm)
    command = gc.get_control_string(var_ent)
    vo.write_instrument(dmm, command)
    vo.serial_poll(dmm, check_bit=5)
    response = vo.read_instrument(dmm)
    response = response[:-1]
    stx.delete(1.0, 'end')
```

```
        stx.insert(1.0, response)

        return


var_ent = gc.create_control(root, label='Command',
                            char_width=60)

stx = gc.create_text_box(root, label='Response',
                         char_width=55, lines=20)

btn = gc.action_button(root, label='Submit',
                       action=perform_query)

root.mainloop()

vo.close_session(rm, dmm)
```

Run **visa_query_serial_poll.py**. To acquire 1500 voltage samples from the DMM, enter the following command into the **Command** entry box. Be sure to include the all-important **OPC* within the command (on the older 34401A model DMM, use *15* rather than *1500*) as well as a space after *DC* and *COUN*.

```
CONF:VOLT:DC 10,0.00001;:SAMP:COUN 1500;:INIT;*OPC;:FETC?
```

Press the **Submit** button. If all goes well, after the 1500 voltage samples are acquired by the DMM, they will be transferred to your computer and displayed in the **Response** box, where you can scroll down to view all of the values.

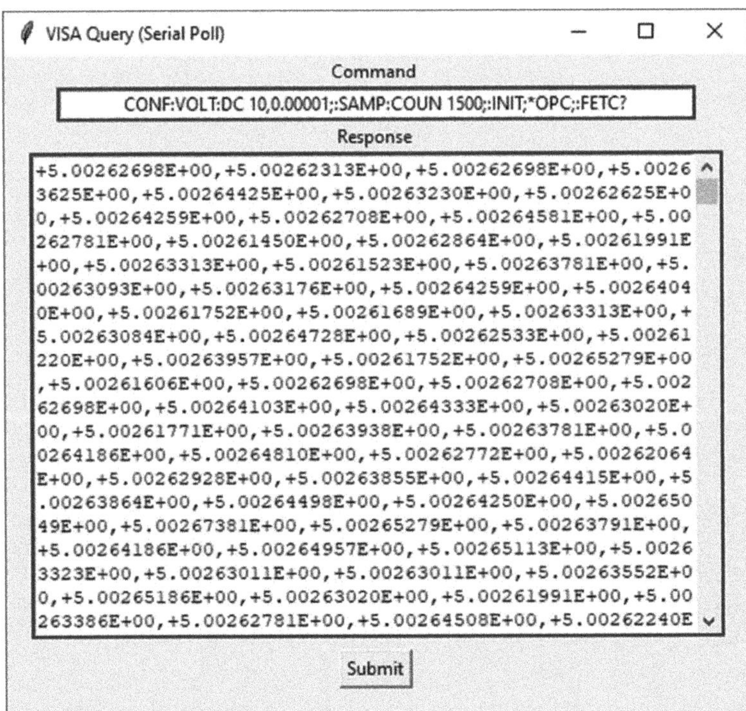

Here is how this program works: First, `status_config_serial_poll()` configures the DMM for the Serial Poll Method. Then, the concatenated command string is retrieved from **Command** and sent to the instrument by `write_instrument()`. The command instructs the multimeter to configure itself for the desired DC Voltage measurement function and the acquisition process is begun by the *INIT* command. The succession of 1500 samples is acquired and temporarily stored in the multimeter's internal memory. The function `serial_poll()` constantly checks the value of the instrument's SBR throughout the entire acquisition process. After the 1500th sample is obtained, **OPC* instructs the instrument to set its SESR's OPC bit (which, in turn, sets the SBR's ESB), and then *FETC?* loads the contents of the internal memory into the instrument's output buffer. At that point, `serial_poll()` detects the setting of ESB, which then triggers the instrument's output buffer to be read by `read_instrument()`.

In reviewing how this program works, a weakness of the Serial Poll Method becomes apparent, namely, the large volume of interface-bus traffic required by this technique. During the 5 s while the 1500 data samples are being gathered, the instrument is polled numerous times by the interface bus so that its status can be continuously monitored. Although effective, the Serial Poll Method is rather inefficient because of its excessive use of the interface bus and processor time.

12.15 Measurement Operation Based on Service Request Method

The Service Request Method provides status reporting with a minimum of interface-bus activity. To configure an IEEE 488.2-compliant instrument for status reporting using the Service Request Method, a new function called `status_config_srq()` is needed. In the Editor, open **visa_operations.py** in **YourNamePythonWork≫Utilities**, then add the following definition for `status_config_srq()`. The constants `EventType.service_request` and `EventMechanism.queue` come from a submodule called `constants` within the `pyvisa` module, so at the top of **visa_operations.py** add the statement `import pyvisa.constants as pycon`:

```
def status_config_srq(instr):

    event_type = pycon.EventType.service_request
    event_mech = pycon.EventMechanism.queue

    instr.clear()

    instr.enable_event(event_type, event_mech)
    instr.write('*CLS;*ESE 1;*SRE 32;*OPC?')
    response = instr.read()

    return response
```

Note the two modifications of this function in comparison to the serial poll configuration function `status_config_serial_poll()`. First, by changing **SRE 0* to **SRE 32* in the command string sent to the instrument, the instrument will assert the SRQ line when the SBR's fifth (ESB) bit is set. The already present **ESE 1* command configures the instrument to set the ESB in response to the setting of the SESR's OPC (Operation Complete) bit. Second, for VISA functions to detect Service Request (SRQ) events during this VISA session, the command `enable_event(event_type,event_mech)` with `event_type` and `event_mech` set equal to `service_request` and `queue`, respectively, is required. With these selections, VISA will create and monitor a queue of service requests.

At the bottom of **visa_operations.py**, modify the test block to the following:

```
if __name__ == '__main__':

    resource_name = \
                'USB0::0x0957::0x0607::MY47022122::INSTR'

    rm, dmm = open_session(resource_name)
    response = status_config_srq(dmm)
    print(response)
    close_session(rm, dmm)
```

Remember to add the statement `import pyvisa.constants as pycon` at the top of your program, then save your work. Run **visa_operations.py**. If your coding is correct, the Shell will display an ASCII character "1".

Next, in **visa_operations.py**, add the following function called `wait_for_srq()`. The command `instr.wait_on_event(event_type,timeout)` sits idly until the instrument denoted by `instr` asserts a SRQ. Upon receipt of the SRQ, the command returns a description of the event, which is assigned to the object `response`. However, there is a limit to the patience of this command. It will only wait up to a total time of `timeout`, which has been defined to be 30000 ms = 30 s.

```
def wait_for_srq(instr):

    event_type = pycon.EventType.service_request
    timeout = 30000

    response = instr.wait_on_event(event_type, timeout)

    return response
```

Finally, in addition to closing the VISA session and resource manager at the conclusion of program execution, the VISA servicing of SRQ events must be disabled before the VISA session is closed. In **visa_operations.py**, add the function `close_session_srq()` as shown here:

```
def close_session_srq(rm, instr):

    event_type = pycon.EventType.service_request
    event_mech = pycon.EventMechanism.queue

    instr.disable_event(event_type, event_mech)
    instr.close()
    rm.close()

    return
```

Save your work as you close **visa_operations.py**.

Now, let's write a program to acquire 1500 voltage samples from a 34410A DMM using the Service Request Method. In the Editor, open **visa_query_serial_poll.py** in **YourNamePython-Work**, and then use **Save As...** to create **visa_query_srq.py** in **YourNamePythonWork**. Most of the required changes are in the `perform_query()` function. Modify that function as shown here (note, by not assigning the return value from `wait_for_srq()` to an object named `response`, that return value is ignored and it will be "garbage-collected"):

```
def perform_query():

    vo.status_config_srq(dmm)
    command = gc.get_control_string(var_ent)
    vo.write_instrument(dmm, command)
    vo.wait_for_srq(dmm)
    response = vo.read_instrument(dmm)
    response = response[:-1]
    stx.delete(1.0, 'end')
    stx.insert(1.0, response)

    return
```

Also, change the argument of `create_window()` to `title='VISA Query (SRQ)'` and, at the end of the program, replace `vo.close_session(rm,dmm)` with `vo.close_session_srq(rm,dmm)`.

Save your work. Here is the complete **visa_query_srq.py** program:

```
import Utilities.visa_operations as vo
import Utilities.gui_creator as gc

root = gc.create_window(title='VISA Query (Service Request)',
                        size='500x450')

resource_name = \
        'USB0::0x0957::0x0607::MY47022122::INSTR'

rm, dmm = vo.open_session(resource_name)

def perform_query():

    vo.status_config_srq(dmm)
    command = gc.get_control_string(var_ent)
    vo.write_instrument(dmm, command)
    vo.wait_for_srq(dmm)
    response = vo.read_instrument(dmm)
    response = response[:-1]
    stx.delete(1.0, 'end')
    stx.insert(1.0, response)

    return


var_ent = gc.create_control(root, label='Command',
                            char_width=60)
```

```
stx = gc.create_text_box(root, label='Response',
                         char_width=55, lines=20)

btn = gc.action_button(root, label='Submit',
                       action=perform_query)

root.mainloop()

vo.close_session_srq(rm, dmm)
```

Run **visa_query_srq.py**. Again, to acquire 1500 voltage samples from the DMM, enter the following command into the **Command** control. Be sure to include the *OPC (on the older 34401A model DMM, use *15* rather than *1500*).

```
CONF:VOLT:DC 10,0.00001;:SAMP:COUN 1500;:INIT;*OPC;:FETC?
```

Press the **Submit** button. Does the program successfully acquire the requested 1500 DC voltage samples from your DMM? Do you understand the operation of this program and how the Service Request Method manages to work with a minimum of interface-bus activity?

12.16 Python Dictionary

In Section 12.17, we'll create an *instrument driver*, that is, user-friendly software that, once created, can be used to easily produce the SCPI commands required to communicate with our DMM. An essential component of this instrument driver will be a set of *lookup tables*, which store various specifications of our instrument. Python provides a handy data structure perfectly suited to this programming need—the *dictionary*—and so let's take a brief detour to learn about this useful tool.

A Python dictionary is a collection of *key–value* pairs enclosed in curly brackets. The form of a dictionary is as follows:

```
dict = {key_1:value_1, key_2:value_2, key_3:value_3,...}
```

As shown, a key and its associated value are connected by a colon, and key–value pairs are separated by commas. A key as well as its value can be any type of Python object, but for our instrument-driver application, a key will be a string and its value will be either a string or a number.

As an example, consider three possible measurements for the multimeter—DC voltage, DC current, and resistance—are represented by the three strings `dc_voltage`, `dc_current`, and `resistance`, respectively. Then, using these strings as keys, the following dictionary called `measurement` stores the SCPI command headers that configure the DMM for each of these measurements:

```
    measurement = {'dc_voltage':'CONF:VOLT:DC',
                   'dc_current':'CONF:CURR:DC',
                   'resistance':'CONF:RES'}
```

In this example, both the key and its associated value are strings.

To access a particular header, for example, the one related to the DC current measurement, one simply uses the square-bracketing indexing syntax we have used previously with lists and arrays:

```
header = measurement['dc_current']
```

As another example, the following dictionary called `voltage_range` stores all of the DMM's allowed voltage ranges:

```
voltage_range = {'100 mV':0.1,
                 '1 V':1.0,
                 '10 V':10.0,
                 '100 V':100.0,
                 '1000 V':1000.0}
```

This time, the key is a string and its associated value is a float.

12.17 Creating an Instrument Driver

An instrument driver is a collection of modular software routines that perform the operations required in the computer control of a programmable instrument. These operations include configuring, triggering, status checking, sending commands to, and receiving data from the instrument. Above, `status_config_serial_poll()` and `status_config_srq()` are examples of configuration functions that would be useful to include as part of the DMM instrument driver. We will now write another configuration function, this time one that prepares the multimeter for taking a desired measurement.

The 34410A DMM is capable of implementing 12 types of measurement functions, including DC and AC voltage, DC and AC current, two- and four-wire resistance (two-wire is the "normal" method for measuring resistance; the more involved four-wire technique is necessary only when measuring very small resistance samples), frequency and period of an AC signal, continuity, diode check, capacitance, and temperature. To gain experience with some of the Python tools available for developing instrument drivers, let's write a driver that offers the choice of configuring the 34410A DMM for a DC voltage, DC current, or two-wire resistance measurement. You, of course, can be more ambitious and write your program to control up to all 12 possible measurement functions.

Referring to the 34410A DMM user manual, we find that our driver must allow a user to select one of the following three possible SCPI commands to configure the instrument for the desired measurement function:

```
CONFigure:VOLTage:DC<Space><Range>,<Resolution>
CONFigure:CURRent:DC<Space><Range>,<Resolution>
CONFigure:RESistance<Space><Range>,<Resolution>
```

Here, the possible values of *<Range>* for the DC voltage measurements are 0.1, 1, 10, 100, and 1000 V, while the possible DC current ranges are 100 μA, 1 mA, 10 mA, 100 mA, 1, and 3 A. For the resistance measurement, the allowed *<Range>* values are 100 ohm, 1 kohm, 10 kohm, 100 kohm, 1 Mohm, 10 Mohm, and 100 Mohm (we'll ignore the 1-Gohm range because it is not available on the older 34401A model). For each of these three measurements, the measurement resolution (i.e., its precision) may be selected as 3, 0.7, 0.3, 0.1, or 0.03 parts per million (ppm), where these "fractional resolution" values are the fraction of <*Range*>. The resolution is determined by the number of power line cycles (NPLC) devoted to the measurement (called the *integration time*), which correspond to 0.02, 0.2, 1, 10, and 100 NPLC, respectively, for the fractional resolutions listed. Once the fractional resolution is selected, <*Resolution*> equals the fractional resolution times <*Range*>. The 34410A DMM can operate at five more precision levels, but to stay compatible with other instruments (e.g., 34401A and Rigol 3058), we will work with just the five given here.

As part of our instrument driver, we will write a function called `dmm_command_string()`, which will allow a user to construct the desired SCPI command string in a user-friendly way. In the Editor, open **visa_operations.py** in **YourNamePythonWork≫Utilities**. Our plan is to write a function that, with the input of its arguments, produces a command string like

```
CONF:VOLT:DC 10,0.00001;:SAMP:COUN 1500;:INIT;*OPC;:FETC?
```

Within **visa_operations.py**, start coding this function with the following definition statement. As shown, `dmm_command_string()` has five arguments, the first four of which are keys to be used in looking up chosen values in dictionaries, while the last argument `sampcount_int` is the integer number of samples to be acquired:

```
def dmm_command_string(measurement_key, range_key,
                       resfrac_key, zero_key,
                       sampcount_int):
```

For the three measurements DC voltage, DC current, and resistance, we need to create dictionaries that encode the following information about the 34410A DMM:

Measurement header: CONF:VOLT:DC, CONF:CURR:DC, CONF:RES
Voltage range: 100 mV, 1 V, 10 V, 100 V, 1000 V
Current range: 100 μA, 1 mA, 10 mA, 100 mA, 1 A, 3 A
Resistance range: 100 ohm, 1 kohm, 10 kohm, 100 kohm, 1 Mohm, 10 Mohm, 100 Mohm
Fractional resolution (NPLC): 3 ppm (0.02 NPLC), 0.7 ppm (0.2 NPLC), 0.3 ppm (1 NPLC), 0.1 ppm (10 NPLC), 0.03 ppm (100 NPLC)

Note, for 34401A or Rigol DM3058 owners:

Fractional resolution (NPLC): 100 ppm (0.02 NPLC), 10 ppm (0.2 NPLC), 3 ppm (1 NPLC), 1 ppm (10 NPLC), 0.3 ppm (100 NPLC)

To include this information in your program, below the function definition statement add the following five dictionaries, noting the choice for each key–value pair:

```
measurement = {'dc_voltage':'CONF:VOLT:DC',
               'dc_current':'CONF:CURR:DC',
               'resistance':'CONF:RES'}

voltage_range = {'100 mV':0.1,
                 '1 V':1.0,
                 '10 V':10.0,
                 '100 V':100.0,
                 '1000 V':1000.0}

current_range = {'0.1 mA':1.0e-4,
                 '1 mA':1.0e-3,
                 '10 mA':0.01,
                 '100 mA':0.1,
                 '1 A':1.0,
                 '3 A':3.0}

resistance_range = {'100 ohm':1.0e2,
                    '1 kohm':1.0e3,
                    '10 kohm':1.0e4,
                    '100 kohm':1.0e5,
                    '1 Mohm':1.0e6,
                    '10 Mohm':1.0e7,
                    '100 Mohm':1.0e8}

fraction_res = {'3 ppm (0.02 NPLC)':3.0,
                '0.7 ppm (0.2 NPLC)':0.7,
                '0.3 ppm (1 NPLC)':0.3,
                '0.1 ppm (10 NPLC)':0.1,
                '0.03 ppm (100 NPLC)':0.03,}

auto_zero = {'On':':ZERO:AUTO ON',
             'Off':':ZERO:AUTO OFF'}
```

Below the dictionaries, include the *if . . . elif* statement shown next that determines the appropriate dictionary and then looks up the value of `range` requested by the argument `range_key`. The calculation below the *if. . . elif* statement calculates the requested `resolution`. Note that the values found in the three `range` dictionaries as well as the `fraction_res` dictionary are floats, not strings (e.g., `10.0`, not `'10'`) because numerical values are required for the calculation that determines `resolution`.

```
if measurement_key == 'dc_voltage':
    range = voltage_range[range_key]
elif measurement_key == 'dc_current':
    range = current_range[range_key]
elif measurement_key == 'resistance':
    range = resistance_range[range_key]
```

```
    fractional_resolution = (fraction_res[resfrac_key]
                             *1.0e-6)

    resolution = range*fractional_resolution
```

Finally, construct the complete command string by concatenating its five components as shown. The operator += appends the string on its right-hand side to the existing string on its left-hand side. The f-string method is used to convert floats to strings with the correct formatting. The additions of `' '`, `','`, and `';'` put spaces, commas, and semicolons in the correct places. Note `' '` has a space between the single quotation marks; i.e., it is equivalent to `'\s'`.

```
    command = measurement[measurement_key] + ' '
    command += f'{range:7.2e}' + ','
    command += f'{resolution:7.2e}' + ';'
    command += auto_zero[zero_key] + ';'
    command += (':SAMP:COUN' + ' '
                + f'{sampcount_int:d}' + ';')

    return command
```

Save your work. The complete definition (except for the dictionary coding) of `dmm_command_string()` is as follows:

```
def dmm_command_string(measurement_key, range_key,
                       resfrac_key, zero_key,
                       sampcount_int):

    measurement = ...

    voltage_range = ...

    current_range = ...

    resistance_range = ...

    fraction_res = ...

    auto_zero = ...


    if measurement_key == 'dc_voltage':
        range = voltage_range[range_key]
    elif measurement_key == 'dc_current':
        range = current_range[range_key]
    elif measurement_key == 'resistance':
        range = resistance_range[range_key]
```

```
    fractional_resolution = (fraction_res[resfrac_key]
                             *1.0e-6)

    resolution = range*fractional_resolution


    command = measurement[measurement_key] + ' '
    command += f'{range:7.2e}' + ','
    command += f'{resolution:7.2e}' + ';'
    command += auto_zero[zero_key] + ';'
    command += (':SAMP:COUN' + ' '
                + f'{sampcount_int:d}' + ';')

    return command
```

To check your coding, at the bottom of **visa_operations.py**, change the test block to read as follows:

```
if __name__ == '__main__':

    command = dmm_command_string('dc_voltage', '10 V',
                                 '0.7 ppm (0.2 NPLC)',
                                 'On', 10)

    print(command)
```

Save your work, then run **visa_operations.py**. If your coding s correct, in the Shell you will find this command string:

```
CONF:VOLT:DC 1.00e+01,7.00e-06;:ZERO:AUTO ON;:SAMP:COUN 10;
```

With that success, modify the test block at the bottom of **visa_operations.py** as follows, so that the command string is written to the DMM.

```
if __name__ == '__main__':

    command = dmm_command_string('dc_voltage', '10 V',
                                 '0.7 ppm (0.2 NPLC)',
                                 'On', 10)

    resource_name = \
            'USB0::0x0957::0x0607::MY47022122::INSTR'

    rm, dmm = open_session(resource_name)

    write_instrument(dmm, command)

    close_session(rm, dmm)
```

Save your work, then run **visa_operations.py**. After running the program, the multimeter will be configured for the DC voltage measurement, but will be left in *remote mode*. You can switch to *local mode* by pressing the instrument's front-panel Shift/Local key. The 34410A DMM can then be triggered (equivalent to sending the INIT command over the interface bus) with the **Trigger** button. A star (*) annunciator will blink on the instrument's front-panel display as it acquires the requested number of voltage samples.

To complete the instrument driver, in **visa_operations.py**, add the following definition for the function `dmm_start_acquisition()`. After configuring a measurement by sending the `command` string created via `dmm_command_string()`, this function will initiate the data-acquisition measurement by sending the SCPI command `':INIT;*OPC;:FETC?'` or `':INIT;:FETC?'`, depending on whether synchronization methods are being used. Setting the argument `sync=False` means that, by default, no synchronization methods are used (remember synchronization methods are necessary when execution of a command will take longer than the timeout value).

```
def dmm_start_acquisition(instr, sync=False):

    if sync == False:
        command = ':INIT;:FETC?'
    elif sync == True:
        command = ':INIT;*OPC;:FETC?'

    instr.write(command)

    return
```

Save your work, then modify the test block at the bottom of **visa_operations.py** as follows.

```
if __name__ == '__main__':

    command = dmm_command_string('dc_voltage', '10 V',
                                 '0.7 ppm (0.2 NPLC)',
                                 'On', 10)

    resource_name= \
            'USB0::0x0957::0x0607::MY47022122::INSTR'

    rm, dmm = open_session(resource_name)

    status_config_srq(dmm)
    write_instrument(dmm, command)
    dmm_start_acquisition(dmm, sync=True)
    wait_for_srq(dmm)
    response = read_instrument(dmm)
    print(response)

    close_session_srq(rm,dmm)
```

Save your work, then run **visa_operations.py**. Your program should acquire ten DC voltage samples under the service request synchronization method, and display the results in the Shell. Try running this program again with `sampcount_int=1500` (or `sampcount_int=15` for 34401A DMM owners).

12.18 Using the Instrument Driver to Write an Application Program

Ultimately, the merit of an instrument driver is measured by the ease with which you can use it to write an *application program* to fulfill some specialized need in your laboratory work. Let's write a GUI-equipped application program called **data_sampler.py** that instructs the DMM to implement the service request synchronization method as it carries out a multisample DC voltage, DC current, or resistance measurement. Our goal is to create an interactive GUI window for this program that looks as shown here:

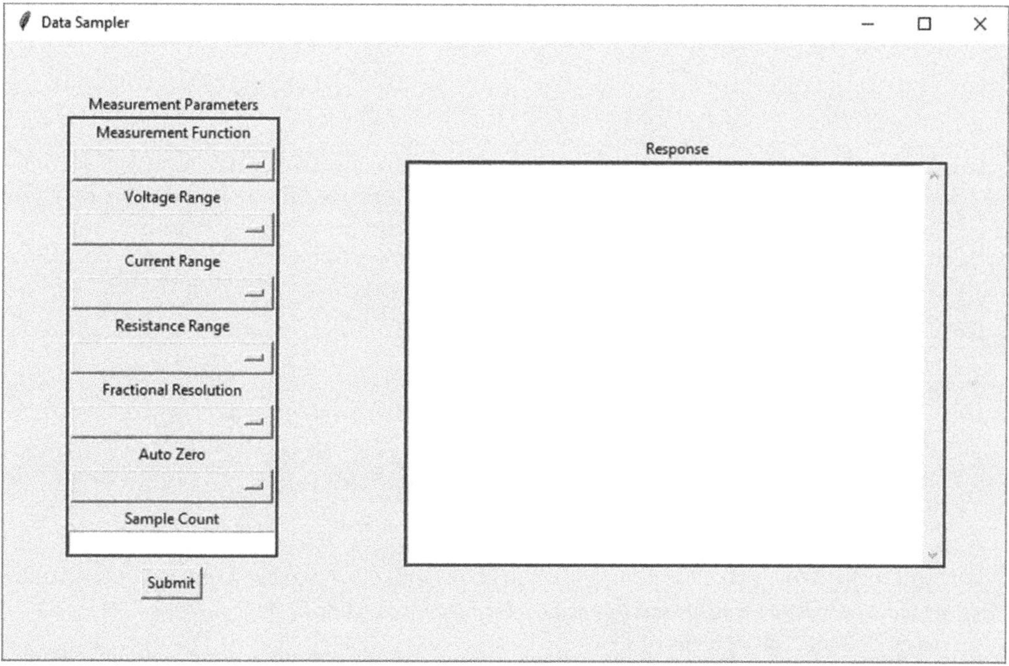

To create the **Measurements Parameters** control panel, open **gui_creator.py py** in **YourNamePythonWork≫Utilities**. The template code that creates each of the option menus is the following. A descriptive label is placed above the option menu. The menu options presented to a user are in the `options` list and the user's selection from that list is stored in the variable `var`:

```
lbl = tk.Label(frame, text='Menu Title Here')
lbl.pack(fill=tk.X)

var = tk.StringVar()

options = [Menu Options List Here]
opm = tk.OptionMenu(frame, var, *options)

opm.config(width=char)
opm.pack(fill=tk.X)
```

In **gui_creator.py py**, code the definition of the function `create_dmm_control()`. A label **Measurement Parameters** is placed above the blue-bordered frame that contains six option menus and an entry box. The six `OptionMenu()` widgets are indexed from `0` to `5`, each programmed with the possible values of the quantity named in its associated label. The final `Entry()` widget receives the integer number of samples to be acquired. The values input to these widgets are stored in seven string variables, which are returned as the seven-element list `var_list`:

```
def create_dmm_control(container):

    char = 20

    lbl = tk.Label(container,
                   text='Measurement Parameters')
    lbl.pack()

    frame = tk.Frame(container, bg='blue', bd=2)
    frame.config(width=char)
    frame.pack()


    lbl_0 = tk.Label(frame,
                     text='Measurement Function')
    lbl_0.pack(fill=tk.X)
    var_0 = tk.StringVar()
    options_0 = ['DC Voltage', 'DC Current', 'Resistance']
    opm_0 = tk.OptionMenu(frame, var_0, *options_0)
    opm_0.config(width=char)
    opm_0.pack(fill=tk.X)

    lbl_1 = tk.Label(frame, text='Voltage Range')
    lbl_1.pack(fill=tk.X)
    var_1 = tk.StringVar()
    options_1 = ['100 mV', '1 V', '10 V', '100 V', '1000 V']
    opm_1 = tk.OptionMenu(frame, var_1, *options_1)
    opm_1.config(width=char)
    opm_1.pack(fill=tk.X)
```

```
    lbl_2 = tk.Label(frame, text='Current Range')
    lbl_2.pack(fill=tk.X)
    var_2 = tk.StringVar()
    options_2 = ['0.1 mA', '1 mA', '10 mA', '100 mA',
                 '1 A', '3 A']
    opm_2 = tk.OptionMenu(frame, var_2, *options_2)
    opm_2.config(width=char)
    opm_2.pack(fill=tk.X)

    lbl_3 = tk.Label(frame, text='Resistance Range')
    lbl_3.pack(fill=tk.X)
    var_3 = tk.StringVar()
    options_3 = ['100 ohm', '1 kohm', '10 kohm',
                 '100 kohm', '1 Mohm', '10 Mohm', '100 Mohm']
    opm_3 = tk.OptionMenu(frame, var_3, *options_3)
    opm_3.config(width=char)
    opm_3.pack(fill=tk.X)


    lbl_4 = tk.Label(frame,
                     text='Fractional Resolution')
    lbl_4.pack(fill=tk.X)
    var_4 = tk.StringVar()
    options_4 = ['3 ppm (0.02 NPLC)',
                 '0.7 ppm (0.2 NPLC)', '0.3 ppm (1 NPLC)',
                 '0.1 ppm (10 NPLC)', '0.03 ppm (100 NPLC)']
    opm_4 = tk.OptionMenu(frame, var_4, *options_4)
    opm_4.config(width=char)
    opm_4.pack(fill=tk.X)

    lbl_5 = tk.Label(frame, text='Auto Zero')
    lbl_5.pack(fill=tk.X)
    var_5 = tk.StringVar()
    options_5 = ['On', 'Off']
    opm_5 = tk.OptionMenu(frame, var_5, *options_5)
    opm_5.config(width=char)
    opm_5.pack(fill=tk.X)

    lbl_6 = tk.Label(frame, text='Sample Count')
    lbl_6.pack(fill=tk.X)
    var_6 = tk.StringVar()
    ent = tk.Entry(frame, textvariable=var_6)
    ent.config(width=char, justify=tk.CENTER)
    ent.pack(fill=tk.X)

    var_list = [var_0, var_1, var_2, var_3, var_4,
                var_5, var_6]

    return var_list
```

Save your work. Next, in **gui_creator.py**, code the definition of the function `get_dmm_control_args(var_list)`. Given `var_list`, this function first retrieves the first element, which corresponds to the selected measurement function. Then, based on this selection, the *if... elif* statement determines the correct `measurement_key` and `range_key` to be used in the lookup dictionaries. The `resfrac_key` and `zero_key` are also retrieved from associated option menus as well as the integer `sampcount_int` value. The five resulting values are collected into a tuple called `args` and returned by the function. The ordering of the elements of `args` corresponds to the order of the arguments for `dmm_command_string()`:

```
def get_dmm_control_args(var_list):

    measurement = var_list[0].get()

    if measurement == 'DC Voltage':
        measurement_key = 'dc_voltage'
        range_key = var_list[1].get()
    elif measurement == 'DC Current':
        measurement_key = 'dc_current'
        range_key = var_list[2].get()
    elif measurement == 'Resistance':
        measurement_key = 'resistance'
        range_key = var_list[3].get()

    resfrac_key = var_list[4].get()
    zero_key = var_list[5].get()
    sampcount_int = int(var_list[6].get())

    args = (measurement_key, range_key, resfrac_key,
            zero_key, sampcount_int)

    return args
```

Save your work as your close in **gui_creator.py py**.

Now, we are ready to code **data_sampler.py**. In the Editor, begin by opening **visa_query_srq.py**. Then, using **File≫Save As...** create **data_sampler.py** and save it in **YourNamePythonWork**. Our GUI plan is this: Place the **Measurement Parameters** control panel in the `controls_box` frame at the left side of our GUI window, and the instrument's **Response** in a scrolled text widget at the right. Thus, in the top lines of the program, add `create_frames()` as shown here:

```
import Utilities.visa_operations as vo
import Utilities.gui_creator as gc

root = gc.create_window(title='Data Sampler',
                        size='800x500')

control_box, right_frame = gc.create_frames(root)
```

```
resource_name = \
          'USB0::0x0957::0x0607::MY47022122::INSTR'

rm, dmm = vo.open_session(resource_name)
```

Then, in the bottom portion of the program, replace `create_control()` with `create_dmm_control()` as shown with the `container=control_box`. Also, for `create_text_box()` make `container=right_frame`:

```
var_list = gc.create_dmm_control(control_box)

stx = gc.create_text_box(right_frame, label='Response',
                         char_width=55, lines=20)

btn = gc.action_button(control_box, label='Submit',
                       action=perform_query)

root.mainloop()

vo.close_session_srq(rm, dmm)
```

Finally, modify `perform_query()` as shown next. With this code, the command string is constructed from information gleaned from the **Measurement Parameters** control panel:

```
def perform_query():

    vo.status_config_srq(dmm)

    args = gc.get_dmm_control_args(var_list)
    command = vo.dmm_command_string(*args)

    vo.write_instrument(dmm, command)
    vo.dmm_start_acquisition(dmm, sync=True)
    vo.wait_for_srq(dmm)
    response = vo.read_instrument(dmm)
    response = response[:-1]
    stx.delete(1.0, 'end')
    stx.insert(1.0, response)

    return
```

Save your work. The complete **data_sampler.py** program is as follows:

```
import Utilities.visa_operations as vo
import Utilities.gui_creator as gc

root = gc.create_window(title='Data Sampler',
                        size='800x500')

control_box, right_frame = gc.create_frames(root)

resource_name = \
          'USB0::0x0957::0x0607::MY47022122::INSTR'
```

```
rm, dmm = vo.open_session(resource_name)

def perform_query():

    vo.status_config_srq(dmm)

    args = gc.get_dmm_control_args(var_list)
    command = vo.dmm_command_string(*args)

    vo.write_instrument(dmm, command)
    vo.dmm_start_acquisition(dmm, sync=True)
    vo.wait_for_srq(dmm)
    response = vo.read_instrument(dmm)
    response = response[:-1]
    stx.delete(1.0, 'end')
    stx.insert(1.0, response)

    return


var_list = gc.create_dmm_control(control_box)

stx = gc.create_text_box(right_frame, label='Response',
                         char_width=55, lines=20)

btn = gc.action_button(control_box, label='Submit',
                       action=perform_query)

root.mainloop()

vo.close_session_srq(rm, dmm)
```

Run **data_sampler.py** and program it to acquire fifty DC voltage samples at 0.7 ppm fractional resolution and autozeroing off. The result will appear as shown here:

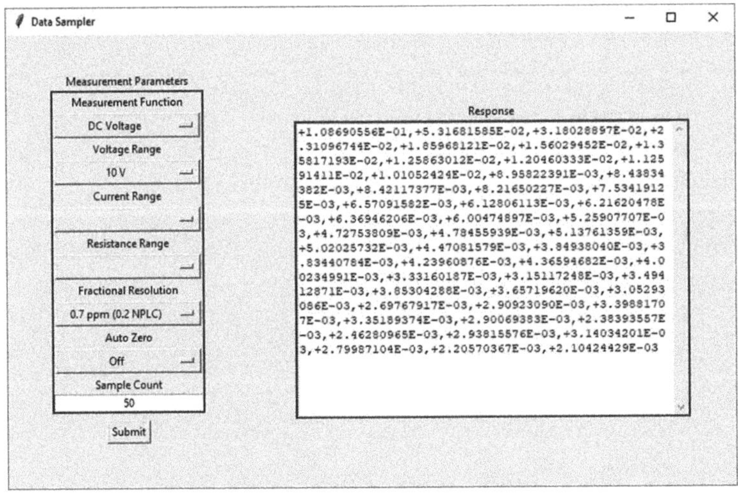

Try rerunning **data_sampler.py**, this time programming it to acquire 1500 samples and then pat yourself on the back for a job well done!

Finally, a useful 34410A DMM instrument-driver utility would perform the following task: Take the instrument's `response` string (data samples delimited by commas with the terminating LF character stripped off) and convert it to a numeric NumPy array, which can then be used in a numeric calculation and/or plotted by Matplotlib routines. That task is accomplished with just these two lines of code:

```
    response = response.split(',')
    numpy_array = np.genfromtxt(response, delimiter=',')
```

First, the Python string method `split()` converts the comma-delimited single string `response` into a list of strings, where we have reused the `response` object to contain this list. Each string in this list represents one of the numbers in the original single string, where the proper locations to slice off each of these number has been identified by the commas. Second, the NumPy function `genfromtxt()` converts the list of strings (each separated from its two neighbors by commas) into a 1D NumPy array called `numpy_array`. Since the elements in this array are floating-point numbers, they can be used in mathematical calculations or plotting routines, and/or written to a spreadsheet data file.

If interested, add the following function to **visa_operations.py** (also add `import numpy as np` at the top of the module):

```
def response_string_to_numpy_array(response):

    response = response.split(',')
    numpy_array = np.genfromtxt(response, delimiter=',')

    return numpy_array
```

Do It Yourself

data_logger_dmm.py Assume that you have a gadget in your laboratory that is providing you with some interesting information about X, which is a slowly varying quantity (say, its variations take place on the time scale of 1 s or longer). X might be the temperature of an object or the intensity of a light source. Additionally, say, the gadget provides this information about X in the form of a "voltage code"; that is, it produces an output voltage V that is some known function of X. Then, with a 34410A DMM and an appropriate GUI-equipped application program, you can monitor X (via measurement of V) as a function of time.

Write the GUI-equipped application program that, when run, continuously obtains a single voltage sample every 1 s over a total time of 50 s. While running, the program provides a real-time plot of V versus t data on the GUI; define time $t = 0$ at the moment that the first voltage sample is acquired. Once all fifty voltage samples are acquired, allow the user to store all of the accumulated data in a CSV spreadsheet file with the t and V data in the spreadsheet's first and second columns, respectively. The GUI of your program should appear as shown below. In this case, a 0.1-Hz triangle wave was input to the DMM.

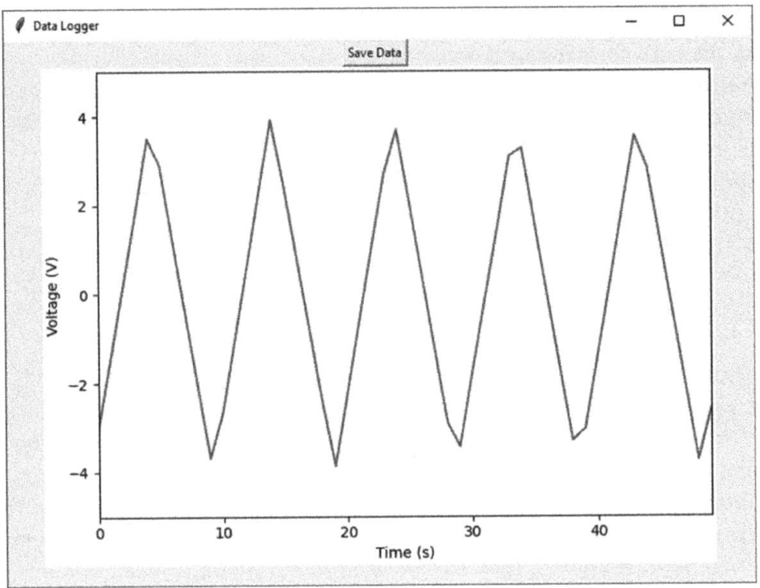

Programming tips: With Chapter 4's **sinewave_animation_blit_gui.py** open in the Editor, use **File≫Save As**. . . to create **data_logger_dmm.py**, then modify the functions `init()` and `animate()` as follows:

```
def init():
    ax.set_xlim(0, (num-1)*delta_t/1000.)
    ax.set_ylim(-a, a)
    ax.set_xlabel('Time (s)')
    ax.set_ylabel('Voltage (V)')

    return line,

def animate(i, t, y):

    t_i = i * delta_t/1000.

    command = 'CONF:VOLT:DC 10,0.00001;:INIT;:FETC?'
    vo.write_instrument(dmm, command)
    response = vo.read_instrument(dmm)
    y_i = float(response[:-1])

    t.append(t_i)
    y.append(y_i)

    line.set_data(t, y)

    return line,
    return
```

Also, assign `delta_t = 1000`, and `num = 50`. Within `FuncAnimation()`, set `interval=delta_t`. Then replace the code related to the **Stop Plot** button with the following function called `save_data()`, along with the creation code for a **Save Data** button. Once the data are acquired, a user can press the **Save Data** button and store the of V versus t values in a CSV spreadsheet:

```
def save_data():

    fname = fd.asksaveasfilename()
    np.savetxt(fname, np.transpose((t, y)), fmt='%.3f',
               delimiter=',')
    vo.close_session_srq(rm, dmm)
    root.quit()

    return

btn = tk.Button(root, text='Save Data', command=save_data)
btn.pack()
```

The required import statements are as follows:

```
import Utilities.visa_operations as vo
import matplotlib.pyplot as plt
import matplotlib.animation as animation
import matplotlib.backends.backend_tkagg as bt
import tkinter as tk
import numpy as np
import tkinter.filedialog as fd
```

Use It!

Configuring LAN/Ethernet Interface on Agilent 34410A DMM Many stand-alone instruments today, including the Agilent 34410A DMM, come with a local area network (LAN) interface for remote control. To utilize the LAN interface in your PyVISA-based code, you will need your instrument's VISA resource name for the DMM, which follows the syntax for a USB-interfaced instrument,

```
TCPIP0IP address::LAN device name::INSTR
```

where `IP address` is the *internet protocol* (IP) address of your instrument, which has the form XXX.XXX.XXX.XXX. If desired, a *static IP address* (supplied by your network manager) can be manually programmed into your instrument using the front-panel **Utility≫Remote I/O** menu. However, if available, *Dynamic Host Configuration Protocol (DHCP)* is a method that automatically assigns an IP address to your instrument. By default, the Agilent 34410A DMM has DHCP activated, so when you connect your DMM to your LAN via an ethernet cable, your DMM will request an IP address from the LAN. The resulting IP address can be found in the **Utility≫Remote I/O** menu.

To establish communication with the DMM, in the **Command Prompt** type

```
ping <IP address>
```

where `<IP address>` is the address you found in the **Utility≫Remote I/O** menu.

Alternately, the NI-VISA download includes the NI-MAX program (for Windows and Mac only). Start NI-MAX, then double-click on **Devices and Interface** in the upper-left corner. Next, right click on **Network Devices** and select **Create New VISA TCP/IP Resource**... In the dialog window that opens, select **Auto-detect of LAN Instrument**, then click **Next.** In the next dialog window, your instrument will be detected and its VISA resource name displayed. Click **Finish**.

Finally, start IDLE and, in the Shell, enter the following sequence of commands:

```
>>> import pyvisa
>>> rm = pyvisa.ResourceManager()
>>> rm.list_resources()
('TCPIP0::10.0.1.36::inst0::INSTR',)
```

For my case, I find that the VISA resource name to use for LAN communication with my DMM is `'TCPIP0::10.0.1.36::inst0::INSTR'`.

Problems

1. **thermocouple_thermometer.py** Thermocouples are widely used as temperature sensors. A thermocouple is constructed by joining the ends of two dissimilar metals, for example, a chromel and an alumel wire for a type K thermocouple. This junction produces a millivolt-level voltage that has a well-documented temperature dependence, where the temperature is measured relative to a "cold junction" reference temperature. Conveniently, this cold junction can be provided by a compact electronic device called a *cold junction compensator (CJC)*, which effectively makes the reference temperature equal to 0°C.

   On its most sensitive scale, the 34410A multimeter can directly measure thermocouple voltages accurately enough to resolve small temperature changes. Connect a type K thermocouple to a CJC (e.g., Omega MCJ Series device) and then connect the plus and minus outputs of the CJC to the HI and LO Voltage Inputs of the 34410A multimeter. Then write a GUI-equipped program that, every 250 ms until a front-panel **Stop** button is pressed, reads the thermocouple voltage, converts this value to the corresponding temperature in Celsius, and then displays this temperature in a GUI indicator. To convert the type K thermocouple voltage to its corresponding temperature, see Problem 2 in Chapter 10.

   Run your program and use it to measure room temperature as well as the temperature of your skin.

2. **thermistor_thermometer.py** As we found in Chapter 10, the resistance of a thermistor can be used as a thermometer. Connect the thermistor you used in Chapter 10 to the HI and LO Inputs of the 34410A multimeter, then write a GUI-equipped program that samples the resistance of the thermistor, every 250 ms until a front-panel **Stop** button is pressed. Within your code, use the Steinhart–Hart equation to convert each resistance sample to the corresponding temperature in Celsius, and then to display this temperature in a GUI indicator.

   Run your program and use it to measure room temperature as well as the temperature of your skin.

APPENDIX

Installing Python and IDLE

A.1 Python Versions

Python is a cross-platform programming language. It runs on all major operating systems and any Python program you write can be run on any computer that has Python installed. Like every programming language, developers are continuously introducing newer versions of Python in order to increase its power and versatility. The code presented in this book was written using Python 3.12 (the current version at the time of writing), but all of the Python commands used are included in Python 3.6 or later.

In this Appendix, the steps for setting up Python on your personal computing system are outlined. If, instead, you are using a shared (e.g., instructional, research, or commercial laboratory) computing system, please consult with the appropriate person (e.g., instructor, research supervisor, work colleague) for instructions on accessing Python.

A.2 Installing Python on Windows

First, check if Python is already present on your system by using the Windows command-line interface called **Command Prompt** (or **PowerShell**). To open a **Command Prompt** window, (depending on your version of Windows) enter *cmd* in the search box next to the bottom-left Windows icon or in the **Start≫Run** input box. Then, in the **Command Prompt** window, type *python -V*, where the lowercase *python* runs Python and the (uppercase) option *-V* requests Python to print its version number. Press *<Enter>* and, if you have Python properly installed, its version number will be printed on a new line. If Python is not installed, you will get an error message (e.g., stating that *python* is not a recognized command).

To install the latest version of Python (which is necessary for this book, if your present version is older than 3.6), go to https://www.python.org/downloads/. The webpage will present you with a yellow **Download** button with the appropriate Python installer for your Windows system. Click on the **Download** button and, after the installer has been downloaded, run the installer. IMPORTANT: In the installer's first window, be sure to check the box labeled **Add python.exe to PATH**. This selection configures Windows with the file directory information it needs to find the Python program. Then, click on **Install Now**, and follow the instruction in the subsequent windows.

Once the Python installation is complete, verify that the version you downloaded is present in your system. Open a new **Command Prompt** window, then type *python -V* and press *<Enter>*. If Python has been properly installed, its version number will be printed on a new line. If the printed version is not the one you just installed, you may have to uninstall an older version of Python that previously existed on your system. However, if other Python programmers use this same system, consult with them before going forward with any uninstall operation.

If you want to know the directory in which Python has been installed, do the following: In the **Command Prompt** window, type *python* and press *<Enter>*. This command will start the Python Interpreter and you will be presented with a Python prompt (>>>), where a Python statement can be entered and then executed. At the Python prompt, enter the following two commands:

```
>>> import sys
>>> sys.executable
```

When you press *<Enter>* after typing the second command, the directory path to Python will be displayed.

If the *python -V* command yields an error message, most likely Windows has not been configured for the correct path to Python, possibly because the **Add python.exe to PATH** box was not check during the installation process. To correct this problem, the path (i.e., the directory containing Python) needs to be added to Windows' **Environmental Variables**. For the procedure on how to do this task manually, use the internet search keywords "How to add Python to PATH on Windows". Perhaps easier, first uninstall Python from your system. Then, reinstall Python, this time remembering to check the **Add python.exe to PATH** box in the first installer window. Windows will be automatically configured with the correct path to Python.

Your computer is now outfitted with the Python programming language Interpreter and the Python Standard Library of programming elements. Also provided are the integrated development environment *IDLE* with which you can write and run Python programs as well as the Preferred Installer Program, or *pip* for short, that is used to download and install Python software from the Python Package Index (PyPI).

A.3 Installing Python on Mac

Python is installed on most macOS systems, but usually is the outdated Python 2 version; do not delete this version of Python as it is used by the macOS. First, check if the newer Python 3 is present on your system by using the macOS command-line interface application called **Terminal**. To open a Terminal window, click on **Terminal.app** in the **Applications≫Utilities** folder. Then, at the prompt, type *python3 -V*, where the lowercase *python3* runs Python 3 and the (uppercase) option *-V* requests Python to print its version number. Press *<Enter>* and, if you have Python 3 properly installed, its version number will be printed on a new line. If Python is not installed, you will get an error message.

To install the latest version of Python 3 (which is necessary for this book, if your present version is older than 3.6), go to https://www.python.org/downloads/. The webpage will present you with a yellow **Download** button with the appropriate Python installer for your macOS system. Click on the **Download** button. After the installer has been downloaded, run the installer, and follow the instruction in the subsequent windows.

Once the Python installation is complete, verify that the version you downloaded is present in your system. Open a new **Terminal** window, then type *python3 -V* and press *<Enter>*. If Python 3 has been properly installed, its version number will be printed on a new line. If the printed version is not the one you just installed, you may have to uninstall an older version of Python that previously existed on your system. However, if other Python programmers use this same system, consult with them before going forward with any uninstall operation.

If you want to know the directory in which Python has been installed, do the following: In the **Terminal** window, type *python3* and press *<Enter>*. This command will start the Python Interpreter and you will be presented with a Python prompt (>>>), where a Python statement can be entered and then executed. At the Python prompt, enter the two commands

```
>>> import sys
>>> sys.executable
```

When you press <*Enter*> after typing the second command, the directory path to Python will be displayed.

Your computer is now outfitted with the Python 3 programming language Interpreter and the Python Standard Library of programming elements. Also provided are the integrated development environment *IDLE* with which you can write and run Python programs as well as the Preferred Installer Program, or *pip* for short, that is used to download and install Python software from the Python Package Index (PyPI).

A.4 Installing Python on Linux

Python is preinstalled on almost every Linux system. If the version of Python on your Linux system is earlier than 3.6, go to https://www.python.org/downloads/. The webpage will present you with a yellow **Download** button with the appropriate Python installer for your Linux system.

A.5 Python Program Creation and IDLE Editor

A Python program comprises a sequence of text-based commands, saved in a file with the extension **.py**. Hence, any text editor can be used to write a Python program. That said, a text editor designed specifically to edit code is the approach taken by all Python programmers. A dedicated code editor can be as basic as a text editor that highlights your code structure with emblematic colors in order to catch common coding errors as well as the ability to run a completed program. Other handy features may include automatic indentation, indicators that measure line numbers and lengths, and keyboard shortcuts for common operations.

A more sophisticated programming tool is called the *integrated development environment* (IDE). An IDE is a text editor, along with a number of other advanced features such as debugging support, automatic code formatting, and ease of virtual environment creation. This last feature—virtual environment creation—is important in the long-term use of an open-source programming language like Python, especially when using a computing system that is being accessed by multiple users (e.g., a computer shared by a research group). Given that Python is continually evolving, changes made in a current version (even as small as switching a variable name's leading letter from uppercase to lowercase) can cause a functioning program written using a prior version of Python to suddenly become nonfunctional when opened in the newer version of Python. This problem is compounded in a system used by several workers as one programmer may innocently update a Python version and accidently "break" a colleague's previously operative program. A virtual environment solves this problem in that a cordoned-off directory is created with the exact versions of software used to create a program, and this directory remains untouched by any other upgrades taking place on the computer. Learning how to program within a virtual environment is definitely a skill needed for a long-term career in Python. Given this book's aim as an introduction to Python-based experimentation, however, the topic of virtual environments will not be covered.

To create Python programs, there is a wide range of available code editors and IDEs, many of which are free, and some of which are exclusively designed for Python programming. Some popular options are Sublime Text, PyCharm, Spyder, Visual Studio Code, Jupiter Notebooks, and IDLE. While a full-featured IDE can greatly aid in efficient code creation in experienced hands, there is a fairly steep learning curve to understanding all of a program's functionality.

Thus, given this book's introductory focus, I've chosen to present programming as being written using IDLE. IDLE is the easy-to-learn programming editor included when you download Python; hence, it is free, cross-platform, and available to all. Although technically an IDE, IDLE has light computing requirements and so can run even on older machines or those with scaled-down resources. In addition, IDLE provides the user with two windows, the IDLE Shell and the IDLE Editor. The IDLE Shell allows one to interactively execute Python commands line by line, while the IDLE Editor is used to write, save, and execute Python programs. We will make extensive use of both of these tools.

If you are experienced in writing Python programs using an editor or IDE other than IDLE, by all means feel free to use that program as you work your way through the book. All of the book's exercises can be carried out with the editor or IDE of your choice.

Index

Programs Written By Readers of this Book